Theoretical and Applied Mathematics

nova
science publishers
www.novapublishers.com

Theoretical and Applied Mathematics

Engineering Problems and the Laplace Transform
Victor Martinez-Luaces, PhD (Editor)
2025. ISBN: 979-8-89530-282-8 (Hardcover)
2025. ISBN: 979-8-89530-298-9 (eBook)

Applications of Fuzzy Theory in Applied Sciences and Computer Applications
Animesh Kumar Sharma (Editor)
2024. ISBN: 979-8-89113-666-3 (Softcover)
2024. ISBN: 979-8-89113-740-0 (eBook)

Applied Computational Hydraulics for Engineers
Necati Ağıralioğlu and Mehmet Özger (Editors)
2024. ISBN: 979-8-89113-415-7 (Hardcover)
2024. ISBN: 979-8-89113-506-2 (eBook)

The Impossible Quintic Made as Simple as Possible
Graeme Sneddon and Sam Kault (Editors)
2023. ISBN: 979-8-88697-908-4 (Softcover)
2023. ISBN: 979-8-88697-972-5 (eBook)

A Closer Look of Nonlinear Reaction-Diffusion Equations
L. Rajendran, R. Swaminathan and M. Chitra Devi (Editors)
2020. ISBN: 978-1-53618-257-6 (Hardcover)
2020. ISBN: 978-1-53618-356-6 (eBook)

A Closer Look at Boundary Value Problems
Mustafa Avci (Editor)
2020. ISBN: 978-1-53617-857-9 (Hardcover)
2020. ISBN: 978-1-53617-861-6 (eBook)

More information about this series can be found at
https://novapublishers.com/product-category/series/Theoretical-and-Applied-Mathematics/

Victor Martinez-Luaces, PhD
Editor

The Laplace Transform and Its Applications in Science

nova
science publishers
www.novapublishers.com

DOI: https://doi.org/10.52305/OOES1283

Library of Congress Cataloging-in-Publication Data

Names: Martinez-Luaces, Victor editor
Title: The Laplace transform and its applications in science / Victor Martinez-Luaces, PhD, editor.
Description: New York : Nova Science Publishers,, [2024] | Series: Mathematics research developments | Includes bibliographical references and index. |
Identifiers: LCCN 2024056223 (print) | LCCN 2024056224 (ebook) | ISBN 9798891137479 hardcover | ISBN 9798895302934 adobe pdf
Subjects: LCSH: Laplace transformation
Classification: LCC QA432 .L3185 2024 (print) | LCC QA432 (ebook) | DDC 515/.723--dc23/eng/20250213
LC record available at https://lccn.loc.gov/2024056223
LC ebook record available at https://lccn.loc.gov/2024056224

Published by Nova Science Publishers, Inc. † New York

Contents

Preface

This book attempts to show the richness and versatility of one of the most important mathematical tools, such as the Laplace transform.

The work is divided into two sections: the first is devoted to the Laplace transform in mathematics, and the second is about the interaction of this mathematical tool with other sciences.

The first part, containing three chapters, is purely mathematical and shows interesting extensions and generalizations of the Laplace transform and its connection with other branches of mathematics such as probability and queuing theory.

The first chapter addresses and analyzes the fundamental topics of the Laplace transform, from the viewpoint of Time Scales, which represents a unification of the theory of differential equations with that of difference equations, unifying the calculus of finite differences with the integral and differential calculus, and offering a formal frame for the study of hybrid systems. This proposal applies the time scales framework to the Laplace transform, including several relevant theorems about its convergence, the bilateral Laplace transform, and the Fourier transform. Important topics like inversion, differentiation, and delay formulas are studied, as well as relationships with other important tools such as the Z transform and the Hilger Dirac delta function. The chapter includes several interesting examples of these topics.

In the second chapter, a general version of the Laplace transform is analyzed in the study of generalized differential equations. After a brief historical introduction, the work proposes a general definition of a fractional derivative and the corresponding fractional integrals. Based on these definitions, it is possible to generalize partial derivatives, the Laplace transform, the convolution of functions, and differential equations, among other concepts. An integral operator with a general kernel that contains others as particular cases is used to define the general version of the Laplace transform, whose properties are analyzed. Finally, this mathematical tool is applied to differential equations and the work ends with a study of the stability of linear systems of generalized differential equations, based on the real part of their eigenvalues.

In the third chapter, a probabilistic interpretation of the Laplace transform is considered. This different approach allows a new interpretation of some important results, with a notable simplification in the process of calculation. Particularly, the Laplace transform of some exponential and potential-exponential functions and order statistics are obtained. It also shows the connections between Laplace transform, Probability, and Queuing Theory, with some interesting examples.

In the second part of the book –which contains four chapters– it can be observed the versatility of the Laplace transform, through its applications in other scientific disciplines, particularly in chemistry, optics, fluid dynamics, and quantum mechanics, among others.

Chapter 4 studies important topics, such as the use of the Laplace and Borel-Laplace transforms in convergent series and factorially divergent series. The work includes interesting applications to other sciences, for instance, those related to the works of Airy and Stokes for the study of the rainbow, and the quantum mechanics anharmonic oscillator, where the Laplace transform is used in studying asymptotic expansions in certain regions of the complex plane, corresponding to the original divergent series. The well-documented historical notes make this chapter even more interesting.

The first part of Chapter 5 analyzes the solution to nonlinear ordinary and partial differential equations –transform related to mass transfer problems with and without chemical reaction– by using Laplace transform. In the second part, the work deals fundamentally with special functions, including Bessel, Gamma and Polygamma, Kelvin, Struve, Volterra, Whittaker, Mittag-Leffler, Mainardi, Wright, Bateman and Havelock, among others. It is shown how formulas can be obtained for their integrals, derivatives, limits, and also other formulas that establish relations between them, with a particular attention to the differentiation with respect to the parameters and to the evaluation of the asymptotic relations, limits of series, integrals and integral representations.

Chapter 6 is about Laplace transform sum rules (LSR), which are utilized in Quantum ChromoDynamics (QCD). These sum rules are used to obtain different constants, masses, and parameters of interest, related to quarks, gluons, hadrons, and various condensates, which are studied throughout the chapter. Beyond LSR, there are other forms of sum rules used in QCD, which are also discussed and compared with experimental results and/or with values calculated by other methods. The results obtained show that LSR and other sum rules are appropriate to study hadron properties and obtain several fundamental parameters, providing a good alternative to lattice QCD numerical simulations. It is worth mentioning that the chapter includes several previous results of the author, in an extensive list that goes from the late 1970s to the present. Finally, the contribution proposes further research.

In Chapter 7 the transfer function of a continuous stirred-tank reactor (CSTR), as well as CSTRs series and parallels, are presented. After this introduction, several inverse problems, which are the main part of the chapter, arise. Specifically, the idea is to analyze whether or not any rational function can be the transfer function of a CSTRs system, and even more, if that CSTRs system is unique or not, and if small changes in the transfer function correspond to small changes in the system. The final result is that the three problems have a negative answer, that is, neither the existence, uniqueness, nor the stability can be assured for the mentioned inverse problem.

It is important to note that the works presented here –including a great variety of topics and approaches– illustrate the enormous range of applications of this mathematical tool created more than two hundred years ago, which still is a fundamental research topic –pure and applied– for mathematicians, scientists, and engineers.

Acknowledgments

I would like to acknowledge the support staff at Nova Science Publishers, who were very helpful at various stages of developing and producing the volume.

I also wish to thank Marjorie Chaves who, as an assistant editor, collaborated throughout the editing process, and oversaw the style and English language in the chapters of this book.

Chapter 1

Laplace and Fourier Transforms on Time Scales

Svetlin G. Georgiev*
Department of Mathematics, Sorbonne University, Paris, France;
Department of Differential Equations, Faculty of Mathematics and Informatics, Sofia University, Sofia, Bulgaria

Abstract

We have shown that the bilateral Laplace transform on time scales converges absolutely and uniformly on $\mathbb{C}_{s,\alpha,\gamma}$, its derivatives are uniformly continuous on $\mathbb{C}_{s,\alpha,\gamma}$, and established an inversion formula for functions that have finitely many poles of finite order. After that, we showed that the bilateral Laplace transform on time scales can be used to define a Fourier transform on time scales which can be used to solve partial dynamic equations. One interesting future application of this bilateral Laplace transform might be in defining analogues for moment-generating functions for probability distributions defined on time scales. Work in that direction for a unilateral transform was initiated in [1, Definition 20], which defined moment-generating functions for time scales with a minimum value of 0. Of particular interest is an investigation of a proper analogue of the Schwartz class of functions for the continuous transform; finding it would allow for a major expansion of the solutions of partial dynamic equations. Another interesting topic for both the bilateral Laplace transform and the Fourier transform defined here would be the time scales analogues of the convolution theorem, which was treated in full generality in [2] for the unilateral Laplace transform.

* Corresponding Author's Email: svetlingeorgiev1@gmail.com

In: The Laplace Transform and Its Applications in Science
Editor: Victor Martinez-Luaces
ISBN: 979-8-89113-747-9

Keywords: time scale, Laplace transform, bilateral Laplace transform, shifting problem, Fourier transform, Dirac delta functional

Mathematics Subject Classification: 39A10, 34A60, 34B37, 34C10, 34K20, 34K30

1. Introduction

The Laplace transform on time scales has historically been a heavily-studied, useful, and interesting object, while the bilateral Laplace transform has not received nearly as much attention. "Bilateral" refers to the region of integration being the entire time scale,

$$\mathcal{L}^b\{f\}(z;s) = \int_{-\infty}^{\infty} f(t)e_{\ominus z}(\sigma(t),s)\Delta t,$$

as opposed to the more common "unilateral" transform which begins the integral at s. Study of the bilateral Laplace transform on time scales was initiated in [3], which established many of its properties, particularly those related to its region of convergence.

We now point out that the regions of convergence under consideration in this article match those appearing in [3], which takes a more geometrical approach to the regions of convergence than we do here. Significantly, the regions of convergence for bilateral transforms in this manuscript are not the most general ones that exist. It has long been known that the stability region of the time scales exponential can be, e.g. disconnected [4], and so it is not surprising that regions of convergence for transforms whose kernel contains time scales exponentials could also be disconnected. The manuscript [5] leverages the techniques in [6] to develop the most general regions of convergence for unilateral and bilateral Laplace transforms. Thus, the regions of convergence under consideration here are, in general, proper subsets of the maximal regions explored in [5]. The tradeoff in this approach is the ease of computation of our smaller regions of convergence as intersections of various circles compared to the more general ones that could be obtained.

A "nice" Fourier transform for time scales has been long-sought and approached in different ways. For instance, [7, Section 7] defines it as

$$\mathcal{F}\{f\}(u) = \int_{-\infty}^{\infty} f(t)e_{2\mathring{\imath}u}(t)\Delta t = \int_{-\infty}^{\infty} f(t)e^{-2\pi iut}\Delta t, \tag{1.1}$$

where $\mathring{\imath}$ is a certain function. The authors of [8] emphasize the benefits of the definition (1.1) for obtaining a nice convolution operation, prove its inversion integral, and explore some applications to probability theory. Fourier transforms on subclasses of time scales that are also groups were studied in [9]. The similar notion of "discrete Fourier transform" has also been developed in [10]. We will define a Fourier transform on time scales different than those above in terms of the bilateral Laplace transform.

It is important to set realistic expectations of what any "Fourier transform on time scales" can accomplish: from [11],

"*...on a general time scale there are no cisoids. Thus, in general, there are no periodic signals. This has a very important consequence: on a general time scale we cannot define Fourier transforms, which is a powerful drawback of the theory.*"

Indeed, a proper analogue of periodic signals has not been defined on a general time scale, excepting those defined on the restricted class of "periodic time scales". However, (1.1) has been shown to be useful for signals processing, and our Fourier transform has operational properties that make it useful for solving certain partial dynamic equations equations.

2. Definition of the Laplace Transform. Properties

Let $\mathbb{T}$ be a time scale with forward jump operator, backward jump operator, delta differentiation operator and nabla differentiation operator σ, ρ, Δ and ∇, respectively. Let also, $\sup \mathbb{T} = \infty$. Take $s \in \mathbb{T}$. If $z \in \mathbb{C}$ and $1 + \mu(t)z \neq 0$ for all $t \in \mathbb{T}^\kappa$, then

$$(\ominus z)(t) = -\frac{z}{1+\mu(t)z}$$

and

$$\begin{aligned} 1+\mu(t)(\ominus z)(t) &= 1-\frac{\mu(t)z}{1+\mu(t)z} \\ &= \frac{1}{1+\mu(t)z} \end{aligned}$$

for any $t \in \mathbb{T}^\kappa$, and hence, $e_{\ominus z}(\cdot, s)$ is well defined on $\mathbb{T}^\kappa$.

Definition 2.1. *Suppose that $f : \mathbb{T} \to \mathbb{R}$ is regulated. Then the Laplace transform of f is defined by*

$$\mathcal{L}(f)(z,s) = \int_s^\infty e^\sigma_{\ominus z}(t,s)\, f(t)\Delta t \tag{2.1}$$

for $z \in \mathbb{C}$ for which $1+\mu(t)z \neq 0$ for any $t \in \mathbb{T}^\kappa$ and the improper integral (2.1) *exists. When $s = 0$, we will write $\mathcal{L}(f)(z)$.*

Remark 2.1. *Let $f : \mathbb{T} \to \mathbb{R}$ be regulated. With $\mathcal{D}\{f\}$ we will denote the set of all complex numbers z for which $1 + \mu(t) z \neq 0$ for any $t \in \mathbb{T}^\kappa$ and the improper integral* (2.1) *exists.*

Theorem 2.1 (Linearity)**.** *Assume that $f, g : \mathbb{T} \to \mathbb{R}$ are regulated. Then, for any constants α and β, we have*

$$\mathcal{L}(\alpha f + \beta g)(z, s) = \alpha \mathcal{L}(f)(z, s) + \beta \mathcal{L}(g)(z, s)$$

for $z \in \mathcal{D}\{f\} \bigcap \mathcal{D}\{g\}$, $t \in \mathbb{T}$.

Proof. Let $z \in \mathcal{D}\{f\} \bigcap \mathcal{D}\{g\}$, $t \in \mathbb{T}$. Then, we have

$$\begin{aligned}
\mathcal{L}(\alpha f + \beta g)(z, s) &= \int_s^\infty e^\sigma_{\ominus z}(t, s)(\alpha f + \beta g)(t)\Delta t \\
&= \int_s^\infty e^\sigma_{\ominus z}(t, s)(\alpha f(t) + \beta g(t))\Delta t \\
&= \int_s^\infty e^\sigma_{\ominus z}(t, s)\alpha f(t)\Delta t + \int_s^\infty e^\sigma_{\ominus z}(t, s)\beta g(t)\Delta t \\
&= \alpha \int_s^\infty e^\sigma_{\ominus z}(t, s) f(t)\Delta t + \beta \int_s^\infty e^\sigma_{\ominus z}(t, s) g(t)\Delta t \\
&= \alpha \mathcal{L}(f)(z, s) + \beta \mathcal{L}(g)(z, s).
\end{aligned}$$

This completes the proof. □

Lemma 2.1. *Let $z \in \mathbb{C}$. Then*

$$\begin{aligned}
e^\sigma_{\ominus z}(t, s) &= \frac{e_{\ominus z}(t, s)}{1 + \mu(t) z} \\
&= -\frac{(\ominus z)(t)}{z} e_{\ominus z}(t, s), \quad t \in \mathbb{T}.
\end{aligned}$$

Proof. For $z \in \mathbb{C}$ and $t \in \mathbb{T}$, we have

$$\begin{aligned} e^{\sigma}_{\ominus z}(t,s) &= e_{\ominus z}(t,s) + \mu(t)\, e^{\Delta}_{\ominus z}(t,s) \\ &= e_{\ominus z}(t,s) + (\ominus z)(t)\,\mu(t)\, e_{\ominus z}(t,s) \\ &= (1 + (\ominus z)(t)\,\mu(t))\, e_{\ominus z}(t,s) \\ &= \left(1 - \frac{\mu(t)\, z}{1+\mu(t)\, z}\right) e_{\ominus z}(t,s) \\ &= \frac{1}{1+\mu(t)\, z} e_{\ominus z}(t,s) \\ &= -\frac{1}{z}\left(-\frac{z}{1+\mu(t)\, z}\right) e_{\ominus z}(t,s) \\ &= -\frac{(\ominus z)(t)}{z} e_{\ominus z}(t,s). \end{aligned}$$

This completes the proof. □

Theorem 2.2. *Assume $f : \mathbb{T} \to \mathbb{C}$ is such that f^{Δ^l}, $l \in \{0, \ldots, k\}$, are regulated. Then*

$$\mathcal{L}\left(f^{\Delta^k}\right)(z,s) = z^k \mathcal{L}(f)(z,s) - \sum_{l=0}^{k-1} z^l f^{\Delta^{k-1-l}}(s) \qquad (2.2)$$

for those $z \in \mathcal{D}\{f\} \bigcap \mathcal{D}\{f^{\Delta}\} \bigcap \ldots \bigcap \mathcal{D}\left\{f^{\Delta^k}\right\}$ satisfying

$$\lim_{t\to\infty}\left(f^{\Delta^l}(t) e_{\ominus z}(t,s)\right) = 0, \quad l \in \{0, \ldots, k-1\}, \qquad (2.3)$$

for any $t \in \mathbb{T}$.

Proof. Let $t \in \mathbb{T}$. We will use induction.

1. Let $k = 1$. Take $z \in D\{f\} \bigcap D\{f^{\Delta}\}$ so that

$$\lim_{t\to\infty}(f(t) e_{\ominus z}(t,s)) = 0.$$

Integrating by parts and using Lemma 2.1 and the condition (2.3), we get

$$\begin{aligned}\mathcal{L}\left(f^{\Delta}\right)(z,s) &= \int_s^{\infty} f^{\Delta}(t)e^{\sigma}_{\ominus z}(t,s)\,\Delta t\\ &= f(t)e_{\ominus z}(t,s)\Big|_{t=s}^{t\to\infty} - \int_s^{\infty} f(t)\left(\ominus z\right)(t)\,e_{\ominus z}(t,s)\,\Delta t\\ &= -f(s) + z\int_s^{\infty} f(t)e^{\sigma}_{\ominus z}(t,s)\,\Delta t\\ &= -f(s) + z\mathcal{L}(f)(z,s).\end{aligned}$$

2. Assume that (2.2) holds for some $k \in \mathbb{N}$ and for those

$$z \in \mathcal{D}\{f\}\bigcap \mathcal{D}\left\{f^{\Delta}\right\}\bigcap \cdots \bigcap \mathcal{D}\left\{f^{\Delta^k}\right\}$$

for which (2.3) hold.

3. We will prove

$$\mathcal{L}\left(f^{\Delta^{k+1}}\right)(z,s) = z^{k+1}\mathcal{L}(f)(z,s) - \sum_{l=0}^{k} z^l f^{\Delta^{k-l}}(s)$$

for those $z \in \mathcal{D}\{f\}\bigcap\mathcal{D}\{f^{\Delta}\}\bigcap\ldots\bigcap\mathcal{D}\left\{f^{\Delta^{k+1}}\right\}$ for which

$$\lim_{t\to\infty}\left(f^{\Delta^l}(t)e_{\ominus z}(t,s)\right) = 0, \quad l \in \{0,\ldots,k\}. \tag{2.4}$$

Really, let $z \in \mathcal{D}\{f\}\bigcap\mathcal{D}\{f^{\Delta}\}\bigcap\ldots\bigcap\mathcal{D}\left\{f^{\Delta^{k+1}}\right\}$. Integrating by parts and using Lemma 2.1 and (2.2), we get

$$\begin{aligned}
\mathcal{L}\left(f^{\Delta^{k+1}}\right)(z,s) &= \int_s^\infty f^{\Delta^{k+1}}(t)e^\sigma_{\ominus z}(t,s)\,\Delta t \\
&= f^{\Delta^k}(t)e_{\ominus z}(t,s)\Big|_{t=s}^{t\to\infty} - \int_s^\infty f^{\Delta^k}(t)\,(\ominus z)\,(t)e_{\ominus z}(t,s)\,\Delta t \\
&= z\int_s^\infty f^{\Delta^k}(t)e^\sigma_{\ominus z}(t,s)\,\Delta t - f^{\Delta^k}(s) \\
&= z\mathcal{L}\left(f^{\Delta^k}\right)(z,s) - f^{\Delta^k}(s) \\
&= z\left(z^k\mathcal{L}(f)(z,s) - \sum_{l=0}^{k-1} z^l f^{\Delta^{k-1-l}}(s)\right) - f^{\Delta^k}(s) \\
&= z^{k+1}\mathcal{L}(f)(z,s) - \sum_{l=0}^{k-1} z^{l+1} f^{\Delta^{k-1-l}}(s) - f^{\Delta^k}(s) \\
&= z^{k+1}\mathcal{L}(f)(z,s) - \sum_{l=1}^{k} z^l f^{\Delta^{k-l}}(s) - f^{\Delta^k}(s) \\
&= z^{k+1}\mathcal{L}(f)(z,s) - \sum_{l=0}^{k} z^l f^{\Delta^{k-l}}(s).
\end{aligned}$$

This completes the proof.

□

Example 2.1. *Let $z \in \mathbb{C}$ be such that*

$$1 + \mu(t)z \neq 0 \quad \textit{and} \quad \lim_{t\to\infty} e_{\ominus z}(t,s) = 0. \tag{2.5}$$

Then

$$\begin{aligned}
\mathcal{L}(1)(z) &= \int_s^\infty e^\sigma_{\ominus z}(t,s)\Delta t \\
&= \int_s^\infty (1+\mu(t)(\ominus z)(t))\, e_{\ominus z}(t,s)\Delta t \\
&= \int_s^\infty \left(1-\frac{z\mu(t)}{1+z\mu(t)}\right) e_{\ominus z}(t,s)\Delta t \\
&= -\frac{1}{z}\int_s^\infty \left(-\frac{z}{1+z\mu(t)}\right) e_{\ominus z}(t,s)\Delta t \\
&= -\frac{1}{z}\int_s^\infty (\ominus z)(t) e_{\ominus z}(t,s)\Delta t \\
&= -\frac{1}{z}\int_s^\infty e^\Delta_{\ominus z}(t,s)\Delta t \\
&= -\frac{1}{z} e_{\ominus z}(t,s)\Big|_{t=s}^{t\to\infty} \\
&= \frac{1}{z}.
\end{aligned}$$

Example 2.2. *Let* $\mathbb{T}=\mathbb{Z}$, $s=0$ *and*

$$f(t)=t^2+1, \quad t\in\mathbb{T}.$$

We will find $\mathcal{L}(f)(z)$ *for* $z\in\mathcal{D}\{f\}$. *We have*

$$\sigma(t)=t+1, \quad \mu(t)=1, \quad t\in\mathbb{T}.$$

We set

$$f_1(t)=t, \quad t\in\mathbb{T}.$$

Let $z\in\mathbb{C}$ *be such that* $z\in D\{f_1\}\bigcap D\{f_1^\Delta\}\bigcap D\{f\}\bigcap D\{f^\Delta\}$ *and* (2.5) *hold. By* (2.2), *we have*

$$\mathcal{L}\left(f_1^\Delta\right)(z) = z\mathcal{L}(f_1)(z) - f_1(0),$$

or

$$\mathcal{L}(1)(z) = z\mathcal{L}(f_1)(z),$$

or

$$\frac{1}{z} = z\mathcal{L}(f_1)(z),$$

or

$$\mathcal{L}(f_1)(z) = \frac{1}{z^2}.$$

Note that

$$\begin{aligned} f^{\Delta}(t) &= \sigma(t) + t \\ &= t + 1 + t \\ &= 2t + 1, \quad t \in \mathbb{T}. \end{aligned}$$

Therefore

$$\begin{aligned} \mathcal{L}\left(f^{\Delta}\right)(z) &= 2\mathcal{L}(f_1)(z) + \mathcal{L}(1)(z) \\ &= \frac{2}{z^2} + \frac{1}{z} \\ &= \frac{z+2}{z^2}. \end{aligned}$$

Now, we use (2.2) *to get*

$$\mathcal{L}\left(f^{\Delta}\right)(z) = z\mathcal{L}(f)(z) - f(0),$$

or

$$\frac{z+2}{z^2} = z\mathcal{L}(f)(z) - 1,$$

or

$$\frac{z+2}{z^2} + 1 = z\mathcal{L}(f)(z),$$

or

$$\frac{z^2+z+2}{z^2} = z\mathcal{L}(f)(z),$$

or

$$\mathcal{L}(f)(z) = \frac{z^2+z+2}{z^3}.$$

Example 2.3. *Let* $\mathbb{T} = \mathbb{N}_0^2$, $s = 0$ *and*

$$f(t) = \frac{2t + 2\sqrt{t} + 3}{(t+1)^2\left(t + 2\sqrt{t} + 2\right)^2}, \quad t \in \mathbb{T}.$$

We will find $\mathcal{L}(f)(z)$ for $z \in \mathcal{D}\{f\}$. Let

$$h(t) = \frac{1}{(t+1)^2}, \quad t \in \mathbb{T}.$$

Take $z \in D\{f\} \bigcap D\{h\}$. We have

$$\begin{aligned} \sigma(t) &= t + 2\sqrt{t} + 1, \\ \mu(t) &= 2\sqrt{t} + 1, \quad t \in \mathbb{T}. \end{aligned}$$

Note that

$$\begin{aligned} e^{\sigma}_{\ominus z}(t, 0) &= e_{-\frac{z}{1+z\mu(t)}}(\sigma(t), 0) \\ &= e^{-\int_0^{\sigma(t)} \mathrm{Log}(1+z\mu(\tau))\Delta\tau} \\ &= e^{-\sum\limits_{\tau\in[0,t]} \mu(\tau)\mathrm{Log}(1+z\mu(\tau))} \\ &= \prod_{\tau\in[0,t]} \frac{1}{(1+z\mu(\tau))^{\mu(\tau)}} \\ &= \prod_{\tau\in[0,t]} \frac{1}{(1+z(1+2\sqrt{\tau}))^{1+2\sqrt{\tau}}}, \quad t \in \mathbb{T}. \end{aligned}$$

Then

$$\begin{aligned} \mathcal{L}(h)(z) &= \int_0^\infty \frac{1}{(1+t)^2} e^{\sigma}_{\ominus z}(t, 0)\Delta t \\ &= \sum_{t\in[0,\infty)} \frac{2\sqrt{t}+1}{(t+1)^2} \prod_{\tau\in[0,t]} \frac{1}{(1+z(1+2\sqrt{\tau}))^{1+2\sqrt{\tau}}}, \quad t \in \mathbb{T}. \end{aligned}$$

Next,

$$\begin{aligned} h^{\Delta}(t) &= -\frac{\sigma(t)+t+2}{(t+1)^2(\sigma(t)+1)^2} \\ &= -\frac{2t+2\sqrt{t}+3}{(t+1)^2\left(t+2\sqrt{t}+2\right)^2} \\ &= -f(t), \quad t \in \mathbb{T}. \end{aligned}$$

Hence and (2.2)*, we obtain*

$$\mathcal{L}\left(h^{\Delta}\right)(z) = z\mathcal{L}(h)(z) - h^{\Delta}(0),$$

or

$$-\mathcal{L}(f)(z) = z\mathcal{L}(h)(z) - \frac{3}{4},$$

or

$$\mathcal{L}(f)(z) = -z \sum_{t\in[0,\infty)} \frac{2\sqrt{t}+1}{(t+1)^2} \prod_{\tau\in[0,t]} \frac{1}{\left(1+z\left(1+2\sqrt{\tau}\right)\right)^{1+2\sqrt{\tau}}} + \frac{3}{4}.$$

Example 2.4. *Let* $\mathbb{T} = 3^{\mathbb{N}_0} \bigcup \{0\}$*,* $s = 0$ *and*

$$f(t) = \begin{cases} \frac{4}{(t+1)(3t+1)(9t+1)}, & t \neq 0, \quad t \in \mathbb{T}, \\ \frac{3}{8}, \quad t = 0. \end{cases}$$

We will find $\mathcal{L}(f)(z)$ *for* $z \in \mathcal{D}\{f\}$*. Let*

$$h(t) = \frac{1}{1+t}, \quad t \in \mathbb{T}.$$

Take $z \in D\{f\} \bigcap D\{h\} \bigcap D\{h^{\Delta}\} \bigcap D\{h^{\Delta^2}\}$*. We have*

$$\begin{aligned} \sigma(t) &= 3t, \\ \mu(t) &= 2t, \quad t \in 3^{\mathbb{N}_0}, \\ \sigma(0) &= 1, \\ \mu(0) &= 1. \end{aligned}$$

Next,

$$\begin{aligned}
e^{\sigma}_{\ominus z}(t,0) &= e^{\int_0^{\sigma(t)} \mathrm{Log}(1+(\ominus z)(\tau)\mu(\tau))\Delta\tau} \\
&= e^{\int_0^{\sigma(t)} \mathrm{Log}\left(1-\frac{z\mu(\tau)}{1+z\mu(\tau)}\right)\Delta\tau} \\
&= e^{\int_0^{\sigma(t)} \mathrm{Log}\frac{1}{1+z\mu(\tau)}\Delta\tau} \\
&= e^{\sum\limits_{\tau\in[0,t]} \mu(\tau)\mathrm{Log}\frac{1}{1+z\mu(\tau)}} \\
&= \prod_{\tau\in[0,t]} \frac{1}{(1+z\mu(\tau))^{\mu(\tau)}} \\
&= \begin{cases} \frac{1}{1+z}\prod\limits_{\tau\in[1,t]}\frac{1}{(1+2z\tau)^{2\tau}}, & t\in 3^{\mathbb{N}_0}, \\ \frac{1}{z+1}, & t=0. \end{cases}
\end{aligned}$$

Hence,

$$\begin{aligned}
\mathcal{L}(h)(z) &= \int_0^{\infty} \frac{1}{t+1} e^{\sigma}_{\ominus z}(t,0)\Delta t \\
&= \sum_{t\in[0,\infty)} \frac{\mu(t)}{t+1} e^{\sigma}_{\ominus z}(t,0) \\
&= \frac{1}{1+z}\left(1+\sum_{t\in[1,\infty)}\frac{2t}{t+1}\prod_{\tau\in[1,t]}\frac{1}{(1+2z\tau)^{2\tau}}\right).
\end{aligned}$$

Note that

$$\begin{aligned}
h^{\Delta}(t) &= -\frac{1}{(1+t)(1+\sigma(t))} \\
&= \begin{cases} -\frac{1}{2}, & t=0, \\ -\frac{1}{(1+t)(1+3t)}, & t\neq 0, \end{cases}
\end{aligned}$$

for $t = 0$ we get

$$\begin{aligned} h^{\Delta^2}(0) &= \frac{h^{\Delta}(\sigma(0)) - h^{\Delta}(0)}{\sigma(0) - 0} \\ &= h^{\Delta}(1) + \frac{1}{2} \\ &= -\frac{1}{8} + \frac{1}{2} \\ &= \frac{3}{8} \\ &= f(0), \end{aligned}$$

for $t \neq 0$ we obtain

$$\begin{aligned} h^{\Delta^2}(t) &= \frac{h^{\Delta}(\sigma(t)) - h^{\Delta}(t)}{\sigma(t) - t} \\ &= \frac{-\frac{1}{(3t+1)(9t+1)} + \frac{1}{(t+1)(3t+1)}}{2t} \\ &= \frac{4}{(t+1)(3t+1)(9t+1)} \\ &= f(t), \quad t \in \mathbb{T}. \end{aligned}$$

Hence and (2.2)*, we get*

$$\begin{aligned} \mathcal{L}\left(h^{\Delta^2}\right)(z) &= \mathcal{L}(f)(z) \\ &= z^2\mathcal{L}(h)(z) - h^{\Delta}(0) - zh(0) \\ &= \frac{z^2}{1+z}\left(1 + \sum_{t\in[1,\infty)} \frac{2t}{1+t} \prod_{\tau\in[1,t]} \frac{1}{(1+2z\tau)^{2\tau}}\right) \\ &\quad + \frac{1}{2} - z. \end{aligned}$$

Theorem 2.3. *Assume $f : \mathbb{T} \to \mathbb{C}$ is regulated. If*

$$F(x) = \int_s^x f(y)\Delta y$$

for $x \in \mathbb{T}$, then

$$\mathcal{L}(F)(z,s) = \frac{1}{z}\mathcal{L}(f)(z,s)$$

for those $z \in \mathcal{D}\{f\}\backslash\{s\}$ satisfying

$$\lim_{x\to\infty}\left(e_{\ominus z}(x,s)\int_s^x f(y)\Delta y\right) = 0. \tag{2.6}$$

Proof. Using integration by parts, we get

$$\begin{aligned}
\mathcal{L}(F)(z,s) &= \int_s^\infty F(y)e_{\ominus z}(\sigma(y),s)\Delta y \\
&= \int_s^\infty F(y)(1+(\ominus z)(y)\mu(y))e_{\ominus z}(y,s)\Delta y \\
&= \int_s^\infty F(y)\frac{1}{1+z\mu(y)}e_{\ominus z}(y,s)\Delta y \\
&= -\frac{1}{z}\int_s^\infty F(y)\frac{-z}{1+z\mu(y)}e_{\ominus z}(y,s)\Delta y \\
&= -\frac{1}{z}\int_s^\infty F(y)(\ominus z)(y)e_{\ominus z}(y,s)\Delta y \\
&= -\frac{1}{z}\int_s^\infty F(y)e^{\Delta}_{\ominus z}(y,s)\Delta y \\
&= -\frac{1}{z}\left(\lim_{y\to\infty}F(y)e_{\ominus z}(y,s) - F(s)e_{\ominus z}(s,s)\right) \\
&\quad +\frac{1}{z}\int_s^\infty F^{\Delta}(y)e^{\sigma}_{\ominus z}(y,s)\Delta y \\
&= \frac{1}{z}\int_s^\infty f(y)e_{\ominus z}(\sigma(y),s)\Delta y \\
&= \frac{1}{z}\mathcal{L}(f)(z,s),
\end{aligned}$$

for $z \in D\{f\}\backslash\{s\}$ for which (2.6) holds. This completes the proof. □

Theorem 2.4. *For all $n \in \mathbb{N}_0$, we have*

$$\mathcal{L}(h_n(x,s))(z) = \frac{1}{z^{n+1}}, \quad x \in \mathbb{T}, \tag{2.7}$$

for $z \in \mathbb{C}\backslash\{s\}$ *such that* $1 + z\mu(x) \neq 0$, $x \in \mathbb{T}$, *and*

$$\lim_{x \to \infty} (h_n(x,s)e_{\ominus z}(x,s)) = 0. \tag{2.8}$$

Proof. To prove our assertion, we will use induction.

1. Let $n = 0$. We have that $h_0(x,s) = 1$, $x \in \mathbb{T}$, and

$$\mathcal{L}(1)(z,s) = \frac{1}{z}$$

for those $z \in \mathbb{C}\backslash\{s\}$ for which (2.5) hold.

2. Assume that (2.7) holds for some $n \in \mathbb{N}_0$ and for $z \in \mathbb{C}\backslash\{s\}$ such that

$$1 + z\mu(x) \neq 0, \quad x \in \mathbb{T}, \tag{2.9}$$

and (2.8) holds.

3. We will prove that

$$\mathcal{L}(h_{n+1}(x,s))(z) = \frac{1}{z^{n+2}}$$

for those $z \in \mathbb{C}\backslash\{s\}$ such that $1 + z\mu(x) \neq 0$, $x \in \mathbb{T}$, and

$$\lim_{x \to \infty} (h_{n+1}(x,s)e_{\ominus z}(x,s)) = 0. \tag{2.10}$$

Really, for such z we have

$$\begin{aligned} \mathcal{L}(h_{n+1}(x,s))(z,s) &= \int_s^\infty h_{n+1}(y,s)e_{\ominus z}(\sigma(y),s)\Delta y \\ &= \int_s^\infty \left(\int_s^y h_n(t,s)\Delta t\right) e_{\ominus z}(\sigma(y),s)\Delta y. \end{aligned}$$

From here, using Theorem 2.3 and (2.7), we obtain

$$\begin{aligned} \mathcal{L}(h_{n+1}(x,s))(z) &= \frac{1}{z}\mathcal{L}(h_n(x,s))(z) \\ &= \frac{1}{z} \cdot \frac{1}{z^{n+1}} \\ &= \frac{1}{z^{n+2}} \end{aligned}$$

for $z \in \mathbb{C}\backslash\{s\}$ such that (2.9) and (2.10) hold. This completes the proof.

□

Theorem 2.5. *Let* $\alpha \in \mathbb{C}$ *and* $1 + \alpha\mu(x) \neq 0$ *for* $x \in \mathbb{T}$. *Then*

$$\mathcal{L}(e_\alpha(x,s))(z,s) = \frac{1}{z-\alpha}, \quad x \in \mathbb{T},$$

provided

$$\lim_{x\to\infty} e_{\alpha\ominus z}(x,s) = 0, \quad z \neq \alpha. \tag{2.11}$$

Proof. Suppose that $z \in \mathbb{C}$ and (2.11) hold. We have

$$\begin{aligned}
\mathcal{L}(e_\alpha(x,s))(z,s) &= \int_s^\infty e_\alpha(y,s)e_{\ominus z}(\sigma(y),s)\Delta y \\
&= \int_s^\infty e_\alpha(y,s)\left(1 + (\ominus z)(y)\mu(y)\right) e_{\ominus z}(y,s)\Delta y \\
&= \int_s^\infty \frac{1}{1+z\mu(y)} e_\alpha(y,s)e_{\ominus z}(y,s)\Delta y \\
&= \int_s^\infty \frac{1}{1+z\mu(y)} e_{\alpha\ominus z}(y,s)\Delta y \\
&= \frac{1}{\alpha - z}\int_s^\infty \frac{\alpha - z}{1+z\mu(y)} e_{\alpha\ominus z}(y,s)\Delta y,
\end{aligned}$$

i.e.,

$$\mathcal{L}(e_\alpha(x,s))(z,s) = \frac{1}{\alpha - z}\int_s^\infty \frac{\alpha - z}{1+z\mu(y)} e_{\alpha\ominus z}(y,s)\Delta y. \tag{2.12}$$

We note that

$$\begin{aligned}
(\alpha \ominus z)(y) &= \alpha(\oplus(\ominus z)(y)) \\
&= \alpha + (\ominus z)(y) + \alpha(\ominus z)(y)\mu(y) \\
&= \alpha - \frac{z}{1+z\mu(y)} - \frac{\alpha z\mu(y)}{1+z\mu(y)} \\
&= \frac{\alpha - z}{1+z\mu(y)}.
\end{aligned}$$

Hence and (2.12), we obtain

$$\begin{aligned}
\mathcal{L}(e_\alpha(x,s))(z,s) &= \frac{1}{\alpha - z}\int_s^\infty \frac{\alpha - z}{1 + z\mu(y)} e_{\alpha\ominus z}(y,s)\Delta y \\
&= \frac{1}{\alpha - z}\int_s^\infty (\alpha \ominus z)(y) e_{\alpha\ominus z}(y,s)\Delta y \\
&= \frac{1}{\alpha - z}\int_s^\infty e^\Delta_{\alpha\ominus z}(y,s)\Delta y \\
&= \frac{1}{\alpha - z}\left(\lim_{y\to\infty} e_{\alpha\ominus z}(y,s) - e_{\alpha\ominus z}(s,s)\right) \\
&= \frac{1}{z - \alpha}.
\end{aligned}$$

This completes the proof. □

Corollary 2.1. *Let* $\alpha \in \mathbb{C}$ *and*

$$1 \pm i\alpha\mu(x) \neq 0, \quad x \in \mathbb{T}.$$

We have

1. $\mathcal{L}(\cos_\alpha(x,s))(z,s) = \dfrac{z}{z^2 + \alpha^2}$,
2. $\mathcal{L}(\sin_\alpha(x,s))(z,s) = \dfrac{\alpha}{z^2 + \alpha^2}$

provided that

$$\lim_{x\to\infty} e_{i\alpha\ominus z}(x,s) = \lim_{x\to\infty} e_{-i\alpha\ominus z}(x,s) = 0, \quad z \neq \pm i\alpha. \tag{2.13}$$

Proof. 1. Suppose that $z \in \mathbb{C}$ and (2.13) hold. By the definition of $\cos_\alpha(x,s)$, we have

$$\cos_\alpha(x,s) = \frac{e_{i\alpha}(x,s) + e_{-i\alpha}(x,s)}{2}.$$

Hence,

$$\begin{aligned}\mathcal{L}(\cos_\alpha(x,s))(z,s) &= \mathcal{L}\left(\frac{e_{i\alpha}(x,s)+e_{-i\alpha}(x,s)}{2}\right)(z,s)\\ &= \frac{1}{2}\mathcal{L}(e_{i\alpha}(x,s)+e_{-i\alpha}(x,s))(z,s)\\ &= \frac{1}{2}\left(\mathcal{L}(e_{i\alpha}(x,s))(z,s)+\mathcal{L}(e_{-i\alpha}(x,s))(z,s)\right)\\ &= \frac{1}{2}\left(\frac{1}{z-i\alpha}+\frac{1}{z+i\alpha}\right)\\ &= \frac{1}{2}\cdot\frac{z-i\alpha+z+i\alpha}{(z-i\alpha)(z+i\alpha)}\\ &= \frac{z}{z^2+\alpha^2}.\end{aligned}$$

2. By the definition of $\sin_\alpha(x,s)$, we have

$$\sin_\alpha(x,s) = \frac{e_{i\alpha}(x,s)-e_{-i\alpha}(x,s)}{2i}.$$

Hence,

$$\begin{aligned}\mathcal{L}(\sin_\alpha(x,s))(z,s) &= \mathcal{L}\left(\frac{e_{i\alpha}(x,s)-e_{-i\alpha}(x,s)}{2i}\right)(z,s)\\ &= \frac{1}{2i}\mathcal{L}(e_{i\alpha}(x,s)-e_{-i\alpha}(x,s))(z,s)\\ &= \frac{1}{2i}\left(\mathcal{L}(e_{i\alpha}(x,s))(z,s)-\mathcal{L}(e_{-i\alpha}(x,s))(z,s)\right)\\ &= \frac{1}{2i}\left(\frac{1}{z-i\alpha}-\frac{1}{z+i\alpha}\right)\\ &= \frac{1}{2i}\cdot\frac{z+i\alpha-z+i\alpha}{(z-i\alpha)(z+i\alpha)}\\ &= \frac{\alpha}{z^2+\alpha^2}.\end{aligned}$$

This completes the proof.

□

Definition 2.2. *Let $f : \mathbb{N}_0 \to \mathbb{R}$ and let $z \in \mathbb{R}$. Then the $\mathcal{Z}-$transform is defined by*

$$\mathcal{Z}(f)(z) = \sum_{t=0}^{\infty} \frac{f(t)}{z^t}$$

provided the series converges.

Theorem 2.6. *Let $\mathbb{T} = \mathbb{N}_0$. Then*

$$(z+1)\mathcal{L}(f)(z) = \mathcal{Z}(f)(z+1)$$

for every $f : \mathbb{T} \to \mathbb{R}$ and every $z \in \mathcal{D}\{f\}$, provided that the series converges.

Proof. For $y \in \mathbb{T} \cap [0, \infty)$ and $z \in D\{f\}$, we have

$$\begin{aligned}
e_{\ominus z}(\sigma(y), 0) &= (1 + (\ominus z)(y)\mu(y))e_{\ominus z}(y, 0) \\
&= \left(1 - \frac{z\mu(y)}{1 + z\mu(y)}\right) e_{\ominus z}(y, 0) \\
&= \frac{1}{1 + z\mu(y)} e_{\ominus z}(y, 0) \\
&= \frac{1}{1+z} e_{\ominus z}(y, 0) \\
&= \frac{1}{1+z} e^{\int_0^y \frac{1}{\mu(y)} \mathrm{Log}(1+(\ominus z)(y)\mu(y))\Delta y} \\
&= \frac{1}{1+z} e^{\int_0^y \mathrm{Log}\frac{1}{1+z}\Delta y} \\
&= \frac{1}{1+z}\left(\frac{1}{1+z}\right)^y \\
&= \frac{1}{(1+z)^{y+1}}.
\end{aligned}$$

Hence, for $z \in D\{f\}$, we obtain

$$\begin{aligned}(z+1)\mathcal{L}(f)(z) &= \int_0^\infty \frac{f(y)}{(1+z)^{y+1}}\Delta y \\ &= (z+1)\sum_{t=0}^\infty \frac{f(t)}{(1+z)^{t+1}} \\ &= \sum_{t=0}^\infty \frac{f(t)}{(1+z)^t} \\ &= \mathcal{Z}(f)(z+1),\end{aligned}$$

provided that the series converges. This completes the proof. □

Now, suppose that $\mathbb{T}_0$ is a time scale such that $0 \in \mathbb{T}_0$ and $\sup \mathbb{T}_0 = \infty$. The usual convolution of two functions f and g on the real interval $[0, \infty)$ is defined by

$$(f \star g)(x) = \int_0^x f(x-y)g(y)ds \quad for \quad x \geq 0.$$

However, this definition does not work for general time scales because $x, y \in \mathbb{T}_0$ does not imply that $x - y \in \mathbb{T}_0$.

Definition 2.3. *Assume that $\alpha \in \mathbb{C}$ and f is one of the functions $e_\alpha(\cdot, 0), \sinh_\alpha(\cdot, 0), \cosh_\alpha(\cdot, 0), \cos_\alpha(\cdot, 0), \sin_\alpha(\cdot, 0)$, or $h_k(\cdot, 0)$, $k \in \mathbb{N}_0$. If g is a regulated function on $\mathbb{T}_0$, then we define the convolution of f with g by*

$$(f \star g)(x) = \int_0^x f(x, \sigma(y))g(y)\Delta y \quad for \quad x \in \mathbb{T}_0.$$

Theorem 2.7. *(Convolution Theorem) Assume that $\alpha \in \mathbb{C}$ and f is one of the functions $e_\alpha(\cdot, 0), \sinh_\alpha(\cdot, 0), \cosh_\alpha(\cdot, 0), \cos_\alpha(\cdot, 0), \sin_\alpha(\cdot, 0)$, or $h_k(\cdot, 0)$, $k \in \mathbb{N}_0$. If g is a regulated function on $\mathbb{T}_0$ such that*

$$\lim_{x \to \infty} e_{\ominus z}(x, 0)(f \star g)(x) = 0,$$

then

$$\mathcal{L}(f \star g)(z) = \mathcal{L}(f)(z)\mathcal{L}(g)(z). \tag{2.14}$$

Proof. 1. Let $f(x,y) = e_\alpha(x,y)$, $x, y \in \mathbb{T}_0$. Here we assume that $\alpha \in \mathbb{C}$ is regressive. Thus, $e_\alpha(x,y), x, y \in \mathbb{T}_0$, is well defined. Then $f(x) = f(x,0)$, $x \in \mathbb{T}_0$. Consider the initial value problem for the dynamic equation

$$\begin{aligned} l^\Delta - \alpha l &= g(x), \quad x \in \mathbb{T}_0, \\ l(0) &= 0. \end{aligned} \tag{2.15}$$

The solution of this problem is given by

$$l(x) = \int_0^x e_\alpha(x, \sigma(y)) g(y) \Delta y, \quad x \in \mathbb{T}_0,$$

which can be rewritten in the form

$$l(x) = (e_\alpha(\cdot, 0) \star g)(x), \quad x \in \mathbb{T}_0.$$

Suppose that $z \in \mathbb{C}$ satisfies suitable assumptions. Now, we apply the Laplace transform to both sides of the equation (2.15) to get

$$\mathcal{L}\left(l^\Delta - \alpha l\right)(z) = \mathcal{L}(g)(z)$$

or

$$\mathcal{L}(l^\Delta)(z) - \alpha \mathcal{L}(l)(z) = \mathcal{L}(g)(z),$$

or

$$z\mathcal{L}(l)(z) - \alpha \mathcal{L}(l)(z) = \mathcal{L}(g)(z),$$

whereupon

$$\mathcal{L}(l)(z) = \frac{1}{z-\alpha} \mathcal{L}(g)(z).$$

Since

$$\mathcal{L}(e_\alpha(\cdot, 0))(z) = \frac{1}{z-\alpha},$$

we conclude that

$$\mathcal{L}(l)(z) = \mathcal{L}(f)(z)\mathcal{L}(g)(z)$$

or we get (2.14).

2. Let $f(x,y) = \cosh_\alpha(x,y), x, y \in \mathbb{T}_0$. Here we assume that $\alpha \in \mathbb{C}$ and $-\mu\alpha^2 \in \mathcal{R}$ so that $\cosh_\alpha(x,y)$, $x, y \in \mathbb{T}_0$, is well defined. Then,

$$\begin{aligned} f(x) &= f(x,0) \\ &= \cosh_\alpha(x,0), \quad x \in \mathbb{T}_0. \end{aligned}$$

We have

$$\cosh_\alpha(x,0) = \frac{e_\alpha(x,0)+e_{-\alpha}(x,0)}{2}.$$

Then

$$\begin{aligned}(f\star g)(x) &= \int_0^x \cosh_\alpha(x,\sigma(y))g(y)\Delta y\\ &= \frac{1}{2}\int_0^x \left(e_\alpha(x,\sigma(y))+e_{-\alpha}(x,\sigma(y))\right)g(y)\Delta y\\ &= \frac{1}{2}\int_0^x e_\alpha(x,\sigma(y))g(y)\Delta y + \frac{1}{2}\int_0^x e_{-\alpha}(x,\sigma(y))g(y)\Delta y\\ &= \frac{1}{2}(e_\alpha(x,\cdot)\star g)(x) + \frac{1}{2}(e_{-\alpha}(x,\cdot)\star g)(x), \quad x\in\mathbb{T}_0.\end{aligned}$$

From here, under suitable assumptions for $z\in\mathbb{C}$, we get

$$\begin{aligned}\mathcal{L}(f\star g)(z) &= \mathcal{L}\left(\frac{1}{2}(e_\alpha(\cdot,0)\star g)+\frac{1}{2}(e_{-\alpha}(\cdot,0)\star g)\right)(z)\\ &= \frac{1}{2}\mathcal{L}(e_\alpha(\cdot,0)\star g)(z)+\frac{1}{2}\mathcal{L}(e_{-\alpha}(\cdot,0)\star g)(z)\\ &= \frac{1}{2}\mathcal{L}(e_\alpha(\cdot,0))(z)\mathcal{L}(g)(z)+\frac{1}{2}\mathcal{L}(e_{-\alpha}(\cdot,0))(z)\mathcal{L}(g)(z)\\ &= \frac{1}{2}\cdot\frac{1}{z-\alpha}\mathcal{L}(g)(z)+\frac{1}{2}\cdot\frac{1}{z+\alpha}\mathcal{L}(g)(z)\\ &= \frac{1}{2}\left(\frac{1}{z-\alpha}+\frac{1}{z+\alpha}\right)\mathcal{L}(g)(z)\\ &= \frac{z}{z^2-\alpha^2}\mathcal{L}(g)(z)\\ &= \mathcal{L}(\cosh_\alpha(\cdot,0))(z)\mathcal{L}(g)(z)\\ &= \mathcal{L}(f)(z)\mathcal{L}(g)(z).\end{aligned}$$

3. Let $f(x,y)=\sinh_\alpha(x,y)$, $x,y\in\mathbb{T}_0$. Suppose that $\alpha\in\mathbb{C}$ and $-\mu\alpha^2\in\mathcal{R}$.

Thus, $\sinh_\alpha(x,y)$, $x,y\in\mathbb{T}_0$, is well defined. We have

$$\begin{aligned} f(x) &= f(x,0) \\ &= \sinh_\alpha(x,0), \quad x\in\mathbb{T}_0. \end{aligned}$$

Then

$$\sinh_\alpha(x,y)=\frac{e_\alpha(x,y)-e_{-\alpha}(x,y)}{2}, \quad x,y\in\mathbb{T}_0,$$

and

$$\begin{aligned} (f\star g)(x) &= \int_0^x \sinh_\alpha(x,\sigma(y))g(y)\Delta y \\ &= \frac{1}{2}\int_0^x \left(e_\alpha(x,\sigma(y))-e_{-\alpha}(x,\sigma(y))\right)g(y)\Delta y \\ &= \frac{1}{2}\int_0^x e_\alpha(x,\sigma(y))g(y)\Delta y-\frac{1}{2}\int_0^x e_{-\alpha}(x,\sigma(y))g(y)\Delta y \\ &= \frac{1}{2}(e_\alpha(\cdot,0)\star g)(x)-\frac{1}{2}(e_{-\alpha}(\cdot,0)\star g)(x), \quad x\in\mathbb{T}_0. \end{aligned}$$

Hence, under suitable assumptions for $z\in\mathbb{C}$, we get

$$\begin{aligned} \mathcal{L}(f\star g)(z) &= \mathcal{L}\left(\frac{1}{2}(e_\alpha(\cdot,0)\star g)(x)-\frac{1}{2}(e_{-\alpha}(\cdot,0)\star g)(x)\right)(z) \\ &= \frac{1}{2}\mathcal{L}(e_\alpha(\cdot,0)\star g)(z)-\frac{1}{2}\mathcal{L}(e_{-\alpha}(\cdot,0)\star g)(z) \\ &= \frac{1}{2}\mathcal{L}(e_\alpha(\cdot,0))(z)\mathcal{L}(g)(z)-\frac{1}{2}\mathcal{L}(e_{-\alpha}(\cdot,0))(z)\mathcal{L}(g)(z) \\ &= \frac{1}{2}\cdot\frac{1}{z-\alpha}\mathcal{L}(g)(z)-\frac{1}{2}\cdot\frac{1}{z+\alpha}\mathcal{L}(g)(z) \\ &= \frac{1}{2}\left(\frac{1}{z-\alpha}-\frac{1}{z+\alpha}\right)\mathcal{L}(g)(z) \\ &= \frac{\alpha}{z^2-\alpha^2}\mathcal{L}(g)(z) \\ &= \mathcal{L}(\sinh_\alpha(\cdot,0))(z)\mathcal{L}(g)(z) \\ &= \mathcal{L}(f)(z)\mathcal{L}(g)(z). \end{aligned}$$

4. Let $f(x,y) = \cos_\alpha(x,y), x, y \in \mathbb{N}_0$. Assume that $\alpha \in \mathbb{C}$ and $\mu\alpha^2 \in \mathcal{R}$. Then, $\cos_\alpha(x,y), x, y \in \mathbb{T}_0$, is well defined. Note that

$$\begin{aligned} f(x) &= f(x,0) \\ &= \cos_\alpha(x,0), \quad x \in \mathbb{T}_0. \end{aligned}$$

We have

$$\cos_\alpha(x,0) = \frac{e_{i\alpha}(x,0) + e_{-i\alpha}(x,0)}{2}, \quad x \in \mathbb{T}_0.$$

Then

$$\begin{aligned} (f \star g)(x) &= \int_0^x \cos_\alpha(x,\sigma(y))g(y)\Delta y \\ &= \frac{1}{2}\int_0^x \left(e_{i\alpha}(x,\sigma(y)) + e_{-i\alpha}(x,\sigma(y))\right) g(y)\Delta y \\ &= \frac{1}{2}\int_0^x e_{i\alpha}(x,\sigma(y))g(y)\Delta y + \frac{1}{2}\int_0^x e_{-i\alpha}(x,\sigma(y))g(y)\Delta y \\ &= \frac{1}{2}(e_{i\alpha}(\cdot,0) \star g)(x) + \frac{1}{2}(e_{-i\alpha}(\cdot,0) \star g)(x), \quad x \in \mathbb{T}_0. \end{aligned}$$

From here, under suitable assumptions for $\alpha \in \mathbb{C}$, we get

$$\begin{aligned}
\mathcal{L}(f \star g)(z) &= \mathcal{L}\left(\frac{1}{2}(e_{i\alpha}(\cdot,0)\star g)+\frac{1}{2}(e_{-i\alpha}(\cdot,0)\star g)\right)(z) \\
&= \frac{1}{2}\mathcal{L}(e_{i\alpha}(\cdot,0)\star g)(z)+\frac{1}{2}\mathcal{L}(e_{-i\alpha}(\cdot,0)\star g)(z) \\
&= \frac{1}{2}\mathcal{L}(e_{i\alpha}(\cdot,0))(z)\mathcal{L}(g)(z)+\frac{1}{2}\mathcal{L}(e_{-i\alpha}(\cdot,0))(z)\mathcal{L}(g)(z) \\
&= \frac{1}{2}\cdot\frac{1}{z-i\alpha}\mathcal{L}(g)(z)+\frac{1}{2}\cdot\frac{1}{z+i\alpha}\mathcal{L}(g)(z) \\
&= \frac{1}{2}\left(\frac{1}{z-i\alpha}+\frac{1}{z+i\alpha}\right)\mathcal{L}(g)(z) \\
&= \frac{z}{z^2+\alpha^2}\mathcal{L}(g)(z) \\
&= \mathcal{L}(\cos_\alpha(\cdot,0))(z)\mathcal{L}(g)(z) \\
&= \mathcal{L}(f)(z)\mathcal{L}(g)(z).
\end{aligned}$$

5. Let $f(x,y)=\sin_\alpha(x,y)$, $x,y\in\mathbb{T}_0$. Assume that $\alpha\in\mathbb{C}$ and $\mu\alpha^2\in\mathcal{R}$. Then $\sin_\alpha(x,y)$, $x,y\in\mathbb{T}_0$, is well defined. Note that

$$\begin{aligned}
f(x) &= f(x,0) \\
&= \sin_\alpha(x,0), \quad x\in\mathbb{T}_0.
\end{aligned}$$

Then

$$\sin_\alpha(x,y)=\frac{e_{i\alpha}(x,y)-e_{-i\alpha}(x,y)}{2i}, \quad x,y\in\mathbb{T}_0.$$

and

$$\begin{aligned}(f \star g)(x) &= \int_0^x \sin_\alpha(x, \sigma(y)) g(y) \Delta y \\ &= \frac{1}{2i} \int_0^x (e_{i\alpha}(x, \sigma(y)) - e_{-i\alpha}(x, \sigma(y))) g(y) \Delta y \\ &= \frac{1}{2i} \int_0^x e_{i\alpha}(x, \sigma(y)) g(y) \Delta y - \frac{1}{2} \int_0^x e_{-i\alpha}(x, \sigma(y)) g(y) \Delta y \\ &= \frac{1}{2i} (e_{i\alpha}(\cdot, 0) \star g)(x) - \frac{1}{2} (e_{-i\alpha}(\cdot, 0) \star g)(x), \quad x \in \mathbb{T}_0.\end{aligned}$$

Hence,

$$\begin{aligned}\mathcal{L}(f \star g)(z) &= \mathcal{L}\left(\frac{1}{2i}(e_{i\alpha}(\cdot, 0) \star g) - \frac{1}{2i}(e_{-i\alpha}(\cdot, 0) \star g)\right)(z) \\ &= \frac{1}{2i} \mathcal{L}(e_{i\alpha}(\cdot, 0) \star g)(z) - \frac{1}{2i} \mathcal{L}(e_{-i\alpha}(\cdot, 0) \star g)(z) \\ &= \frac{1}{2i} \mathcal{L}(e_{i\alpha}(\cdot, 0))(z) \mathcal{L}(g)(z) - \frac{1}{2i} \mathcal{L}(e_{-i\alpha}(\cdot, 0))(z) \mathcal{L}(g)(z) \\ &= \frac{1}{2i} \cdot \frac{1}{z - i\alpha} \mathcal{L}(g)(z) - \frac{1}{2i} \cdot \frac{1}{z + i\alpha} \mathcal{L}(g)(z) \\ &= \frac{1}{2i} \left(\frac{1}{z - i\alpha} - \frac{1}{z + i\alpha}\right) \mathcal{L}(g)(z) \\ &= \frac{\alpha}{z^2 + \alpha^2} \mathcal{L}(g)(z) \\ &= \mathcal{L}(\sin_\alpha(\cdot, 0))(z) \mathcal{L}(g)(z) \\ &= \mathcal{L}(f)(z) \mathcal{L}(g)(z).\end{aligned}$$

6. Let $f(x, y) = h_k(x, y), k \in \mathbb{N}_0, x, y \in \mathbb{T}_0$. Consider the initial value problem for the dynamic equation

$$\begin{aligned}l^{\Delta^{k+1}}(x) &= g(x), \quad x \in \mathbb{T}_0, \\ l^{\Delta^j}(0) &= 0, \quad j \in \{0, 1, \ldots, k\}.\end{aligned} \tag{2.16}$$

The solution of the equation (2.16) is given by

$$l(x) = \int_0^x h_k(x, \sigma(y)) g(y) \Delta y, \quad x \in \mathbb{T}_0.$$

or

$$l(x) = (h_k(x, \sigma(y)) \star g)(x).$$

Under suitable assumptions for $z \in \mathbb{C}$, taking the Laplace transform of both sides of (2.16), gives

$$\mathcal{L}\left(l^{\Delta^{k+1}}\right)(z) = \mathcal{L}(g)(z),$$

whereupon

$$z^{k+1}\mathcal{L}(l)(z) = \mathcal{L}(g)(z)$$

or

$$\begin{aligned} \mathcal{L}(l)(z) &= \frac{1}{z^{k+1}}\mathcal{L}(g)(z) \\ &= \mathcal{L}(h_k(\cdot, 0))(z)\mathcal{L}(g)(z) \\ &= \mathcal{L}(f)(z)\mathcal{L}(g)(z). \end{aligned}$$

This completes the proof.

□

Theorem 2.8. *Assume that f and g are each one of the functions $e_\alpha(\cdot, 0)$, $\cosh_\alpha(\cdot, 0)$, $\sinh_\alpha(\cdot, 0)$, $\cos_\alpha(\cdot, 0)$, $\sin_\alpha(\cdot, 0)$, or $h_k(\cdot, 0)$, not both $h_k(\cdot, 0)$, $k \in \mathbb{N}_0$. Then*

$$f \star g = g \star f.$$

Proof. 1. Let $f(x, y) = e_\alpha(x, y)$ and $g(x, y) = e_\beta(x, y)$, $x, y \in \mathbb{T}_0$. Assume that $\alpha, \beta \in \mathcal{R}$. Thus, $e_\alpha(x, y)$ and $e_\beta(x, y)$, $x, y \in \mathbb{T}_0$, are well defined. Let also,

$$\begin{aligned} p(x) &= (e_\alpha(\cdot, 0) \star e_\beta(\cdot, 0))(x), \\ q(x) &= (e_\beta(\cdot, 0) \star e_\alpha(\cdot, 0))(x), \quad x \in \mathbb{T}_0. \end{aligned}$$

Note that

$$p(0) = q(0) = 0.$$

Also, p and q are solutions to the initial value problems

$$\begin{aligned} p^{\Delta} - \alpha p &= e_{\beta}(x, 0), \quad x \in \mathbb{T}_0, \\ p(0) &= 0, \end{aligned}$$

and

$$\begin{aligned} q^{\Delta} - \beta q &= e_{\alpha}(x, 0), \quad x \in \mathbb{T}_0, \\ q(0) &= 0, \end{aligned}$$

respectively. Then

$$\begin{aligned} p^{\Delta}(0) &= \alpha p(0) + e_{\beta}(0, 0) \\ &= 1, \\ q^{\Delta}(0) &= \beta q(0) + e_{\alpha}(0, 0) \\ &= 1. \end{aligned}$$

We claim that p and q are solutions to the initial value problem

$$\begin{aligned} &m^{\Delta^2} - (\alpha + \beta) m^{\Delta} + \alpha\beta m = 0, \ x \in \mathbb{T}_0, \\ &m(0) = 0, \ m^{\Delta}(0) = 1. \end{aligned} \tag{2.17}$$

We have

$$p^{\Delta}(x) = \alpha p(x) + e_{\beta}(x, 0), \quad x \in \mathbb{T}_0,$$

from where, after differentiating, we obtain

$$p^{\Delta^2}(x) = \alpha p^{\Delta}(x) + \beta e_{\beta}(x, 0), \quad x \in \mathbb{T}_0.$$

Hence,

$$p^{\Delta^2}(x) - (\alpha + \beta) p^{\Delta}(x) + \alpha\beta p(x)$$

$$
\begin{aligned}
&= \alpha p^{\Delta}(x) + \beta e_{\beta}(x,0) - (\alpha+\beta)p^{\Delta}(x) + \alpha\beta p(x) \\
&= \beta e_{\beta}(x,0) - \beta p^{\Delta}(x) + \alpha\beta p(x) \\
&= \beta e_{\beta}(x,0) - \beta(\alpha p(x) + e_{\beta}(x,0)) + \alpha\beta p(x) \\
&= \beta e_{\beta}(x,0) - \alpha\beta p(x) - \beta e_{\beta}(x,0) + \alpha\beta p(x) \\
&= 0, \quad x \in \mathbb{T}_0.
\end{aligned}
$$

Also,

$$q^{\Delta}(x) = \beta q(x) + e_{\alpha}(x,0), \quad x \in \mathbb{T}_0,$$

whereupon, differentiating, we obtain

$$q^{\Delta^2}(x) = \beta q^{\Delta}(x) + \alpha e_{\alpha}(x,0), \quad x \in \mathbb{T}_0.$$

Hence,

$$q^{\Delta^2}(x) - (\alpha+\beta)q^{\Delta}(x) + \alpha\beta q(x)$$

$$
\begin{aligned}
&= \beta q^{\Delta}(x) + \alpha e_{\alpha}(x,0) - (\alpha+\beta)q^{\Delta}(x) + \alpha\beta q(x) \\
&= \alpha e_{\alpha}(x,0) - \alpha q^{\Delta}(x) + \alpha\beta q(x) \\
&= \alpha e_{\alpha}(x,0) - \alpha(\beta q(x) + e_{\alpha}(x,0)) + \alpha\beta q(x) \\
&= \alpha e_{\alpha}(x,0) - \alpha\beta q(x) - \alpha e_{\alpha}(x,0) + \alpha\beta q(x) \\
&= 0, \quad x \in \mathbb{T}_0.
\end{aligned}
$$

Since the problem (2.17) has unique solution, we conclude that

$$p(x) = q(x), \quad x \in \mathbb{T}_0.$$

2. Next, we consider

$$
\begin{aligned}
&(e_\alpha(\cdot,0)\star\cosh_\beta(\cdot,0))(x)\\
&= \left(e_\alpha(\cdot,0)\star\frac{e_\beta(\cdot,0)+e_{-\beta}(\cdot,0)}{2}\right)(x)\\
&= \frac{1}{2}((e_\alpha(\cdot,0)\star(e_\beta(\cdot,0))+e_{-\beta}(\cdot,0))(x))\\
&= \frac{1}{2}((e_\alpha(\cdot,0)\star e_\beta(\cdot,0))(x)+(e_\alpha(\cdot,0)\star e_{-\beta}(\cdot,0))(x))\\
&= \frac{1}{2}((e_\beta(\cdot,0)\star e_\alpha(\cdot,0))(x)+(e_{-\beta}(\cdot,0)\star e_\alpha(\cdot,0))(x))\\
&= \frac{1}{2}((e_\beta(\cdot,0)+e_{-\beta}(\cdot,0))\star e_\alpha(\cdot,0))(x)\\
&= (\cosh_\beta(\cdot,0)\star e_\alpha(\cdot,0))(x), \quad x\in\mathbb{T}_0.
\end{aligned}
$$

3. Let

$$
\begin{aligned}
z(x) &= (e_\alpha(\cdot,0)\star h_k(\cdot,0))(x),\\
q(x) &= (h_k(\cdot,0)\star e_\alpha(\cdot,0))(x), \quad x\in\mathbb{T}_0.
\end{aligned}
$$

We have that z is the solution to the initial value problem

$$
\begin{aligned}
z^\Delta-\alpha z &= h_k(x,0), \quad x\in\mathbb{T}_0,\\
z(0) &= 0.
\end{aligned}
$$

Differentiating this equation i times gives

$$
z^{\Delta^{i+1}}(x)-\alpha z^{\Delta^i}(x)=h_{k-i}(x,0), \quad x\in\mathbb{T}_0, \quad i\in\{1,2,\ldots,k\}.
$$

Also,

$$
\begin{aligned}
z^{\Delta^i}(0) &= 0, \quad i\in\{0,\ldots,k\},\\
z^{\Delta^{k+1}}(0) &= 1.
\end{aligned}
$$

Thus, we get that z is the solution to the initial value problem

$$\begin{aligned} z^{\Delta^{k+2}} - \alpha z^{\Delta^{k+1}} &= 0, \quad x \in \mathbb{T}_0, \\ z^{\Delta^i}(0) &= 0, \quad i \in \{0, \ldots, k\}, \\ z^{\Delta^{k+1}}(0) &= 1. \end{aligned}$$

Since $h_k(0,0) = 0$, $k \in \mathbb{N}$, we obtain that

$$q^{\Delta^i}(0) = 0, \quad i \in \{0, \ldots, k\}.$$

Also, q is the solution of the initial value problem

$$\begin{aligned} q^{\Delta^{k+1}} &= e_\alpha(x, 0), \quad x \in \mathbb{T}_0, \\ q^{\Delta^i}(0) &= 0, \quad i \in \{0, \ldots, k\}. \end{aligned}$$

Hence, differentiating the last equation, we obtain

$$\begin{aligned} q^{\Delta^{k+2}}(x) &= e_\alpha^\Delta(x, 0) \\ &= \alpha e_\alpha(x, 0) \\ &= \alpha q^{\Delta^{k+1}}(x), \quad x \in \mathbb{T}_0, \end{aligned}$$

and

$$\begin{aligned} q^{\Delta^{k+1}}(0) &= e_\alpha(0, 0) \\ &= 1. \end{aligned}$$

Consequently q is the solution of the initial value problem

$$\begin{aligned} q^{\Delta^{k+2}} - \alpha q^{\Delta^{k+1}} &= 0, \quad x \in \mathbb{T}_0, \\ q^{\Delta^i}(0) &= 0, \\ q^{\Delta^{k+1}}(0) &= 1, \quad i \in \{0, \ldots, k\}. \end{aligned}$$

Therefore z and q are solutions to the same initial value problem. Hence, they must coincide. The other combinations for f and g we leave to the reader as en exercise. This completes the proof.

□

Theorem 2.9. *Let $\alpha, \beta \in \mathcal{R}$. Then*

$$\mathcal{L}\left(e_\alpha(\cdot,0)\sin_{\frac{\beta}{1+\alpha\mu}}(\cdot,0)\right)(z) = \frac{\beta}{(z-\alpha)^2+\beta^2}$$

provided

$$\begin{aligned} \lim_{x\to\infty}\left(e_\alpha(x,0)\sin_{\frac{\beta}{1+\mu\alpha}}(x,0)\right) &= 0, \\ \lim_{x\to\infty}\left(e_\alpha(x,0)\left(\sin_{\frac{\beta}{1+\mu\alpha}}(x,0)\right)^\Delta\right) &= 0. \end{aligned}$$

Proof. Let

$$p(x) = e_\alpha(x,0)\sin_{\frac{\beta}{1+\mu\alpha}}(x,0), \quad x \in \mathbb{T}_0.$$

Then

$$p^\Delta(x) = e_\alpha^\Delta(x,0)\sin_{\frac{\beta}{1+\mu\alpha}}(x,0) + e_\alpha(\sigma(x),0)\sin^\Delta_{\frac{\beta}{1+\mu\alpha}}(x,0)$$

$$= \alpha e_\alpha(x,0)\sin_{\frac{\beta}{1+\mu\alpha}}(x,0) + (1+\alpha\mu(x))e_\alpha(x,0)\frac{\beta}{1+\alpha\mu(x)}\cos_{\frac{\beta}{1+\mu\alpha}}(x,0)$$

$$= \alpha e_\alpha(x,0)\sin_{\frac{\beta}{1+\mu\alpha}}(x,0) + \beta e_\alpha(x,0)\cos_{\frac{\beta}{1+\mu\alpha}}(x,0), \quad x \in \mathbb{T}_0.$$

We differentiate the last equation and we get

$$p^{\Delta^2}(x) = \alpha e_\alpha^\Delta(x,0)\sin_{\frac{\beta}{1+\mu\alpha}}(x,0) + \alpha e_\alpha(\sigma(x),0)\sin^\Delta_{\frac{\beta}{1+\mu\alpha}}(x,0)$$

$$+\beta e_\alpha^\Delta(x,0)\cos_{\frac{\beta}{1+\mu\alpha}}(x,0) + \beta e_\alpha(\sigma(x),0)\cos^\Delta_{\frac{\beta}{1+\mu\alpha}}(x,0)$$

$$= \alpha^2 e_\alpha(x,0)\sin_{\frac{\beta}{1+\mu\alpha}}(x,0) + \alpha(1+\alpha\mu(x))e_\alpha(x,0)\frac{\beta}{1+\alpha\mu(x)}\cos_{\frac{\beta}{1+\mu\alpha}}(x,0)$$

$$+\alpha\beta e_\alpha(x,0)\cos_{\frac{\beta}{1+\mu\alpha}}(x,0) - \beta(1+\alpha\mu(x))e_\alpha(x,0)\frac{\beta}{1+\alpha\mu(x)}\sin_{\frac{\beta}{1+\mu\alpha}}(x,0)$$

$$= \alpha^2 e_\alpha(x,0)\sin_{\frac{\beta}{1+\mu\alpha}}(x,0) + \alpha\beta e_\alpha(x,0)\cos_{\frac{\beta}{1+\mu\alpha}}(x,0)$$

$$+\alpha\beta e_\alpha(x,0)\cos_{\frac{\beta}{1+\mu\alpha}}(x,0) - \beta^2 e_\alpha(x,0)\sin_{\frac{\beta}{1+\mu\alpha}}(x,0)$$

$$= (\alpha^2-\beta^2)p(x) + 2\alpha\beta e_\alpha(x,0)\cos_{\frac{\beta}{1+\mu\alpha}}(x,0), \quad x \in \mathbb{T}_0.$$

In this way, we obtain the system

$$\begin{aligned} p^\Delta(x) &= \alpha p(x) + \beta e_\alpha(x,0)\cos_{\frac{\beta}{1+\mu\alpha}}(x,0) \\ p^{\Delta^2}(x) &= (\alpha^2-\beta^2)p(x) + 2\alpha\beta e_\alpha(x,0)\cos_{\frac{\beta}{1+\mu\alpha}}(x,0), \quad x \in \mathbb{T}_0. \end{aligned}$$

Hence,

$$p^{\Delta^2}(x) - 2\alpha p^\Delta(x) = -(\alpha^2+\beta^2)p(x), \quad x \in \mathbb{T}_0,$$

or

$$p^{\Delta^2}(x) + 2\alpha p^\Delta(x) + (\alpha^2+\beta^2)p(x) = 0, \quad x \in \mathbb{T}_0.$$

Also,

$$\begin{aligned} p(0) &= e_\alpha(0,0)\sin_{\frac{\beta}{1+\mu\alpha}}(0,0) \\ &= 0, \\ p^\Delta(0) &= \alpha e_\alpha(0,0)\sin_{\frac{\beta}{1+\mu\alpha}}(0,0) + \beta e_\alpha(0,0)\cos_{\frac{\beta}{1+\mu\alpha}}(0,0) \\ &= \beta. \end{aligned}$$

Consequently we obtain the following initial value problem

$$\begin{aligned} p^{\Delta^2}(x) - 2\alpha p^{\Delta}(x) + (\alpha^2+\beta^2)p(x) &= 0, \quad x \in \mathbb{T}_0, \\ p(0) &= 0, \\ p^{\Delta}(0) &= \beta. \end{aligned} \tag{2.18}$$

Now, we apply the Laplace transform to both sides of the dynamic equation (2.18) and we obtain

$$\mathcal{L}\left(p^{\Delta^2}(\cdot) - 2\alpha p^{\Delta}(\cdot) + (\alpha^2+\beta^2)p(\cdot)\right)(z) = 0$$

or

$$\mathcal{L}\left(p^{\Delta^2}(\cdot)\right)(z) - 2\alpha\mathcal{L}\left(p^{\Delta}(\cdot)\right)(z) + (\alpha^2+\beta^2)\mathcal{L}(p(\cdot))(z) = 0,$$

or

$$z^2\mathcal{L}(p)(z) - p^{\Delta}(0) - zp(0) - 2\alpha z\mathcal{L}(p)(z) + 2\alpha p(0) + (\alpha^2+\beta^2)\mathcal{L}(p)(z) = 0,$$

or

$$z^2\mathcal{L}(p)(z) - \beta - 2\alpha z\mathcal{L}(p)(z) + (\alpha^2+\beta^2)\mathcal{L}(p)(z) = 0,$$

or

$$\left(z^2 - 2\alpha z + \alpha^2 + \beta^2\right)\mathcal{L}(p)(z) = \beta,$$

or

$$\left((z-\alpha)^2 + \beta^2\right)\mathcal{L}(p)(z) = \beta,$$

whereupon

$$\mathcal{L}(p)(z) = \frac{\beta}{(z-\alpha)^2+\beta^2},$$

which completes the proof. □

Definition 2.4. *Let* $a \in \mathbb{T}, \quad a > 0$. *Define the step function* u_a *by*

$$u_a = \begin{cases} 0 & if \quad x \in \mathbb{T} \cap (-\infty, a) \\ 1 & if \quad x \in \mathbb{T} \cap [a, \infty). \end{cases}$$

Theorem 2.10. *Let* $a \in \mathbb{T}_0, \quad a > 0$. *Then*

$$\mathcal{L}(u_a(\cdot))(z) = \frac{e_{\ominus z}(a,0)}{z}$$

for those $z \in \mathcal{D}\{u_a\}$ *such that*

$$\lim_{x\to\infty} e_{\ominus z}(x,0) = 0.$$

Proof. We have

$$\begin{aligned}
\mathcal{L}(u_a(\cdot))(z) &= \int_0^\infty u_a(y)e_{\ominus z}(\sigma(y),0)\Delta y \\
&= \int_a^\infty e_{\ominus z}(\sigma(y),0)\Delta y \\
&= \int_a^\infty (1+\mu(y)(\ominus z)(y))e_{\ominus z}(y,0)\Delta y \\
&= \int_a^\infty \frac{1}{1+\mu(y)z}e_{\ominus z}(y,0)\Delta y \\
&= -\frac{1}{z}\int_a^\infty \frac{-z}{1+\mu(y)z}e_{\ominus z}(y,0)\Delta y \\
&= -\frac{1}{z}\int_a^\infty (\ominus z)(y)e_{\ominus z}(y,0)\Delta y \\
&= -\frac{1}{z}\int_a^\infty e^{\Delta}_{\ominus z}(y,0)\Delta y \\
&= -\frac{1}{z}e_{\ominus z}(y,0)\Big|_{y=a}^{y=\infty} \\
&= -\frac{1}{z}\left(\lim_{y\to\infty} e_{\ominus z}(y,0) - e_{\ominus z}(a,0)\right) \\
&= \frac{1}{z}e_{\ominus z}(a,0).
\end{aligned}$$

This completes the proof. □

Theorem 2.11. *Let* $a \in \mathbb{T}_0, \quad a > 0$. *Assume that* f *is one of the functions* $e_\alpha(\cdot,0)$, $\cos_\alpha(\cdot,0), \sin_\alpha(\cdot,a), \sinh_\alpha(\cdot,0), \cosh_\alpha(\cdot,0)$. *If* $1+z\mu(x)\neq 0, 1+\alpha\mu(x)\neq 0$, $x\in\mathbb{T}_0$, *and*

$$\lim_{x\to\infty} e_{\alpha\ominus z}(x,a) = \lim_{x\to\infty} e_{i\alpha\ominus z}(x,a) = \lim_{x\to\infty} e_{-i\alpha\ominus z}(x,a) = 0,$$

then

$$\mathcal{L}(u_a(\cdot)f(\cdot,a))(z) = e_{\ominus z}(a,0)\mathcal{L}(f(\cdot,a))(z).$$

Proof. 1. Let $f(x,a) = e_\alpha(x,a)$, $x \in \mathbb{T}_0$. Then

$$\begin{aligned} e_\alpha(x,a)e_{\ominus z}(\sigma(x),0) &= (1+\mu(x)(\ominus z)(x))e_\alpha(x,a)e_{\ominus z}(x,0) \\ &= \frac{1}{1+z\mu(x)}e_\alpha(x,a)e_{\ominus z}(x,0)\frac{e_{\ominus z}(0,a)}{e_{\ominus z}(0,a)} \\ &= \frac{1}{1+z\mu(x)}e_\alpha(x,a)e_{\ominus z}(x,a)\frac{1}{e_{\ominus z}(0,a)} \\ &= \frac{1}{1+z\mu(x)}e_{\alpha\ominus z}(x,a)e_{\ominus z}(a,0) \\ &= \frac{1}{\alpha-z}\cdot\frac{\alpha-z}{1+z\mu(x)}e_{\alpha\ominus z}(x,a)e_{\ominus z}(a,0) \\ &= \frac{1}{\alpha-z}(\alpha\ominus z)(x)e_{\alpha\ominus z}(x,a)e_{\ominus z}(a,0) \\ &= \frac{1}{\alpha-z}e^{\Delta}_{\alpha\ominus z}(x,a)e_{\ominus z}(a,0), \quad x \in \mathbb{T}_0. \end{aligned}$$

Thus,

$$\begin{aligned} \mathcal{L}(u_a(\cdot)f(\cdot,a))(z) &= \int_0^\infty u_a(x)f(x,a)e_{\ominus z}(\sigma(x),0)\Delta x \\ &= \int_a^\infty e_\alpha(x,a)e_{\ominus z}(\sigma(x),0)\Delta x \\ &= \frac{1}{\alpha-z}e_{\ominus z}(a,0)\int_a^\infty e^{\Delta}_{\alpha\ominus z}(x,a)\Delta x \\ &= \frac{1}{\alpha-z}e_{\ominus z}(a,0)e_{\alpha\ominus z}(x,a)\Big|_{x=a}^{x=\infty} \\ &= \frac{1}{\alpha-z}e_{\ominus z}(a,0)\left(\lim_{x\to\infty} e_{\alpha\ominus z}(x,a) - e_{\alpha\ominus z}(a,0)\right) \\ &= \frac{1}{z-\alpha}e_{\ominus z}(a,0) \\ &= e_{\ominus z}(a,0)\mathcal{L}(e_\alpha(\cdot,0))(z). \end{aligned}$$

2. Let $f(x,a) = \cos_\alpha(x,a), x \in \mathbb{T}_0$. Then

$$f(x,a) = \frac{1}{2}\left(e_{i\alpha}(x,a) + e_{-\imath\alpha}(x,a)\right), \quad x \in \mathbb{T}_0.$$

Hence,

$$\begin{aligned}
\mathcal{L}(u_a(\cdot)f(\cdot,a))(z) &= \mathcal{L}\left(u_a(\cdot,a)\frac{1}{2}(e_{i\alpha}(\cdot,a) + e_{-\imath\alpha}(\cdot,a))\right)(z) \\
&= \mathcal{L}\left(\frac{1}{2}u_a(\cdot)e_{i\alpha}(\cdot,a) + \frac{1}{2}u_a(\cdot)e_{-i\alpha}(\cdot,a)\right)(z) \\
&= \frac{1}{2}\mathcal{L}(u_a(\cdot)e_{i\alpha}(\cdot,a))(z) + \frac{1}{2}\mathcal{L}(u_a(\cdot)e_{-i\alpha}(\cdot,a))(z) \\
&= \frac{1}{2}e_{\ominus z}(a,0)\mathcal{L}(e_{i\alpha}(\cdot,0))(z) + \frac{1}{2}e_{\ominus z}(a,0)\mathcal{L}(e_{-i\alpha}(\cdot,0))(z) \\
&= e_{\ominus z}(a,0)\mathcal{L}\left(\frac{1}{2}e_{i\alpha}(\cdot,0)\right)(z) + e_{\ominus z}(a,0)\mathcal{L}\left(\frac{1}{2}e_{-i\alpha}(\cdot,0)\right)(z) \\
&= e_{\ominus z}(a,0)\mathcal{L}\left(\frac{1}{2}(e_{i\alpha}(\cdot,0) + e_{-i\alpha}(\cdot,0))\right)(z) \\
&= e_{\ominus z}(a,0)\mathcal{L}(\cos_\alpha(\cdot,0))(z).
\end{aligned}$$

The other possibilities for f we leave to the reader as exercise. This completes the proof.

□

Definition 2.5. *Let $a, b, \alpha \in \mathbb{T}$ and $f : \mathbb{T} \to \mathbb{R}$ be continuous. If $\delta_\alpha : [a,b] \to \mathbb{R}$, satisfies the following conditions*

$$\int_a^b f(x)\delta_\alpha(x)\Delta x = \begin{cases} f(\alpha) & if \quad \alpha \in [a,b) \\ 0 & otherwise, \end{cases}$$

then δ_α will be called Dirac delta function. If $\alpha = 0$, we set $\delta(\cdot) = \delta_0(\cdot)$.

Theorem 2.12. *Let $\alpha \in \mathbb{T}_0, \quad \alpha \geq 0$. Then*

$$\mathcal{L}(\delta_\alpha(\cdot))(z) = e^\sigma_{\ominus z}(\alpha, 0).$$

Proof. We have

$$\begin{aligned}\mathcal{L}(\delta_\alpha(\cdot))(z) &= \int_0^\infty \delta_\alpha(y) e_{\ominus z}(\sigma(y),0)\Delta y \\ &= e_{\ominus z}(\sigma(\alpha),0).\end{aligned}$$

This completes the proof. □

3. Decay of the Exponential Function

For $h > 0$, set

$$\begin{aligned}\mathbb{C}_h &= \left\{ z \in \mathbb{C} : z \neq -\frac{1}{h} \right\}, \\ \mathbb{Z}_h &= \left\{ z \in \mathbb{C} : -\frac{\pi}{h} < \text{Im}(z) \leq \frac{\pi}{h} \right\},\end{aligned}$$

and $\mathbb{C}_0 = \mathbb{C}$, $\mathbb{Z}_0 = \mathbb{Z}$. For $h > 0$, the Hilger real part and imaginary part of a complex number z are defined as follows.

$$\begin{aligned}\text{Re}_h(z) &= \frac{1}{h}\left(|1 + zh| - 1\right), \\ \text{Im}_h(z) &= \text{Arg}(1 + hz),\end{aligned}$$

where $\text{Arg}(\cdot)$ denotes the principal argument function, i.e., $\text{Arg} : \mathbb{C} \to (-\pi, \pi]$, and $\text{Re}_0(z) = \text{Re}(z)$, $\text{Im}_0(z) = \text{Im}(z)$. For any fixed complex number z and $h \in [0, \infty)$, define

$$u(h) = \text{Re}_h(z).$$

Proposition 3.1. *The function $u : [0, \infty) \to \mathbb{R}$ is a nondecreasing function.*

Proof. We have

$$\begin{aligned} u(h) &= \frac{|1+hz|-1}{h} \\ &\geq \frac{\mathrm{Re}(1+hz)-1}{h} \\ &= \frac{1+h\mathrm{Re}(z)-1}{h} \\ &= \mathrm{Re}(z) \\ &= u(0). \end{aligned}$$

Therefore

$$\mathrm{Re}_h(z) \geq \mathrm{Re}(z).$$

Let $h_2 > h_1 > 0$. Then

$$\begin{aligned} u(h_2) - u(h_1) &= \mathrm{Re}_{h_2}(z) - \mathrm{Re}_{h_1}(z) \\ &= \frac{|1+h_2 z|-1}{h_2} - \frac{|1+h_1 z|-1}{h_1} \\ &= \frac{|h_1 + h_1 h_2 z| - h_1 - |h_2 + h_1 h_2 z| + h_2}{h_1 h_2} \\ &\geq \frac{-|h_1 - h_2| + h_2 - h_1}{h_1 h_2} \\ &= 0. \end{aligned}$$

This completes the proof. □

The set of regressive functions and the set of positively regressive functions are denoted by $\mathcal{R}(\mathbb{T}, \mathbb{C})$ and $\mathcal{R}^+(\mathbb{T}, \mathbb{C})$, respectively, and $\mathcal{R}^-(\mathbb{T}, \mathbb{C})$ is defined similarly. For simplicity, we denote by $\mathcal{R}_c(\mathbb{T}, \mathbb{C})$ the set of complex regressive constants, and similarly, we define the sets $\mathcal{R}_c^+(\mathbb{T}, \mathbb{C})$ and $\mathcal{R}_c^-(\mathbb{T}, \mathbb{C})$. Let us set

$$m_\lambda(t,s) = \int_s^t \frac{1}{1+\mu(\tau)\lambda} \Delta\tau \quad \text{for} \quad s,t \in \mathbb{T} \quad \text{and} \quad \lambda \in \mathcal{R}_c(\mathbb{T}, \mathbb{R}).$$

Lemma 3.1. *Let $s \in \mathbb{T}$ and $\lambda \in \mathcal{R}_c^+([s,\infty),\mathbb{R})$. Then*

1. *$m_\lambda^\Delta(t,s) \geq 0$ for all $t \in (s,\infty)$, where the differentiation is with respect to the first argument.*

2. $\lim_{t\to\infty} m_\lambda(t,s) = \infty$.

Proof. 1. Since $\lambda \in \mathcal{R}_c^+([s,\infty),\mathbb{R})$, we have

$$\begin{aligned} m_\lambda^\Delta(t,s) &= \frac{1}{1+\lambda\mu(t)} \\ &> 0, \quad t \in [s,\infty). \end{aligned}$$

2. We will consider two cases.

(a) Let $\sup \mu(\mathbb{T}) < \infty$. Then

$$\begin{aligned} 1+\mu(t)\lambda &\leq 1+\mu(t)|\lambda| \\ &\leq 1+|\lambda|\sup\mu(\mathbb{T}), \quad t \in [s,\infty). \end{aligned}$$

Hence,

$$\begin{aligned} m_\lambda(t,s) &= \int_s^t \frac{\Delta\tau}{1+\lambda\mu(\tau)} \\ &\geq \int_s^t \frac{\Delta\tau}{1+|\lambda|\sup\mu(\mathbb{T})} \\ &= \frac{t-s}{1+|\lambda|\sup\mu(\mathbb{T})} \\ &\to \infty, \quad \text{as} \quad t\to\infty. \end{aligned}$$

(b) Suppose that $\sup \mu(\mathbb{T}) = \infty$. Then, using that $\lambda \in \mathcal{R}_c^+([s,\infty),\mathbb{R})$, we get that $\lambda \geq 0$. If $\lambda = 0$, then

$$\begin{aligned} m_\lambda(t,s) &= t-s \\ &\to \infty, \quad \text{as} \quad t\to\infty. \end{aligned}$$

Assume that $\lambda > 0$. Because $\mu(\mathbb{T}) = \infty$, there exists an increasing sequence $\{\xi_k\}_{k\in\mathbb{N}} \subset [s,\infty)$ such that $\{\mu(\xi_k)\}_{k\in\mathbb{N}}$ is an increasing and divergent sequence. Observe that

$$\lim_{k\to\infty} \frac{\mu(\xi_k)}{1+\lambda\mu(\xi_k)} = \frac{1}{\lambda}.$$

Hence,

$$\begin{aligned} m_\lambda(t,s) &= \int_s^t \frac{\Delta\tau}{1+\lambda\mu(\tau)} \\ &\geq \sum_{\sigma(\xi_k)\leq t, \xi_k\geq s} \int_{\xi_k}^{\sigma(\xi_k)} \frac{\Delta\tau}{1+\lambda\mu(\tau)} \\ &= \sum_{\sigma(\xi_k)\leq t, \xi_k\geq s} \frac{\mu(\xi_k)}{1+\lambda\mu(\xi_k)} \\ &\to \infty, \quad \text{as} \quad t\to\infty. \end{aligned}$$

This completes the proof.

□

For $h \geq 0$, define

$$\mathbb{R}_h = \mathbb{C}_h \bigcap \mathbb{R} = \left\{\lambda \in \mathbb{R} : \lambda \neq -\frac{1}{h}\right\}$$

and

$$\Psi_h(z,\lambda) = \begin{cases} \frac{1}{h}\log\left|\frac{1+h\lambda}{1+hz}\right| & \text{if} \quad h>0 \\ \\ \lambda - \mathrm{Re}(z) & \text{if} \quad h=0. \end{cases}$$

Lemma 3.2. *Let $h \geq 0$ and $\lambda \in \mathbb{R}_h$.*

1. *For all $z \in \mathbb{C}_h$, we have*

$$\Psi_h(z,\lambda) = \Psi_h\left(Re_h(z),\lambda\right).$$

2. *Suppose $1+h\lambda > 0$ and $x_1, x_2 \in \mathbb{R}$, $1+hx_1 > 0$, $1+hx_2 > 0$. If $x_1 < x_2$, then*

$$\Psi_h(x_1,\lambda) > \Psi_h(x_2,\lambda).$$

Proof. 1. If $h = 0$, then

$$\mathrm{Re}_0(z) = \mathrm{Re}(z)$$

and

$$\begin{aligned}\Psi_0(z,\lambda) &= \lambda - \mathrm{Re}(z)\\ &= \lambda - \mathrm{Re}_0(z)\\ &= \Psi_0(\mathrm{Re}_0(z),\lambda).\end{aligned}$$

Now, suppose that $h > 0$. Then

$$\begin{aligned}\Psi_h(\mathrm{Re}_h(z),\lambda) &= \frac{1}{h}\log\left|\frac{1+h\lambda}{1+h\mathrm{Re}_h(z)}\right|\\ &= \frac{1}{h}\log\left|\frac{1+h\lambda}{1+|1+zh|-1}\right|\\ &= \frac{1}{h}\log\left|\frac{1+h\lambda}{|1+zh|}\right|\\ &= \frac{1}{h}\log\left|\frac{1+h\lambda}{1+zh}\right|\\ &= \Psi_h(z,\lambda).\end{aligned}$$

2. If $h = 0$, then

$$\begin{aligned}\Psi_0(x_1,\lambda) &= \lambda - x_1\\ &> \lambda - x_2\\ &= \Psi_0(x_2,\lambda).\end{aligned}$$

Suppose that $h > 0$. Then

$$\begin{aligned}\Psi_h(x_1,\lambda) &= \frac{1}{h}\log\left|\frac{1+h\lambda}{1+hx_1}\right| \\ &= \frac{1}{h}\log\left(\frac{1+h\lambda}{1+hx_1}\right) \\ &> \frac{1}{h}\log\left(\frac{1+h\lambda}{1+hx_2}\right) \\ &= \Psi_h(x_2,\lambda).\end{aligned}$$

This completes the proof. □

Define $\mu_* : \mathbb{T} \to \mathbb{R}_0^+$ by

$$\mu_*(s) = \inf_{\tau\in[s,\infty)} \mu(\tau), \quad s \in \mathbb{T}.$$

For $h > 0$ and $\lambda \in \mathbb{R}$, define

$$\mathbb{C}_h(\lambda) = \{z \in \mathbb{C}_h : \mathrm{Re}_h(z) > \lambda\}\,.$$

Theorem 3.1 (Decay of the Exponential Function).
Let $s \in \mathbb{T}$ and $\lambda \in \mathcal{R}_c^+([s,\infty),\mathbb{R})$. Then, for any $z \in \mathbb{C}_{\mu_*(s)}(\lambda)$, we have the following properties.

1.

$$|e_{\lambda\ominus z}(t,s)| \leq e_{\lambda\ominus Re_{\mu_*(s)}(z)}(t,s), \quad t \in [s,\infty).$$

2.

$$\lim_{t\to\infty} e_{\lambda\ominus Re_{\mu_*(s)}(z)}(t,s) = 0.$$

3.

$$\lim_{t\to\infty} e_{\lambda\ominus z}(t,s) = 0.$$

Proof. 1. Set

$$\psi_h(z) = \begin{cases} \frac{1}{h}\log|1+hz| & \text{if} \quad h > 0 \\ \mathrm{Re}(z) \quad \text{if} \quad h = 0 \end{cases}$$

for $z \in \mathbb{C}_h$. Then

$$|e_f(t,s)| = \left|e^{\int_s^t \psi_{\mu(\tau)}(f(\tau))\Delta\tau}\right|, \quad t \in [s,\infty),$$

and

$$\begin{aligned} \psi_{\mu(\tau)}((\lambda \ominus z)(\tau)) &= \begin{cases} \frac{1}{\mu(\tau)} \log|1+\mu(\tau)((\lambda\ominus z)(\tau))| & \text{if} \quad \mu(\tau)>0 \\ \text{Re}(\lambda - z) \quad \text{if} \quad \mu(\tau)=0 \end{cases} \\ &= \begin{cases} \frac{1}{\mu(\tau)} \log\left|\frac{1+\lambda\mu(\tau)}{1+z\mu(\tau)}\right| & \text{if} \quad \mu(\tau)>0 \\ \lambda - \text{Re}(z) \quad \text{if} \quad \mu(\tau)=0, \end{cases} \end{aligned}$$

$\tau \in [s,\infty)$. Therefore

$$\psi_{\mu(\tau)}((\lambda\ominus z)(\tau)) = \Psi_{\mu(\tau)}(z,\lambda), \quad \tau \in [s,\infty).$$

Hence, for $\lambda \in \mathcal{R}_c^+([s,\infty),\mathbb{R})$ and $z \in \mathbb{C}_{\mu_*(s)}(\lambda)$, using Lemma 3.2, and $\text{Re}_{\mu_*(s)}(z) > \lambda$, we find

$$\begin{aligned} 1+\mu(\tau)\text{Re}_{\mu(\tau)}(z) &\geq 1+\mu(\tau)\text{Re}_{\mu_*(s)}(z) \\ &\geq 1+\lambda\mu(\tau) \\ &> 0, \end{aligned}$$

and

$$\begin{aligned} |e_{\lambda\ominus z}(t,s)| &= e^{\int_s^t \Psi_{\mu(\tau)}(\text{Re}_{\mu(\tau)}(z,\lambda))\Delta\tau} \\ &\leq e^{\int_s^t \Psi_{\mu(\tau)}(\text{Re}_{\mu_*(\tau)}(z,\lambda))\Delta\tau} \\ &= e_{\lambda\ominus\text{Re}_{\mu_*(\tau)}(z,\lambda)}(t,s) \end{aligned}$$

for any $t \in [s,\infty)$.

2. Observe that

$$\begin{aligned} \left(\lambda\ominus\text{Re}_{\mu_*(s)}(z)\right)(t) &= \frac{\lambda - \text{Re}_{\mu_*(s)}(z)}{1+\mu(t)\text{Re}_{\mu_*(s)}(z)} \\ &= \frac{\lambda - \text{Re}_{\mu_*(s)}(z)}{1+\lambda\mu(t)+\mu(t)\left(\text{Re}_{\mu_*(s)}(z)-\lambda\right)}, \quad t \in [s,\infty), \end{aligned}$$

and

$$
\begin{aligned}
1+\mu(t)\left(\lambda\ominus \mathrm{Re}_{\mu*(s)}(z)\right)(t) &= 1+\mu(t)\frac{\lambda-\mathrm{Re}_{\mu*(s)}(z)}{1+\lambda\mu(t)+\mu(t)(\mathrm{Re}_{\mu*(s)}(z)-\lambda)}\\
&= \frac{1+\lambda\mu(t)}{1+\lambda\mu(t)+\mu(t)(\mathrm{Re}_{\mu*(s)}(z)-\lambda)}, \quad t\in[s,\infty).
\end{aligned}
$$

Therefore

$$\lambda\ominus \mathrm{Re}_{\mu*(s)}(z)<0$$

and

$$\lambda\ominus \mathrm{Re}_{\mu*(s)}(z)\in \mathcal{R}_c^+([s,\infty),\mathbb{R}).$$

Now, applying Lemma 3.1, we have

$$
\begin{aligned}
e_{\lambda\ominus \mathrm{Re}_{\mu*(s)}(z)}(t,s) &\le e^{\int_s^t\left(\lambda\ominus \mathrm{Re}_{\mu*(s)}(z)\right)(\tau)\Delta\tau}\\
&= e^{-\left(\mathrm{Re}_{\mu*(s)}(z)-\lambda\right)\int_s^t\frac{\Delta\tau}{1+\mu(\tau)\mathrm{Re}_{\mu*(s)}(z)}}\\
&= e^{-\left(\mathrm{Re}_{\mu*(s)}(z)-\lambda\right)m_{\mathrm{Re}_{\mu*(s)}(z)}(t,s)}\\
&\to 0, \quad \text{as} \quad t\to\infty.
\end{aligned}
$$

3. By 1 and 2, we obtain

$$
\begin{aligned}
|e_{\lambda\ominus z}(t,s)| &\le e_{\lambda\ominus \mathrm{Re}_{\mu*(s)}(z)}(t,s)\\
&\to 0, \quad \text{as} \quad t\to\infty.
\end{aligned}
$$

This completes the proof.

□

Definition 3.1 (Exponential Order). *A function $f\in \mathcal{C}_{rd}(\mathbb{T})$ has exponential order α on $[s,\infty)$, if*

1. $\alpha\in \mathcal{R}_c^+([s,\infty),\mathbb{R})$,
2. *there exists a constant K such that*

$$|f(t)|\le Ke_\alpha(t,s), \quad t\in[s,\infty).$$

Example 3.1. *$e_\alpha(\cdot, s)$, $s \in \mathbb{T}$, $\alpha \in \mathcal{R}_c^+([s,\infty),\mathbb{R})$ is a function of exponential order α.*

Example 3.2. *Let $\epsilon > 0$ and $n \in \mathbb{N}_0$. Then, for all $s, t \in \mathbb{T}$, $t \geq s$, we have*

$$\begin{aligned} h_n(t,s) &= \frac{1}{\epsilon^n}\epsilon^n h_n(t,s) \\ &\leq \frac{1}{\epsilon^n}\sum_{k=0}^{\infty}\epsilon^k h_k(t,s) \\ &= \frac{1}{\epsilon^n}e_\epsilon(t,s). \end{aligned}$$

Thus, $h_n(\cdot, s)$ is a function of exponential order ϵ on $[s,\infty)$.

Lemma 3.3. *Let $s \in \mathbb{T}$ and $f \in \mathcal{C}_{rd}([s,\infty))$ be a function of exponential order α. Then*

$$\lim_{t\to\infty} f(t)e_{\ominus z}(t,s) = 0$$

for any $z \in \mathbb{C}_{\mu_*(s)}(\alpha)$.

Proof. Since f is of exponential order α, there exists a constant $K > 0$ such that

$$|f(t)| \leq K e_\alpha(t,s), \quad t \in [s,\infty).$$

Hence and Theorem 3.1, we get

$$\begin{aligned} |f(t)e_{\ominus z}(t,s)| &= |f(t)|\,|e_{\ominus z}(t,s)| \\ &\leq K e_\alpha(t,s) e_{\ominus \mathrm{Re}_{\mu_*(s)}(z)}(t,s) \\ &= K e_{\alpha \ominus \mathrm{Re}_{\mu_*(s)}(z)}(t,s) \\ &\to 0, \quad \text{as} \quad t \to \infty. \end{aligned}$$

This completes the proof. □

4. Convergence of the Laplace Transform

Theorem 4.1 (Absolute Convergence of the Laplace Transform). *Let $s \in \mathbb{T}$ and $f \in \mathcal{C}_{rd}([s,\infty))$ be of exponential order α. Then the Laplace transform $\mathcal{L}(f)(\cdot, s)$ exists on $\mathbb{C}_{\mu_*(s)}(\alpha)$ and converges absolutely.*

Proof. Since $f \in \mathcal{C}_{rd}([s,\infty))$ is of exponential order α on $[s,\infty)$, there exists a constant $K>0$ such that

$$|f(t)| \le K e_\alpha(t,s), \quad t \in [s,\infty).$$

Note that

$$\mathbf{Re}_{\mu(t)}(z) \ge \mathbf{Re}_{\mu_*(s)}(z), \quad t \in [s,\infty).$$

Thus,

$$\begin{aligned} |1+\mu(t)z| &= 1+\mu(t)\mathbf{Re}_{\mu(t)}(z) \\ &\ge 1+\mu(t)\mathbf{Re}_{\mu_*(s)}(z), \quad t \in [s,\infty). \end{aligned}$$

Hence, using Theorem 3.1, we obtain

$$\begin{aligned} \left|\int_s^t f(\tau)e_{\ominus z}(\sigma(\tau),s)\Delta\tau\right| &\le \int_s^t |f(\tau)e_{\ominus z}(\sigma(\tau),s)|\Delta\tau \\ &\le K\int_s^t |e_\alpha(\tau,s)e_{\ominus z}(\sigma(\tau),s)|\Delta\tau \\ &= K\int_s^t \frac{1}{|1+\mu(\tau)z|}|e_{\alpha\ominus z}(\tau,s)|\Delta\tau \\ &\le K\int_s^t \frac{e_{\alpha\ominus \mathbf{Re}_{\mu_*(s)}(z)}(\tau,s)}{1+\mu(\tau)\mathbf{Re}_{\mu_*(s)}(z)}\Delta\tau \\ &= \frac{K}{\alpha-\mathbf{Re}_{\mu_*(s)}(z)}\int_s^t \frac{\alpha-\mathbf{Re}_{\mu_*(s)}(z)}{1+\mu(\tau)\mathbf{Re}_{\mu_*(s)}(z)}e_{\alpha\ominus \mathbf{Re}_{\mu_*(s)}(z)}(\tau,s)\Delta\tau \\ &= \frac{K}{\alpha-\mathbf{Re}_{\mu_*(s)}(z)}\int_s^t (\alpha\ominus\mathbf{Re}_{\mu_*(s)}(z))e_{\alpha\ominus \mathbf{Re}_{\mu_*(s)}(z)}(\tau,s)\Delta\tau \\ &= \frac{K}{\mathbf{Re}_{\mu_*(s)}(z)-\alpha}\left(1-e_{\alpha\ominus \mathbf{Re}_{\mu_*(s)}(z)}(t,s)\right) \\ &\to \frac{K}{\mathbf{Re}_{\mu_*(s)}(z)-\alpha}, \quad \text{as} \quad t\to\infty, \quad z \in \mathbb{C}_{\mu_*(s)}(\alpha). \end{aligned}$$

Thus,

$$\begin{aligned} \left|\int_s^\infty f(\tau)e_{\ominus z}(\sigma(\tau),s)\Delta\tau\right| &\le \int_s^\infty |f(\tau)e_{\ominus z}(\sigma(\tau),s)|\Delta\tau \\ &\le \frac{K}{\mathbf{Re}_{\mu_*(s)}(z)-\alpha}, \quad z \in \mathbb{C}_{\mu_*(s)}(\alpha). \end{aligned} \tag{4.1}$$

Consequently the Laplace transform of the function f converges absolutely on $\mathbb{C}_{\mu_*(s)}(z)$. This completes the proof. □

Corollary 4.1. *Let $f \in \mathcal{C}_{rd}([s,\infty))$ be of some exponential order. Then, by* (4.1), *we have that*

$$\lim_{Re(z)\to\infty} \mathcal{L}(f)(z,s) = 0.$$

Therefore

$$\lim_{|z|\to\infty} \mathcal{L}(f)(z,s) = 0$$

in the case when $\mu_*(s) > 0$.

Theorem 4.2 (Uniform Convergence of the Laplace Transform). *Let $s \in \mathbb{T}$ and $f \in \mathcal{C}_{rd}([s,\infty))$ be of exponential order α. Then, the Laplace transform $\mathcal{L}(f)$ converges uniformly in the half plane $\mathbb{C}_{\mu_*(s)}(\beta)$, where $\beta > \alpha$.*

Proof. Since f is of exponential order α on $[s,\infty)$, there exists a constant $K > 0$ such that

$$|f(t)| \le K e_\alpha(t,s), \quad t \in [s,\infty).$$

As in the proof of Theorem 4.1, we have, for $t \ge s$,

$$\int_t^\infty |f(\tau) e_{\ominus z}(\sigma(\tau),s)|\Delta\tau \le \frac{K}{\mathrm{Re}_{\mu_*(s)}(z) - \alpha} e_{\alpha\ominus \mathrm{Re}_{\mu_*(s)}(z)}(t,s).$$

Thus,

$$\int_t^\infty |f(\tau) e_{\ominus z}(\sigma(\tau),s)|\Delta\tau \le \frac{K}{\beta - \alpha} e_{\alpha\ominus \mathrm{Re}_{\mu_*(s)}(z)}(t,s), \quad t \in [s,\infty). \tag{4.2}$$

Because

$$1 + \mu(\tau)\mathrm{Re}_{\mu_*(s)}(z) \ge 1 + \mu(\tau)\alpha, \quad \tau \in [s,t),$$

we have

$$m_{\mathrm{Re}_{\mu_*(s)}(z)}(t,s) \le m_\alpha(t,s), \quad z \in \mathbb{C}_{\mu_*(s)}(\alpha), \quad t \in [s,\infty).$$

Hence and Lemma 3.1, 2, we arrive at

$$\begin{aligned} e_{\alpha\ominus \mathrm{Re}_{\mu_*(s)}(z)}(t,s) &\le e^{\int_s^t \left(\alpha \ominus \mathrm{Re}_{\mu_*(s)}(z)\right)(\tau)\Delta\tau} \\ &= e^{-\left(\mathrm{Re}_{\mu_*(s)}(z) - \alpha\right)\int_s^t \frac{\Delta\tau}{1+\mu(\tau)\mathrm{Re}_{\mu_*(s)}(z)}} \\ &= e^{-\left(\mathrm{Re}_{\mu_*(s)}(z) - \alpha\right) m_{\mathrm{Re}_{\mu_*(s)}(z)}(t,s)} \\ &\to 0, \quad \text{as} \quad t \to \infty. \end{aligned}$$

Therefore, for any $\epsilon > 0$, there exists an $r \in [s, \infty)$ such that

$$\left| \int_t^\infty f(\tau) e_{\ominus z}(\sigma(\tau), s) \Delta\tau \right| \le \epsilon, \quad t \in [r, \infty), \quad z \in \mathbb{C}_{\mu_*(s)}(\beta).$$

This completes the proof. □

5. Inversion of the Laplace Transform

Theorem 5.1. *Let $s \in \mathbb{T}$ and $f \in C_{rd}([s, \infty))$ be of exponential order α. Consider $\mathcal{L}(f)(\cdot, s)$ on $\mathbb{C}_{\mu_*(s)}(\beta)$ for some $\beta > \alpha$ and suppose that it has finitely many regressive poles of finite order $\{z_1, z_2, \ldots, z_n\}$ and $\widetilde{F_{\mathbb{R}}}(z)$ is the Laplace transform of the function $\tilde{f}$ on $\mathbb{R}$ that corresponds to the transform $\mathcal{L}(f)(z, s)$ of f on $\mathbb{T}$. If*

$$\int_{c-i\infty}^{c+i\infty} \left| \widetilde{F_{\mathbb{R}}}(z) \right| |dz| < \infty,$$

then

$$f(t) = \sum_{i=1}^{n} Res_{z=z_i} e_z(t, s) \mathcal{L}(f)(z, s)$$

has Laplace transform $\mathcal{L}(f)(z, s)$ for all $z \in \mathbb{C}_{\mu_*(s)}(\beta) \bigcap \mathbb{C}_0(c)$.

Proof. Without loss of generality, suppose that $s = 0$. By Theorem 4.2, we have that $\mathcal{L}(f)(\cdot)$ converges uniformly on $\mathbb{C}_{\mu_*(s)}(\beta) \bigcap \mathbb{C}_0(c)$ and hence, it is analytic in this region. Next, by Corollary 4.1, we have that

$$\lim_{|z| \to \infty} \mathcal{L}(f)(z) = 0.$$

Let C be the collection of the Laplace transforms over $\mathbb{R}$, D be the collection of the Laplace transforms over $\mathbb{T}$, i.e., $C = \left\{ \widetilde{F_{\mathbb{R}}}(z) \right\}$, $D = \{\mathcal{L}(f)(z)\}$, where

$$\widetilde{F_{\mathbb{R}}}(z) = G(z) e^{-z\tau},$$

$$\mathcal{L}(f)(z) = G(z) e_{\ominus z}(\tau, 0)$$

for G a rational function and for $\tau \in \mathbb{T}$ a constant. Let $C_{p-c0}(\mathbb{R}, \mathbb{R})$ denotes the space of piecewise continuous functions of exponential order, $C_{prd-ez}(\mathbb{T}, \mathbb{R})$ denotes the space of piecewise right-dense continuous functions of double exponential type in which the exponential function coincides with a Hilger exponential function. Each of $\theta, \gamma, \theta^{-1}, \gamma^{-1}$ map functions involving the continuous exponential to the time scale exponential and vice versa. For example, γ maps the function

$$\widetilde{F}_{\mathbb{R}}(z) = \frac{e^{-za}}{z}$$

to the function

$$\mathcal{L}(f)_{(}z) = \frac{e_{\ominus z}(a, 0)}{z},$$

while γ^{-1} maps $\mathcal{L}(f)(z)$ back to $\widetilde{F}_{\mathbb{R}}(z)$ in the obvious manner. If the representation of $\mathcal{L}(f)(z)$ is independent of the exponential, that is $\tau = 0$, then γ and its inverse γ^{-1} will act as the identity. For example,

$$\gamma\left(\frac{1}{1+z^2}\right) = \gamma^{-1}\left(\frac{1}{1+z^2}\right) = \frac{1}{1+z^2}.$$

The map θ will set the continuous exponential function to the time scale exponential function in the following manner: if we write $\widetilde{f} \in C_{p-c0}(\mathbb{R}, \mathbb{R})$ as

$$\widetilde{f}(t) = \sum_{i=1}^{n} \text{Res}_{z=z_i} e^{zt} \widetilde{F}_{\mathbb{R}}(z),$$

then

$$\theta\left(\widetilde{f}(t)\right) = \sum_{i=1}^{n} \text{Res}_{z=z_i} e_z(t, 0) \mathcal{L}(f)(z).$$

To go from $\widetilde{F}_{\mathbb{R}}(z)$ to $\mathcal{L}(f)(z)$, we simply switch expressions involving the continuous exponential in $\widetilde{F}_{\mathbb{R}}$ with the time scale exponential giving $\mathcal{L}(f)(z)$ as was done for γ and its inverse θ^{-1} will then act on $g \in C_{prd-ez}(\mathbb{T}, \mathbb{R})$,

$$g(t) = \sum_{i=1}^{n} \text{Res}_{z=z_i} e_z(t, 0) G_{\mathbb{T}}(z)$$

as

$$\theta^{-1}(g(t)) = \sum_{i=1}^{n} \text{Res}_{z=z_i} e^{zt} \widetilde{G}_{\mathbb{R}}(z).$$

For example, for the unit step function $\widetilde{u}_a(t)$ on $\mathbb{R}$, we know from the continuous result that we may write the step function as

$$\widetilde{u}_a(t) = \text{Res}_{z=0} e^{zt} \frac{e^{-az}}{z},$$

so that if $a \in \mathbb{T}$, then

$$\theta\left(\widetilde{u}_a(t)\right) = \mathrm{Res}_{z=0} e_z(t,0)\frac{e_{\ominus z}(a,0)}{z}.$$

Now, for a given time scale Laplace transform $\mathcal{L}(f)(z)$, we begin by mapping to $\widetilde{F}_{\mathbb{R}}(z)$ via γ^{-1}. We have that the inverse of $\widetilde{F}_{\mathbb{R}}(z)$ exists for all z with $z \in \mathbb{C}_{\mu_*(s)}(\beta) \bigcap \mathbb{C}_0(c)$ and it is given by

$$\widetilde{f}(t) = \sum_{i=1}^{n} \mathrm{Res}_{z=z_i} e^{zt} \widetilde{F}_{\mathbb{R}}(z).$$

Applying θ to $\widetilde{f}(t)$ to retrieve the time scale function, we get

$$f(t) = \sum_{i=1}^{n} \mathrm{Res}_{z=z_i} e_z(t,0)\mathcal{L}(f)(z),$$

whereby

$$\left(\gamma \circ \widetilde{F}_{\mathbb{R}} \circ \theta^{-1}\right)(f(t)) = \mathcal{L}(f)(z).$$

This completes the proof. □

Theorem 5.2. *If the functions $f, g : \mathbb{T} \to \mathbb{R}$ satisfy Theorem 5.1 and have the same Laplace transform, then $f = g$ a.e.*

Proof. Let $s \in \mathbb{T}$. Suppose that

$$\int_s^\infty e^\sigma_{\ominus z}(t,s) f(t)\Delta t = \int_s^\infty e^\sigma_{\ominus z}(t,s) g(t)\Delta t.$$

Hence, $h = f - g$ has Laplace transform zero and $h \in \ker\mathcal{L}$. If we denote by $\mathcal{L}^{-1}$ the inversion of $\mathcal{L}$, then

$$\begin{aligned}\mathcal{L}^{-1} \circ \mathcal{L}(h) &= \mathcal{L}^{-1}(0)\\ &= 0\\ &= h.\end{aligned}$$

Therefore $f = g$ a.e. This completes the proof. □

6. The Laplace Transform of Power Series

Suppose that $s \in \mathbb{T}$.

Theorem 6.1. *Assume that there exist $M, \alpha > 0$ such that*

$$|a_k| \le M\alpha^k, \quad k \in \mathbb{N}_0.$$

Then, for any $z \in \mathbb{C}_{\mu_*(s)}(\alpha)$, we have

$$\begin{aligned} \mathcal{L}\left(\sum_{k=0}^{\infty} a_k h_k(\cdot, s)\right)(z, s) &= \sum_{k=0}^{\infty} a_k \mathcal{L}(h_k(\cdot, s))(z, s) \\ &= \sum_{k=0}^{\infty} \frac{a_k}{z^{k+1}}. \end{aligned}$$

Proof. Let

$$f(t) = \sum_{k=0}^{\infty} a_k h_k(t, s), \quad t \in [s, \infty),$$

which is a convergent series. Since $h_k(\cdot, s)$ is a continuous function on $[s, \infty)$, we have that f is a continuous function on $[s, \infty)$. Next,

$$\begin{aligned} |f(t)| &= \left|\sum_{k=0}^{\infty} a_k h_k(t, s)\right| \\ &\le \sum_{k=0}^{\infty} |a_k| h_k(t, s) \\ &\le M \sum_{k=0}^{\infty} \alpha^k h_k(t, s) \\ &= M e_\alpha(t, s), \quad t \in [s, \infty). \end{aligned}$$

Thus, f is of exponential order α. Observe that

$$|e_{\ominus z}(t, s)| \le e_{\ominus \mathrm{Re}_{\mu_*(s)}(z)}(t, s), \quad t \in [s, \infty), \quad z \in \mathbb{C}_{\mu_*(s)}(\alpha),$$

and

$$\begin{aligned}
\left| f(t) - \sum_{k=0}^{n-1} a_k h_k(t,s) \right| &= \left| \sum_{k=n}^{\infty} a_k h_k(t,s) \right| \\
&\le \sum_{k=n}^{\infty} |a_k| h_k(t,s) \\
&\le M \sum_{k=n}^{\infty} \alpha^k h_k(t,s) \\
&= M \left(\sum_{k=0}^{\infty} \alpha^k h_k(t,s) - \sum_{k=0}^{n-1} \alpha^k h_k(t,s) \right) \\
&= M \left(e_\alpha(t,s) - \sum_{k=0}^{n-1} \alpha^k h_k(t,s) \right), \quad t \in [s,\infty).
\end{aligned}$$

Hence,

$$\begin{aligned}
&\left| \mathcal{L}(f)(z,s) - \sum_{k=0}^{n-1} a_k \mathcal{L}(h_k(\cdot,s))(z,s) \right| \\
&= \left| \mathcal{L}\left(f - \sum_{k=0}^{n-1} a_k h_k(\cdot,s) \right)(z,s) \right| \\
&\le \mathcal{L}\left(\left| f - \sum_{k=0}^{n-1} a_k h_k(\cdot,s) \right| \right) (\mathrm{Re}_{\mu*(s)}(z), s) \\
&\le M\mathcal{L}\left(e_\alpha(\cdot,s) - \sum_{k=0}^{n-1} \alpha^k h_k(\cdot,s) \right) (\mathrm{Re}_{\mu*(s)}(z), s) \\
&= M \left(\frac{1}{\mathrm{Re}_{\mu*(s)}(z) - \alpha} - \frac{1}{\mathrm{Re}_{\mu*(s)}(z)} \sum_{k=0}^{n-1} \left(\frac{\alpha}{\mathrm{Re}_{\mu*(s)}(z)} \right)^k \right) \\
&= M \left(\frac{1}{\mathrm{Re}_{\mu*(s)}(z) - \alpha} - \frac{1}{\mathrm{Re}_{\mu*(s)}(z)} \frac{1 - \left(\frac{\alpha}{\mathrm{Re}_{\mu*(s)}(z)} \right)^n}{1 - \frac{\alpha}{\mathrm{Re}_{\mu*(s)}(z)}} \right) \\
&= \frac{M}{\mathrm{Re}_{\mu*(s)}(z) - \alpha} \left(\frac{\alpha}{\mathrm{Re}_{\mu*(s)}(z)} \right)^n \\
&\to 0, \quad \text{as} \quad n \to \infty, \quad z \in \mathbb{C}_{\mu*(s)}(\alpha).
\end{aligned}$$

This completes the proof. □

7. Differentiation of the Laplace Transform

Assume that there exist $a, b \in \mathbb{T}$, $b > a$, and a bounded domain $D \subset \mathbb{C}$ such that $s, t \in [a, b]$ and for $z \in D$ we have $z \in \mathcal{R}_c(\mathbb{T}, \mathbb{C})$. Then, for any $t, s \in [a, b]$, $t \geq s$, and for all $z \in D$, we have

$$\begin{aligned}
\frac{d}{dz} e_z(t, s) &= \frac{d}{dz} \sum_{k=0}^{\infty} h_k(t, s) z^k \\
&= \sum_{k=1}^{\infty} k h_k(t, s) z^{k-1} \\
&= \sum_{k=0}^{\infty} (k+1) h_{k+1}(t, s) z^k \\
&= \sum_{k=0}^{\infty} \left. \sum_{l=0}^{k} h_{k+1}(t, s) \right) z^k \\
&= \sum_{k=0}^{\infty} \left. \sum_{l=0}^{k} \int_s^t h_{k-l}(t, \sigma(\tau)) h_k(\tau, s) \Delta\tau \right) z^k \\
&= \int_s^t \sum_{k=0}^{\infty} \left. \sum_{l=0}^{k} h_{k-l}(t, \sigma(\tau)) h_l(\tau, s) \right) z^k \Delta\tau \\
&= \int_s^t \left. \sum_{k=0}^{\infty} h_k(t, \sigma(\tau)) z^k \right) \left(\sum_{k=0}^{\infty} h_k(\tau, s) z^k \right) \Delta\tau \\
&= \int_s^t e_z(t, \sigma(\tau)) e_z(\tau, s) \Delta\tau \\
&= \int_s^t \frac{e_z(t, \tau) e_z(\tau, s)}{1 + z\mu(\tau)} \Delta\tau \\
&= m_z(t, s) e_z(t, s).
\end{aligned}$$

Because $a, b \in \mathbb{T}$ and $D \subset \mathbb{C}$ was arbitrarily chosen, we conclude that

$$\frac{d}{dz} e_z(t, s) = m_z(t, s) e_z(t, s), \quad z \in \mathcal{R}_c(\mathbb{T}, \mathbb{C}), \quad t \in [s, \infty).$$

Now, we differentiate with respect to z the equation

$$e_z^{\Delta}(t,s) = z e_z(t,s), \quad e_z(s,s) = 1, \quad t \geq s,$$

and we get

$$\left(\frac{d}{dz} e_z(t,s)\right)^{\Delta} = e_z(t,s) + z\frac{d}{dz} e_z(t,s), \quad \frac{d}{dz} e_z(s,s) = 0, \quad t \geq s.$$

Hence, using the variation of parameters, we find

$$\begin{aligned} \frac{d}{dz} e_z(t,s) &= \int_s^t e_z(t,\sigma(\tau)) e_z(\tau,s)\Delta\tau \\ &= e_z(t,s) m_z(t,s), \quad t \in [s,\infty), \quad z \in \mathcal{R}_c(\mathbb{T},\mathbb{C}). \end{aligned}$$

Also,

$$\begin{aligned} \frac{d}{dz} e_{\ominus z}(t,s) &= \frac{d}{dz}\left(\frac{1}{e_z(t,s)}\right) \\ &= -\frac{m_z(t,s) e_z(t,s)}{(e_z(t,s))^2} \\ &= -\frac{m_z(t,s)}{e_z(t,s)} \\ &= -m_z(t,s) e_{\ominus z}(t,s), \quad t,s \in [a,b]. \end{aligned}$$

Theorem 7.1. *Let $s \in \mathbb{T}$ and $f \in \mathcal{C}_{rd}(\mathbb{T})$ be of exponential order α with a constant K. Then, for any $k \in \mathbb{N}$ and $z \in \mathbb{C}_{\mu_*(s)}(\alpha)$, $\frac{\mathrm{d}^k}{\mathrm{d}z^k}\mathcal{L}(f)(z,s)$ exists with*

$$\frac{\mathrm{d}^k}{\mathrm{d}z^k}\mathcal{L}(f)(z,s) = \mathcal{L}(f u_k(z,\sigma(\cdot),s))(z,s),$$

where

$$u_n(z,t,s) = \begin{cases} 1 & \text{if} \quad n = 0 \\ \frac{d}{dz} u_{n-1}(z,t,s) - u_{n-1}(z,t,s) m_z(t,s) & \text{if} \quad n \in \mathbb{N}, \end{cases}$$

$t \in [s,\infty)$.

Proof. We will prove that for $k \in \mathbb{N}$, the integrals

$$\int_s^\infty f(t) \frac{\mathrm{d}^k}{\mathrm{d}z^k} e_{\ominus z}(\sigma(t), s) \Delta t$$

are uniformly convergent with respect to $z \in \mathbb{C}_{\mu_*(s)}(\alpha)$. It suffices to prove that for $\ell, r \in \mathbb{N}_0$, the integrals

$$\int_s^\infty f(t) \left(m_z(\sigma(t), s)\right)^\ell \frac{\mathrm{d}^r}{\mathrm{d}z^r} \Big[m_z(\sigma(t), s)\Big] e_{\ominus z}(\sigma(t), s) \Delta t$$

are uniformly convergent with respect to $z \in \mathbb{C}_{\mu_*(s)}(\alpha)$. For $t \geq s$ and $r \in \mathbb{N}$, we compute

$$\frac{\mathrm{d}^r}{\mathrm{d}z^r} m_z(t, s) = (-1)^r r! \int_s^t \frac{(\mu(\tau))^r}{(1 + z\mu(\tau))^{r+1}} \Delta \tau.$$

Set $p_1(x) = \dfrac{x}{1 + \gamma x}$ for $x, \nu \in \mathbb{R}$ such that $x, 1 + \gamma x > 0$. Then $p_1'(x) = \dfrac{1}{(1 + \gamma x)^2} > 0$ for $x > 0$ and $1 + \gamma x > 0$. Since $\lim\limits_{x \to \infty} p_1(x) = \dfrac{1}{\gamma}$, there exists a constant $P_1 > 0$ such that $p_1(x) \leq P_1$ for $x > 0$ and $1 + \gamma x > 0$. Hence,

$$\begin{aligned}
\left| \frac{\mathrm{d}^r}{\mathrm{d}z^r} m_z(t, s) \right| &\leq r! \int_s^t \frac{(\mu(\tau))^r}{|1 + z\mu(\tau)|^{r+1}} \Delta \tau \\
&= r! \int_s^t \frac{(\mu(\tau))^r}{(1 + \mu(\tau) \mathrm{Re}_{\mu(\tau)}(z))^{r+1}} \Delta \tau \\
&\leq r! \int_s^t \frac{(\mu(\tau))^r}{(1 + \mu(\tau) \mathrm{Re}_{\mu_*(s)}(z))^r (1 + \mu(\tau) \mathrm{Re}_{\mu(\tau)}(z))} \Delta \tau \\
&\leq r! \int_s^t \frac{(\mu(\tau))^r}{(1 + \mu(\tau)\alpha)^r (1 + \mu(\tau) \mathrm{Re}_{\mu(\tau)}(z))} \Delta \tau \\
&\leq P_1^r r! \int_s^t \frac{1}{1 + \mu(\tau) \mathrm{Re}_{\mu(\tau)}(z)} \Delta \tau \\
&= P_1^r r! m_{\mathrm{Re}_\mu(z)}(t, s), \quad t \geq s,
\end{aligned}$$

and

$$\begin{aligned}
|m_z(t,s)| &= \left|\int_s^t \frac{1}{1+\mu(\tau)z}\Delta\tau\right| \\
&\leq \int_s^t \frac{1}{|1+\mu(\tau)z|}\Delta\tau \\
&= \int_s^t \frac{1}{1+\mu(\tau)\mathrm{Re}_{\mu(\tau)}(z)}\Delta\tau \\
&= m_{\mathrm{Re}_\mu(z)}(t,s), \quad t \geq s.
\end{aligned}$$

Therefore we have to estimate the integral

$$J_1 = \int_s^\infty f(t)(m_z(\sigma(t),s))^\ell e_{\ominus z}(\sigma(t),s)\Delta t.$$

Using the Pötzsche chain rule, we get

$$\begin{aligned}
&\left(\left(m_{\mathrm{Re}_\mu(z)}(t,s)\right)^\ell\right)^\Delta = \\
&= \ell\left(\int_0^1 \left(m_{\mathrm{Re}_\mu(z)}(t,s) + h\mu(t)m^\Delta_{\mathrm{Re}_\mu(z)}(t,s)\right)^{\ell-1} \mathrm{d}h\right) m^\Delta_{\mathrm{Re}_\mu(z)}(t,s) \\
&= \frac{\ell}{1+\mu(t)\mathrm{Re}_\mu(z)} \int_0^1 \left(m_{\mathrm{Re}_\mu(z)}(t,s) + h\mu(t)m^\Delta_{\mathrm{Re}_\mu(z)}(t,s)\right)^{\ell-1} \mathrm{d}h \\
&\leq \frac{\ell}{1+\mu(t)\mathrm{Re}_\mu(z)} \left(m_{\mathrm{Re}_\mu(z)}(t,s) + \mu(t)m^\Delta_{\mathrm{Re}_\mu(z)}(t,s)\right)^{\ell-1} \\
&= \frac{\ell}{1+\mu(t)\mathrm{Re}_\mu(z)} \left(m_{\mathrm{Re}_\mu(z)}(\sigma(t),s)\right)^{\ell-1}, \quad t \geq s.
\end{aligned}$$

Then, for $s_1 \geq s$, after multiple applications of integration by parts, we find

$$\left|\int_s^{s_1} f(t)(m_z(\sigma(t),s))^\ell e_{\ominus z}(\sigma(t),s)\Delta t\right|$$

$$\begin{aligned}
&\leq \int_s^{s_1} |f(t)e_{\ominus z}(\sigma(t),s)||m_z(\sigma(t),s)|^\ell \Delta t \\
&\leq K\int_s^{s_1} \frac{|e_{\alpha\ominus z}(t,s)|}{|1+\mu(t)z|}\left(m_{\mathrm{Re}_\mu(z)}(\sigma(t),s)\right)^\ell \Delta t \\
&\leq K\int_s^{s_1} \frac{e_{\alpha\ominus \mathrm{Re}_{\mu*(s)}(z)}}{1+\mu(t)\mathrm{Re}_{\mu*(s)}(z)}\left(m_{\mathrm{Re}_{\mu*(s)}(z)}(\sigma(t),s)\right)^\ell \Delta t \\
&= \frac{K}{\alpha-\mathrm{Re}_{\mu*(s)}(z)}\int_s^{s_1} e^\Delta_{\alpha\ominus \mathrm{Re}_{\mu*(s)}(z)}(t,s)\left(m_{\mathrm{Re}_{\mu*(s)}(z)}(\sigma(t),s)\right)^\ell \Delta t \\
&= \frac{K}{\alpha-\mathrm{Re}_{\mu*(s)}(z)}e_{\alpha\ominus \mathrm{Re}_{\mu*(s)}(z)}(s_1,s)\left(m_{\mathrm{Re}_{\mu*(s)}(z)}(s_1,s)\right)^\ell \\
&\quad+\frac{K}{\mathrm{Re}_{\mu*(s)}(z)-\alpha}\int_s^{s_1}\left(\left(m_{\mathrm{Re}_{\mu*(s)}(z)}(t,s)\right)^\ell\right)^\Delta e_{\alpha\ominus \mathrm{Re}_{\mu*(s)}(z)}(t,s)\Delta t \\
&\leq \frac{K}{\alpha-\mathrm{Re}_{\mu*(s)}(z)}e_{\alpha\ominus \mathrm{Re}_{\mu*(s)}(z)}(s_1,s)\left(m_{\mathrm{Re}_{\mu*(s)}(z)}(s_1,s)\right)^\ell \\
&\quad+\frac{K\ell}{\mathrm{Re}_{\mu*(s)}(z)-\alpha}\int_s^{s_1}\frac{\left(m_{\mathrm{Re}_{\mu*(s)}(z)}(\sigma(t),s)\right)^{\ell-1}}{1+\mu(t)\mathrm{Re}_{\mu*(s)}(z)}e_{\alpha\ominus \mathrm{Re}_{\mu*(s)}(z)}(t,s)\Delta t \\
&= \frac{K}{\alpha-\mathrm{Re}_{\mu*(s)}(z)}e_{\alpha\ominus \mathrm{Re}_{\mu*(s)}(z)}(s_1,s)\left(m_{\mathrm{Re}_{\mu*(s)}(z)}(s_1,s)\right)^\ell \\
&\quad+\frac{K\ell}{\left(\mathrm{Re}_{\mu*(s)}(z)-\alpha\right)\left(\alpha-\mathrm{Re}_{\mu*(s)}(z)\right)}\int_s^{s_1}\left(m_{\mathrm{Re}_{\mu*(s)}(z)}(\sigma(t),s)\right)^{\ell-1} \\
&\quad e^\Delta_{\alpha\ominus \mathrm{Re}_{\mu*(s)}(z)}(t,s)\Delta t \\
&= \frac{K}{\alpha-\mathrm{Re}_{\mu*(s)}(z)}e_{\alpha\ominus \mathrm{Re}_{\mu*(s)}(z)}(s_1,s)\left(m_{\mathrm{Re}_{\mu*(s)}(z)}(s_1,s)\right)^\ell \\
&\quad+\frac{K\ell}{\left(\mathrm{Re}_{\mu*(s)}(z)-\alpha\right)\left(\alpha-\mathrm{Re}_{\mu*(s)}(z)\right)}\left(m_{\mathrm{Re}_{\mu*(s)}(z)}(s_1,s)\right)^{\ell-1}e_{\alpha\ominus \mathrm{Re}_{\mu*(s)}(z)}(s_1,s) \\
&\quad+\frac{K\ell}{\left(\mathrm{Re}_{\mu*(s)}(z)-\alpha\right)^2}\int_s^{s_1}\left(\left(m_{\mathrm{Re}_{\mu*(s)}(z)}(t,s)\right)^{\ell-1}\right)^\Delta e_{\alpha\ominus \mathrm{Re}_{\mu*(s)}(z)}(t,s)\Delta t \\
&\leq \\
&\vdots \\
&\leq \frac{K}{\alpha-\mathrm{Re}_{\mu*(s)}(z)}e_{\alpha\ominus \mathrm{Re}_{\mu*(s)}(z)}(s_1,s)\left(m_{\mathrm{Re}_{\mu*(s)}(z)}(s_1,s)\right)^\ell
\end{aligned}$$

$$+\frac{K\ell}{\left(\mathrm{Re}_{\mu_*(s)}(z)-\alpha\right)\left(\alpha-\mathrm{Re}_{\mu_*(s)}(z)\right)}\quad m_{\mathrm{Re}_{\mu_*(s)}(z)}(s_1,s)\qquad e_{\alpha\ominus\mathrm{Re}_{\mu_*(s)}(z)}(s_1,s)$$

$$+\cdots-\frac{K\ell!}{\left(\mathrm{Re}_{\mu_*(s)}(z)-\alpha\right)^{\ell+1}}\int_s^{s_1}e^{\Delta}_{\alpha\ominus\mathrm{Re}_{\mu_*(s)}(z)}(t,s)\Delta t$$

$$\begin{aligned}
=&\quad\frac{K}{\alpha-\mathrm{Re}_{\mu_*(s)}(z)}e_{\alpha\ominus\mathrm{Re}_{\mu_*(s)}(z)}(s_1,s)\left(m_{\mathrm{Re}_{\mu_*(s)}(z)}(s_1,s)\right)^{\ell}\\
&+\frac{K\ell}{\left(\mathrm{Re}_{\mu_*(s)}(z)-\alpha\right)\left(\alpha-\mathrm{Re}_{\mu_*(s)}(z)\right)}\left(m_{\mathrm{Re}_{\mu_*(s)}(z)}(s_1,s)\right)^{\ell-1}e_{\alpha\ominus\mathrm{Re}_{\mu_*(s)}(z)}(s_1,s)\\
&+\cdots-\frac{K\ell!}{\left(\mathrm{Re}_{\mu_*(s)}(z)-\alpha\right)^{\ell+1}}\left(e_{\alpha\ominus\mathrm{Re}_{\mu_*(s)}(z)}(s_1,s)-1\right).
\end{aligned}$$

Therefore

$$|J_1|\le\frac{K\ell!}{\left(\mathrm{Re}_{\mu_*(s)}(z)-\alpha\right)^{\ell+1}}.$$

Hence,

$$\begin{aligned}
\frac{d}{dz}\mathcal{L}(f)(z,s)&=\frac{d}{dz}\int_s^{\infty}f(\tau)e_{\ominus z}(\sigma(\tau),s)\Delta\tau\\
&=\int_s^{\infty}f(\tau)\frac{d}{dz}e_{\ominus z}(\sigma(\tau),s)\Delta\tau\\
&=-\int_s^{\infty}f(\tau)m_z(\sigma(\tau,s)e_{\ominus z}(\sigma(\tau),s)\Delta\tau\\
&=-\mathcal{L}\left(m_z(\sigma(\cdot),s)f\right)(z,s)\\
&=\mathcal{L}(u_1(z,\sigma(\cdot),s))(z,s).
\end{aligned}$$

for all $z\in\mathbb{C}_{\mu_*(s)}(\alpha)$. Assume that

$$\frac{d^k}{dz^k}\mathcal{L}(f)(z,s)=\mathcal{L}(u_{k-1}(z,\sigma(\cdot),s)f)(z,s)$$

for all $z\in\mathbb{C}_{\mu_*(s)}(\alpha)$. We will prove that

$$\frac{d^{k+1}}{dz^{k+1}}\mathcal{L}(f)(z,s)=\mathcal{L}(u_k(z,\sigma(\cdot),s)f)(z,s)$$

for all $z\in\mathbb{C}_{\mu_*(s)}(\alpha)$. Really,

$$\frac{d^{k+1}}{dz^{k+1}}\mathcal{L}(f)(z,s)\quad=\quad\frac{d}{dz}\mathcal{L}(u_{k-1}(z,\sigma(\cdot),s)f)(z,s)$$

$$= \frac{d}{dz}\int_s^\infty u_{k-1}(z,\sigma(t),s)f(t)e_{\ominus z}(\sigma(t),s)\Delta t$$

$$= \int_s^\infty \left(\frac{d}{dz}u_{k-1}(z,\sigma(t),s) - m_z(\sigma(t),s)u_{k-1}(z,\sigma(t),s)\right) f(t)e_{\ominus z}(\sigma(t),s)\Delta t$$

$$= \int_s^\infty u_k(z,\sigma(t),s)f(t)e_{\ominus z}(\sigma(t),s)\Delta t$$

$$= \mathcal{L}(u_k(z,\sigma(\cdot),s)f)(z,s)$$

for all $z \in \mathbb{C}_{\mu_*(s)}(\alpha)$. This completes the proof. □

8. The Dirac Delta Functional

Definition 8.1. *The Hilger Dirac delta function is defined as follows.*

$$\delta_\tau^H(t) = \begin{cases} \frac{\delta[t-\tau]}{\mu(t)} & \text{if} \quad \mu(t) > 0 \\ \delta(t-\tau) & \text{otherwise,} \end{cases}$$

where $\delta[\cdot]$ denotes the Kronecker delta, i.e., $\delta[0] = 1$ and $\delta[t] = 0$ for $t \neq 0$, $\delta(\cdot)$ is the Dirac delta function.

Let $f : \mathbb{T} \to \mathbb{R}$ be a given function having unit area. With $\mathcal{C}_c^\infty(\mathbb{T})$ we denote the $\mathcal{C}^\infty(\mathbb{T})$ functions with compact support.

Definition 8.2. *For $g^\sigma \in \mathcal{C}_c^\infty(\mathbb{T})$ and for all $\epsilon > 0$, define the time scale Dirac delta functional as follows.*

$$\langle \delta_a(t), g^\sigma \rangle = \begin{cases} \int_0^\infty \delta_a^H(x) g^\rho(\sigma(x))\Delta x & \text{if} \quad a \quad \text{is} \quad \text{right-scattered} \\ \lim_{\epsilon\to 0}\int_0^\infty \frac{1}{\epsilon} f\left(\frac{x}{\epsilon}\right) g(\sigma(x))\Delta x & \text{if} \quad a \quad \text{is} \quad \text{right-dense.} \end{cases}$$

Below, we will list some of the properties of the Dirac delta functional.

1. Suppose that a is right-dense. Consider

$$f(x) = \begin{cases} \frac{1}{\epsilon} & \text{if} \quad a \le x \le a+\epsilon \\ 0 & \text{otherwise.} \end{cases}$$

Then

$$\begin{aligned}\langle \delta_a(x), g^\sigma \rangle &= \lim_{\epsilon\to 0}\int_0^\infty \frac{1}{\epsilon} f\left(\frac{x}{\epsilon}\right) g(\sigma(x))\Delta x\\ &= \lim_{\epsilon\to 0}\frac{1}{\epsilon}\int_a^{a+\epsilon} g(\sigma(x))\Delta x\\ &= g(\sigma(a))\\ &= g(a),\end{aligned}$$

where $g^\sigma \in \mathcal{C}_c^\infty(\mathbb{T})$.

2. Let a is right-scattered. Then

$$\begin{aligned}\langle \delta_a(x), g^\sigma \rangle &= \int_0^\infty \delta_a^H(x) g^\rho(\sigma(x))\Delta x\\ &= g^\rho(\sigma(a))\\ &= g(a),\end{aligned}$$

where $g^\sigma \in \mathcal{C}_c^\infty(\mathbb{T})$.

3. We have

$$\begin{aligned}\mathcal{L}(\delta)(z) &= \int_0^\infty e^\sigma_{\ominus z}(t,0)\delta(t)\Delta t\\ &= 1.\end{aligned}$$

4. We have

$$\begin{aligned}\mathcal{L}(\delta^\Delta)(z) &= \int_0^\infty \delta^\Delta(t) e^\sigma_{\ominus z}(t,0)\Delta t\\ &= \delta(t)e_{\ominus z}(t,0)\Big|_{t=0}^{t=\infty} - \int_0^\infty (\ominus z)(t) e_{\ominus z}(t,0)\delta(t)\Delta t\\ &= z\int_0^\infty e^\sigma_{\ominus z}(t,0)\delta(t)\Delta t + 1\\ &= z\mathcal{L}(\delta)(z) + 1\\ &= z+1.\end{aligned}$$

5. Let H be the Heaviside function. Then

$$\begin{aligned}\langle H^{\Delta}, g^{\sigma}\rangle &= \int_0^{\infty} H^{\Delta}(t) g^{\sigma}(t)\Delta t \\ &= H(t)g(t)\Big|_{t=0}^{t=\infty} - \int_0^{\infty} H(t) g^{\Delta}(t)\Delta t \\ &= -\int_0^{\infty} g^{\Delta}(t) \\ &= g(0) \\ &= \langle \delta, g^{\sigma}\rangle,\end{aligned}$$

where $g^{\sigma} \in \mathcal{C}_c^{\infty}(\mathbb{T})$.

9. The Bilateral Laplace Transform

Suppose that $\mathbb{T}$ is a time scale with forward jump operator and delta differentiation operator σ and Δ, respectively, and $\inf \mathbb{T} = -\infty$, $\sup \mathbb{T} = \infty$. For $s \in \mathbb{T}$, define

$$\begin{aligned}\mu_*(s) &= \inf_{t\in[s,\infty)} \mu(t), \\ \mu^*(s) &= \sup_{t\in(-\infty,s]} \mu(t), \\ \overline{\mu}(s) &= \inf_{t\in(-\infty,s]} \mu(t).\end{aligned}$$

For $s, t \in \mathbb{T}$ and $\lambda \in \mathcal{R}(\mathbb{T})$, set

$$M_{\lambda}(t,s) = \int_t^s \frac{1}{1+\lambda\mu(\tau)}\Delta\tau.$$

Lemma 9.1. *Let $s \in \mathbb{T}$, $\lambda \in \mathcal{R}^+((-\infty, s])$. Then*

1. *$M_{\lambda}^{\Delta}(t,s) < 0$ for all $t \in (-\infty, s)$, where the differentiation is with respect to t.*

2. $\lim_{t\to-\infty} M_{\lambda}(t,s) = \infty$.

Proof. 1. By the definition of the function M_λ, it follows

$$\begin{aligned} M_\lambda^\Delta(t,s) &= -\frac{1}{1+\lambda\mu(t)} \\ &< 0, \quad t \in (-\infty, s]. \end{aligned}$$

2. We will consider two cases.

(a) Let $\sup_{t\in(-\infty,s]} \mu(t) < \infty$. Then

$$\begin{aligned} 1+\lambda\mu(t) &\le 1+|\lambda|\mu(t) \\ &\le 1+|\lambda| \sup_{t\in(-\infty,s]} \mu(t), \quad t \in (-\infty, s]. \end{aligned}$$

Hence,

$$\begin{aligned} M_\lambda(t,s) &= \int_t^s \frac{\Delta\tau}{1+\lambda\mu(\tau)} \\ &\ge \int_t^s \frac{\Delta\tau}{1+|\lambda| \sup_{t\in(-\infty,s]} \mu(t)} \\ &= \frac{s-t}{1+|\lambda| \sup_{t\in(-\infty,s]} \mu(t)} \\ &\to \infty, \quad \text{as} \quad t \to -\infty. \end{aligned}$$

(b) Suppose that $\sup_{t\in(-\infty,s]} \mu(t) = \infty$. Because $\lambda \in \mathcal{R}^+((-\infty, s])$, we have that $\lambda \ge 0$. If $\lambda = 0$, then

$$\begin{aligned} M_\lambda(t,s) &= s-t \\ &\to \infty, \quad \text{as} \quad t \to -\infty. \end{aligned}$$

Assume that $\lambda > 0$. Since $\sup_{t\in(-\infty,s]} \mu(t) = \infty$, there exists a decreasing and divergent sequence $\{\xi_k\}_{k\in\mathbb{N}} \subset (-\infty, s]$ of right-scattered points such that $\{\mu(\xi_k)\}_{k\in\mathbb{N}}$ is an increasing and divergent sequence. Hence, using that

$$\lim_{k\to\infty} \frac{\mu(\xi_k)}{1+\lambda\mu(\xi_k)} = \frac{1}{\lambda},$$

we get

$$\begin{aligned} M_\lambda(t,s) &= \int_t^s \frac{\Delta\tau}{1+\lambda\mu(\tau)} \\ &\geq \sum_{\sigma(\xi_k)\geq t, \xi_k\leq s} \int_{\xi_k}^{\sigma(\xi_k)} \frac{\Delta\tau}{1+\lambda\mu(\tau)} \\ &= \sum_{\sigma(\xi_k)\geq t, \xi_k\leq s} \frac{\mu(\xi_k)}{1+\lambda\mu(\xi_k)} \\ &\to \infty, \quad \text{as} \quad t\to-\infty. \end{aligned}$$

This completes the proof.

□

Theorem 9.1 (Decay of the Exponential Function). *Let $s\in\mathbb{T}$, $\lambda\in\mathcal{R}^+((-\infty,s])$ and*

$$\overline{\mathbb{C}}_{\mu^*(s)}(\lambda) = \left\{z\in\mathbb{C} : Re_{\mu^*(s)}(z) < \lambda\right\}.$$

Then, for any $z\in\overline{\mathbb{C}}_{\mu^*(s)}(\lambda)$, we have the following properties.

1.

$$|e_{\lambda\ominus z}(t,s)| \leq e_{\lambda\ominus Re_{\mu^*(s)}(z)}(t,s), \quad t\in(-\infty,s],$$

2.

$$\lim_{t\to-\infty} e_{\lambda\ominus Re_{\mu^*(s)}(z)}(t,s) = 0,$$

3. $\lim_{t\to-\infty} e_{\lambda\ominus z}(t,s) = 0.$

Proof. Let $\Psi_h(z,\lambda)$ be the function, defined in Section 3.

1. We have

$$\begin{aligned} |e_{\lambda\ominus z}(t,s)| &= e^{\int_s^t \Psi_{\mu(\tau)}\left(\mathrm{Re}_{\mu(\tau)}(z),\lambda\right)\Delta\tau} \\ &\leq e^{\int_s^t \Psi_{\mu(\tau)}\left(\mathrm{Re}_{\mu^*(s)}(z),\lambda\right)\Delta\tau} \\ &= e_{\lambda\ominus \mathrm{Re}_{\mu^*(s)}(z)}(t,s), \quad t\in(-\infty,s]. \end{aligned}$$

2. For $z \in \overline{\mathbb{C}}_{\mu^*(s)}(\lambda)$, we have

$$\begin{aligned}\lambda \ominus \mathbf{Re}_{\mu^*(s)}(z) &= \frac{\lambda - \mathbf{Re}_{\mu^*(s)}(z)}{1+\mu(t)\mathbf{Re}_{\mu^*(s)}(z)}\\ &> \frac{\lambda - \mathbf{Re}_{\mu^*(s)}(z)}{1+\lambda\mu(t)}\\ &> 0, \quad t \in (-\infty, s],\end{aligned}$$

and

$$\begin{aligned}1+\mu(t)\left(\lambda \ominus \mathbf{Re}_{\mu^*(s)}(z)\right) &= \frac{1+\lambda\mu(t)}{1+\mu(t)\mathbf{Re}_{\mu^*(s)}(z)}\\ &> \frac{1+\lambda\mu(t)}{1+\lambda\mu(t)}\\ &= 1, \quad t \in (-\infty, s].\end{aligned}$$

Thus,

$$\lambda \ominus \mathbf{Re}_{\mu^*(s)}(z) \in \mathcal{R}^+((-\infty, s])$$

and

$$\begin{aligned}e_{\lambda\ominus \mathbf{Re}_{\mu^*(s)}(z)}(t,s) &= e^{\left(\lambda - \mathbf{Re}_{\mu^*(s)}(z)\right)\int_s^t \frac{1}{1+\mu(\tau)\mathbf{Re}_{\mu^*(s)}(z)}\Delta\tau}\\ &= e^{\left(\mathbf{Re}_{\mu^*(s)}(z)-\lambda\right)M_\lambda(t,s)}\\ &\to 0, \quad \text{as} \quad t \to -\infty.\end{aligned}$$

3. By 1 and 2, we obtain

$$\begin{aligned}|e_{\lambda\ominus z}(t,s)| &\le e_{\lambda\ominus \mathbf{Re}_{\mu^*(s)}(z)}(t,s)\\ &\to 0, \quad \text{as} \quad t \to -\infty.\end{aligned}$$

This completes the proof.

□

Definition 9.1. *Suppose that $s \in \mathbb{T}$, $f : \mathbb{T} \to \mathbb{R}$ is regulated. Then the bilateral Laplace transform of f is defined by*

$$\mathcal{L}^b(f)(z,s) = \int_{-\infty}^{\infty} f(t) e^{\sigma}_{\ominus z}(t,s)\Delta t$$

for $z \in \mathbb{C}$ for which $1 + \mu(t)z \neq 0$ for any $t \in \mathbb{T}^\kappa$ and the improper integral exists.

Definition 9.2 (Double Exponential Order)**.** *Let $s \in \mathbb{T}$. A function $f \in \mathcal{C}_{rd}(\mathbb{T})$ has double exponential order (α, γ) on $\mathbb{T}$ if*

1. *$\alpha \in \mathcal{R}^+([s,\infty))$, $\gamma \in \mathcal{R}^+((-\infty, s])$,*
2. *there exist positive constants K_α, K_γ such that*

$$\begin{aligned} |f(t)| &\leq K_\alpha e_\alpha(t,s), \quad t \in [s,\infty), \\ |f(t)| &\leq K_\gamma e_\gamma(t,s), \quad t \in (-\infty, s]. \end{aligned}$$

Example 9.1. *Let $\beta > 0$. Consider the function*

$$f(t) = e_\beta(t,s), \quad t \in \mathbb{T}.$$

If $s \leq t$, then $\alpha = \beta$. If $s > t$, then $0 \leq f(t) \leq 1$ and $e_{\ominus\beta}(t,s) > 1$. Thus,

$$f(t) \leq e_{\ominus\beta}(t,s).$$

Therefore f is of double exponential order $(\beta, \ominus\beta)$.

Lemma 9.2. *Let $s \in \mathbb{T}$ and $f \in \mathcal{C}_{rd}(\mathbb{T})$ be a function of double exponential order (α, γ). Then*

1. *$\lim_{t\to\infty} f(t)e_{\ominus z}(t,s) = 0$ for any $z \in \mathbb{C}_{\mu_*(s)}(\alpha)$.*
2. *$\lim_{t\to-\infty} f(t)e_{\ominus z}(t,s) = 0$ for any $z \in \overline{\mathbb{C}}_{\mu^*(s)}(\gamma)$.*

Proof. Since f is of double exponential order (α,γ), there exist positive constants K_α and K_γ such that

$$\begin{aligned} |f(t)| &\leq K_\alpha e_\alpha(t,s), \quad t \in [s,\infty), \\ |f(t)| &\leq K_\gamma e_\gamma(t,s), \quad t \in (-\infty, s]. \end{aligned}$$

1. The first statement follows by Lemma 3.3.

2. By Theorem 9.1, we have

$$\begin{aligned} |f(t)e_{\ominus z}(t,s)| &\leq K_\gamma \left|e_{\gamma\ominus z}(t,s)\right| \\ &\leq K_\gamma e_{\gamma\ominus \mathrm{Re}_{\mu^*(s)}(z)}(t,s) \\ &\to 0, \quad \text{as} \quad t\to-\infty. \end{aligned}$$

This completes the proof.

□

For $z \in \mathbb{C}$, denote

$$\overline{\overline{\mu}}(s) = \mu^*(s) \quad \text{if} \quad \mathrm{Re}_{\overline{\mu}(s)}(z) \leq 0$$

and

$$\overline{\overline{\mu}}(s) = \overline{\mu}(s) \quad \text{if} \quad \mathrm{Re}_{\overline{\mu}(s)}(z) \geq 0.$$

Definition 9.3. *Let $s \in \mathbb{T}$, $\alpha \in \mathcal{R}^+([s,\infty))$, $\gamma \in \mathcal{R}^+((-\infty,s])$. We say that (s,α,γ) is an admissible triple if*

$$\begin{aligned} \mathbb{C}_{s,\alpha,\gamma} &= \Big\{ z\in\mathbb{C} : Re_{\mu^*(s)}(z) < \gamma, \quad Re_{\mu_*(s)}(z) > \alpha, \\ &\qquad 1+\overline{\overline{\mu}}(s)Re_{\overline{\overline{\mu}}(s)}(z) \neq 0 \Big\}. \end{aligned}$$

Theorem 9.2 (Absolute Convergence of the Bilateral Laplace Transform). *Let (s,α,γ) be an admissible triple, $f \in C_{rd}(\mathbb{T})$ be of double exponential order (α,γ). Then $\mathcal{L}^b(f)(\cdot,s)$ exists on $\mathbb{C}_{s,\alpha,\gamma}$ and converges absolutely.*

Proof. Since f is of double exponential order (α,γ), there exist positive constants K_α, K_γ such that

$$\begin{aligned} |f(t)| &\leq K_\alpha e_\alpha(t,s), \quad t\in[s,\infty), \\ |f(t)| &\leq K_\gamma e_\gamma(t,s), \quad t\in(-\infty,s]. \end{aligned}$$

Next, for $t \leq s$, we have

$$\begin{aligned} |1+\mu(t)z| &= 1+\mu(t)\mathrm{Re}_{\mu(t)}(z) \\ &\geq 1+\overline{\overline{\mu}}(s)\mathrm{Re}_{\overline{\overline{\mu}}(s)}(z). \end{aligned}$$

For $z \in \mathbb{C}_{s,\alpha,\gamma}$, we find

$$\begin{aligned} \left|\mathcal{L}^b(f)(z,s)\right| &= \left|\int_s^\infty f(\tau)e_{\ominus z}(\sigma(\tau),s)\Delta\tau + \int_{-\infty}^s f(\tau)e_{\ominus z}(\sigma(\tau),s)\Delta\tau\right| \\ &\leq \int_s^\infty |f(\tau)e_{\ominus z}(\sigma(\tau),s)|\,\Delta\tau + \int_{-\infty}^s |f(\tau)e_{\ominus z}(\sigma(\tau),s)|\,\Delta\tau \\ &= J_1+J_2. \end{aligned}$$

By the proof of Theorem 4.1, we have

$$J_1 \leq \frac{K}{\mathrm{Re}_{\mu_*(s)}(z)-\alpha}, \quad z \in \mathbb{C}_{s,\alpha,\gamma}.$$

Consider the integral

$$J_2 = \int_t^s |f(\tau)e_{\ominus z}(\sigma(\tau),s)|\,\Delta\tau, \quad t \leq s.$$

Using Theorem 9.1, 1, 2, we obtain

$$\begin{aligned} J_2 &\leq K_\gamma \int_t^s \frac{|e_{\gamma\ominus z}(\tau,s)|}{|1+\mu(\tau)z|}\Delta\tau \\ &= K_\gamma \int_t^s \frac{|e_{\gamma\ominus z}(\tau,s)|}{1+\mu(\tau)\mathrm{Re}_{\mu(\tau)}(z)}\Delta\tau \\ &\leq K_\gamma \int_t^s \frac{e_{\gamma\ominus \mathrm{Re}_{\mu(\tau)}(z)}(\tau,s)}{1+\mu(\tau)\mathrm{Re}_{\mu^*(s)}(z)}\Delta\tau \\ &= K_\gamma \int_t^s \frac{1}{\gamma-\mathrm{Re}_{\mu^*(s)}(z)} e^{\Delta}_{\gamma\ominus \mathrm{Re}_{\mu^*(s)}(z)}(\tau,s)\Delta\tau \\ &\leq \frac{K_\gamma}{\gamma-\mathrm{Re}_{\mu^*(s)}(z)}\left(1-e_{\gamma\ominus \mathrm{Re}_{\mu^*(s)}(z)}(t,s)\right), \quad t \leq s. \end{aligned}$$

Therefore

$$J_2 \leq \frac{K_\gamma}{\gamma-\mathrm{Re}_{\mu^*(s)}(z)}, \quad t \leq s,$$

and

$$\left|\mathcal{L}^b(f)(z,s)\right| \le \frac{K_\alpha}{\mathrm{Re}_{\mu_*(s)}(z)-\alpha} + \frac{K_\gamma}{\gamma - \mathrm{Re}_{\mu^*(s)}(z)}.$$

Consequently the bilateral Laplace transform of the function f converges absolutely on $\mathbb{C}_{s,\alpha,\gamma}$. This completes the proof. □

Corollary 9.1. *Let (s,α,γ) be an admissible triple, $f \in \mathcal{C}_{rd}(\mathbb{T})$ be of double exponential order (α,γ). Then*

$$\lim_{|z|\to\infty} \mathcal{L}^b(f)(z,s) = 0.$$

Proof. Note that $|z| \to \infty$ implies $\mathrm{Re}_{\mu_*(s)}(z) \to \infty$, $\mathrm{Re}_{\mu^*(s)}(z) \to \infty$ and $\mathrm{Re}_{\mu(s)}(z) \to \overline{\infty}$. This completes the proof. □

Theorem 9.3 (Uniform Convergence of the Bilateral Laplace Transform). *Let (s,α,γ)be an admissible triple, $f \in \mathcal{C}_{rd}(\mathbb{T})$ be of double exponential order (α,γ). Then $\mathcal{L}^b(f)(\cdot,s)$ converges uniformly in $\mathbb{C}_{s,\alpha,\gamma}$.*

Proof. Let $\epsilon > 0$ be arbitrarily chosen. By the proof of Theorem 4.2, we have that there exists an $r \in [s,\infty)$ such that

$$\left|\int_t^\infty f(\tau)e_{\ominus z}(\sigma(\tau),s)\Delta\tau\right| \le \epsilon, \quad t \in [r,\infty), \quad z \in \mathbb{C}_{s,\alpha,\gamma}.$$

Let $t \in (-\infty, s]$ and $a \le t$. By the proof of Theorem 9.2, we get

$$\begin{aligned}\left|\int_a^t f(\tau)e_{\ominus z}(\sigma(\tau),s)\Delta\tau\right| &\le \frac{K_\gamma}{\gamma - \mathrm{Re}_{\mu^*(s)}(z)}\left(e_{\gamma\ominus \mathrm{Re}_{\mu^*(s)}(z)}(t,s) - e_{\gamma\ominus \mathrm{Re}_{\mu^*(s)}(z)}(a,s)\right)\\ &\le \frac{K_\gamma}{\gamma - \mathrm{R}_{\mu^*(s)}(z)} e_{\gamma\ominus \mathrm{Re}_{\mu^*(s)}(z)}(t,s), \quad t \le s.\end{aligned}$$

Thus, there exists an $r_1 \in (-\infty, s]$, such that

$$\left|\int_{-\infty}^t f(\tau)e_{\ominus z}(\sigma(\tau),s)\Delta\tau\right| \le \epsilon, \quad t \in (-\infty, r_1], \quad z \in \mathbb{C}_{s,\alpha,\gamma}.$$

This completes the proof. □

Theorem 9.4. *Let (s,α,γ) be an admissible triple and $f \in \mathcal{C}_{rd}([s,\infty))$ be of double exponential order (α,γ). Consider $\mathcal{L}^b(f)(\cdot,s)$ on $\mathbb{C}_{s,\alpha,\gamma}$ and suppose that it has finitely many regressive poles of finite order $\{z_1, z_2, \ldots, z_n\}$ and $\widetilde{F}^b_{\mathbb{R}}(z)$ is* *the bilateral Laplace transform of the function $\widetilde{f}$ on $\mathbb{R}$ that corresponds to the transform $\mathcal{L}^b(f)(z,s)$ of f on $\mathbb{T}$. If*

$$\int_{c-i\infty}^{c+i\infty} \left|\widetilde{F}_{\mathbb{R}}^b(z)\right| |dz| < \infty,$$

then

$$f(t) = \sum_{i=1}^{n} Res_{z=z_i} e_z(t,s)\mathcal{L}^b(f)(z,s)$$

has bilateral Laplace transform $\mathcal{L}^b(f)(z,s)$ for all $z \in \mathbb{C}_{s,\alpha,\gamma} \bigcap \mathbb{C}_0(c)$.

Theorem 9.5. *If the functions $f, g : \mathbb{T} \to \mathbb{R}$ satisfy Theorem 9.4 and have the same bilateral Laplace transform, then $f = g$ a.e.*

The proofs of Theorem 9.4 and Theorem 9.5 repeat the proof of Theorem 5.1 and Theorem 5.2 and therefore we omit them.

10. The Fourier Transform

Suppose that $\mathbb{T}$ is a time scale with forward jump operator and delta differentiation operator σ and Δ, respectively. Let also, $\inf \mathbb{T} = -\infty$, $\sup \mathbb{T} = \infty$ and $s \in \mathbb{T}$. Denote

$$\begin{aligned} \mu_*(s) &= \inf_{t\in[s,\infty)} \mu(t), \quad \mu^*(s) = \sup_{t\in(-\infty,s]} \mu(t), \quad \overline{\mu}(s) = \inf_{t\in(-\infty,s]} \mu(t), \\ \overline{\overline{\mu}}(s) &= \mu^*(s) \quad \text{if} \quad \mathrm{Re}_{\overline{\mu}(s)}(x) \le 0, \\ \overline{\overline{\mu}}(s) &= \overline{\mu}(s) \quad \text{if} \quad \mathrm{Re}_{\overline{\mu}(s)}(x) \ge 0, \end{aligned}$$

where $x \in \mathbb{R}$.

Definition 10.1. *Suppose that $f : \mathbb{T} \to \mathbb{R}$ is regulated. Then the Fourier transform of the function f is defined by*

$$\mathcal{F}(f)(x,s) = \int_{-\infty}^{\infty} f(t) e_{\ominus ix}^{\sigma}(t,s)\Delta t$$

for $x \in \mathbb{R}$ for which $1 + ix\mu(t) \neq 0$ for any $t \in \mathbb{T}^\kappa$ and the improper integral exists.

Definition 10.2. *Let* $\alpha \in \mathcal{R}^+([s,\infty))$, $\gamma \in \mathcal{R}^+((-\infty,s])$. *We say that* (s,α,γ) *is a real admissible triple if*

$$R_{s,\alpha,\gamma} = \Big\{ x \in \mathbb{R} : Re_{\mu^*(s)}(ix) < \gamma, \quad Re_{\mu_*(s)}(ix) > \alpha, \\ 1 + \overline{\overline{\mu}}(s) Re_{\overline{\mu}(s)}(ix) \neq 0 \Big\} \neq \emptyset.$$

If $f \in \mathcal{C}_{rd}(\mathbb{T})$, then the triple (s,α,γ) is a real admissible triple and f is of double exponential order (α,γ), then by Theorem 9.3 and Theorem 9.4, it follows that $\mathcal{F}(f)(\cdot,s)$ exists on $R_{s,\alpha,\gamma}$ and converges absolutely and uniformly on $R_{s,\alpha,\gamma}$. Below we will list some of the properties of the Fourier transform.

Theorem 10.1. *Let* $f, g : \mathbb{T} \to \mathbb{R}$, $\alpha, \beta \in \mathbb{C}$. *Then*

$$\mathcal{F}(\alpha f + \beta g)(x,s) = \alpha \mathcal{F}(f)(x,s) + \beta \mathcal{F}(g)(x,s)$$

for those $x \in \mathbb{R}$ *for which* $1 + x\mu(t) \neq 0$, $t \in \mathbb{T}^\kappa$, *and the respective integrals exist.*

Proof. We have

$$\begin{aligned} \mathcal{F}(\alpha f + \beta g)(x,s) &= \int_{-\infty}^{\infty} (\alpha f + \beta g)(t) e^{\sigma}_{\ominus ix}(t,s)\Delta t \\ &= \alpha \int_{-\infty}^{\infty} f(t) e^{\sigma}_{\ominus ix}(t,s)\Delta t + \beta \int_{-\infty}^{\infty} g(t) e^{\sigma}_{\ominus ix}(t,s)\Delta t \\ &= \alpha \mathcal{F}(f)(x,s) + \beta \mathcal{F}(g)(x,s). \end{aligned}$$

This completes the proof. □

Theorem 10.2. *Let* $f : \mathbb{T} \to \mathbb{R}$. *Then*

$$\mathcal{F}\left(e^{\sigma}_{y}(\cdot,s) f(\cdot)\right)(x,s) = \mathcal{F}(f)(z,s),$$

where $z = \frac{x+iy}{1+\mu y}$, *for those* $x, y \in \mathbb{R}$ *for which* $1 + \mu(t) y \neq 0$ *for any* $t \in \mathbb{T}^\kappa$ *and the respective integrals exist.*

Proof. Note that

$$\begin{aligned} iz &= i\frac{x+iy}{1+\mu y} \\ &= -\frac{x+iy}{i(1+\mu y)} \end{aligned}$$

and

$$\begin{aligned}
\ominus iz &= \frac{-iz}{1+i\mu z} \\
&= \frac{\frac{x+iy}{i(1+\mu y)}}{1-\mu\frac{x+iy}{i(1+\mu y)}} \\
&= \frac{x+iy}{i+i\mu y-\mu x-i\mu y} \\
&= \frac{x+iy}{i-\mu x} \\
&= \frac{i(x+iy)}{-1-i\mu x} \\
&= \frac{y-ix}{1+i\mu x} \\
&= y\ominus ix.
\end{aligned}$$

Hence,

$$\begin{aligned}
\mathcal{F}\left(e_y^\sigma(\cdot,s)f(\cdot)\right)(x,s) &= \int_{-\infty}^{\infty} e_{\ominus ix}^\sigma(t,s)e_y^\sigma(t,s)f(t)\Delta t \\
&= \int_{-\infty}^{\infty} e_{y\ominus ix}^\sigma(t,s)f(t)\Delta t \\
&= \mathcal{F}(f)(z,s).
\end{aligned}$$

This completes the proof. □

Theorem 10.3. *Let $f:\mathbb{T}\to\mathbb{R}$. For any $k\in\mathbb{N}$, we have*

$$\mathcal{F}\left(f^{\Delta^k}\right)(x,s)=(ix)^k\mathcal{F}(f)(x,s)$$

for those $x\in\mathbb{R}$ for which $1+x\mu(t)\neq 0$, $t\in\mathbb{T}^\kappa$, and the respective integrals exist and

$$\lim_{t\to\pm\infty} f^{\Delta^l}(t)e_{\ominus ix}(t,s)=0,\quad l\in\{0,\ldots,k-1\}.$$

Proof. We will use the principle of mathematical induction.

1. For $k = 1$, we have

$$\begin{aligned}
\mathcal{F}\left(f^{\Delta}\right)(x,s) &= \int_{-\infty}^{\infty} f^{\Delta}(t)e^{\sigma}_{\ominus ix}(t,s)\Delta t \\
&= \lim_{t\to\infty} f(t)e_{\ominus ix}(t,s) - \lim_{t\to-\infty} f(t)e_{\ominus ix}f(t)e_{\ominus ix}(t,s) \\
&\quad - \int_{-\infty}^{\infty} (\ominus ix)f(t)e_{\ominus ix}(t,s)\Delta t \\
&= ix\int_{-\infty}^{\infty} f(t)e^{\sigma}_{\ominus ix}(t,s)\Delta t \\
&= ix\mathcal{F}(f)(x,s).
\end{aligned}$$

2. Assume that

$$\mathcal{F}\left(f^{\Delta^k}\right)(x,s) = (ix)^k\mathcal{F}(f)(x,s)$$

for some $k \in \mathbb{N}$.

3. We will prove that

$$\mathcal{F}\left(f^{\Delta^{k+1}}\right)(x,s) = (ix)^{k+1}\mathcal{F}(f)(x,s).$$

Really, we have

$$\begin{aligned}
\mathcal{F}\left(f^{\Delta^{k+1}}\right)(x,s) &= ix\mathcal{F}\left(f^{\Delta^k}\right)(x,s) \\
&= (ix)^{k+1}\mathcal{F}(f)(x,s).
\end{aligned}$$

This completes the proof.

□

Theorem 10.4. *Let $f : \mathbb{T} \to \mathbb{R}$. Then*

$$\overline{\mathcal{F}(f)(x,s)} = \mathcal{F}(f)(-x,s)$$

for those $x \in \mathbb{R}$ for which $1 \pm x\mu(t) \neq 0$, $t \in \mathbb{T}^{\kappa}$, and the respective integrals exist.

Proof. Let $t \in \mathbb{T}^\kappa$ and $x \in \mathbb{R}$ be such that $1 \pm x\mu(t) \neq 0$. Then, we have

$$\begin{aligned}
(\ominus(ix))(t) &= -\frac{ix}{1+i\mu(t)x} \\
&= -\frac{ix(1-i\mu(t)x)}{(1+i\mu(t)x)(1-i\mu(t)x)} \\
&= -\frac{ix+\mu(t)x^2}{1+(\mu(t))^2x^2}, \\
1+\mu(t)(\ominus(ix))(t) &= 1-\mu(t)\frac{ix+\mu(t)x^2}{1+(\mu(t))^2x^2} \\
&= \frac{1+(\mu(t))^2x^2-i\mu(t)x-(\mu(t))^2x^2}{1+(\mu(t))^2x^2} \\
&= \frac{1-i\mu(t)x}{1+(\mu(t))^2x^2} \\
&= \frac{1}{\sqrt{1+(\mu(t))^2x^2}} \quad \frac{1}{\sqrt{1+(\mu(t))^2x^2}} - i\frac{\mu(t)x}{\sqrt{1+(\mu(t))^2x^2}}\Bigg)
\end{aligned}$$

and

$$(\ominus(i(-x)))(t) = \frac{1}{\sqrt{1+(\mu(t))^2x^2}} \quad \frac{1}{\sqrt{1+(\mu(t))^2x^2}} + i\frac{\mu(t)x}{\sqrt{1+(\mu(t))^2x^2}}\Bigg).$$

Let

$$r(t) = \frac{1}{\sqrt{1+(\mu(t))^2x^2}}, \quad \cos\theta(t) = \frac{1}{\sqrt{1+(\mu(t))^2x^2}}.$$

Then

$$1+\mu(t)(\ominus(ix))(t) = r(t)e^{-i\theta(t)}$$

and

$$1+\mu(t)(\ominus(i(-x)))(t) = r(t)e^{i\theta(t)}.$$

Also,

$$\begin{aligned}
\log(1+\mu(t)(\ominus(ix))(t)) &= \ln(r(t)) - i(\theta+2k\pi), \\
\log(1+\mu(t)(\ominus(i(-x)))(t)) &= \ln(r(t)) + i(\theta+2k\pi), \quad k \in \mathbb{Z}.
\end{aligned}$$

From here and from the definition of the Fourier transform, we have

$$
\begin{aligned}
\mathcal{F}(f)(x,s) &= \int_{-\infty}^{\infty} f(t) e^{\sigma}_{\ominus(ix)}(t,s)\Delta t \\
&= \int_{-\infty}^{\infty} e^{\int_s^{\sigma(t)} \frac{1}{\mu(\tau)} \log(1+\mu(\tau)(\ominus(ix))(\tau))\Delta\tau} f(\tau)\Delta\tau \\
&= \int_{-\infty}^{\infty} e^{\int_s^{\sigma(t)} \frac{1}{\mu(\tau)} (\ln r(\tau) - i(\theta(\tau)+2k\pi))\Delta\tau} f(t)\Delta t
\end{aligned}
$$

and

$$
\begin{aligned}
\overline{\mathcal{F}(f)(x,s)} &= \overline{\int_{-\infty}^{\infty} e^{\int_s^{\sigma(t)} \frac{1}{\mu(\tau)} (\ln r(\tau) - i(\theta(\tau)+2k\pi))\Delta\tau} f(t)\Delta t} \\
&= \int_{-\infty}^{\infty} \overline{e^{\int_s^{\sigma(t)} \frac{1}{\mu(\tau)} (\ln r(\tau) - i(\theta(\tau)+2k\pi))\Delta\tau} f(t)}\Delta t \\
&= \int_{-\infty}^{\infty} \overline{e^{\int_s^{\sigma(t)} \frac{1}{\mu(\tau)} (\ln r(\tau) - i(\theta(\tau)+2k\pi))\Delta\tau}} f(t)\Delta t \\
&= \int_{-\infty}^{\infty} e^{\overline{\int_s^{\sigma(t)} \frac{1}{\mu(\tau)} (\ln r(\tau) - i(\theta(\tau)+2k\pi))\Delta\tau}} f(t)\Delta t \\
&= \int_{-\infty}^{\infty} e^{\int_s^{\sigma(t)} \overline{\frac{1}{\mu(\tau)} (\ln r(\tau) - i(\theta(\tau)+2k\pi))}\Delta\tau} f(t)\Delta t \\
&= \int_{-\infty}^{\infty} e^{\int_s^{\sigma(t)} \frac{1}{\mu(\tau)} (\ln r(\tau) + i(\theta(\tau)+2k\pi))\Delta\tau} f(t)\Delta t \\
&= \int_{-\infty}^{\infty} e^{\int_s^{\sigma(t)} \frac{1}{\mu(\tau)} \log(1+\mu(\tau)(\ominus(i(-x)))(\tau))\Delta\tau} f(\tau)\Delta\tau \\
&= \int_{-\infty}^{\infty} f(t) e^{\sigma}_{\ominus(i(-x))}(t,s)\Delta t \\
&= \mathcal{F}(f)(-x,s).
\end{aligned}
$$

This completes the proof. □

Theorem 10.5. *Let $a \in \mathbb{T}$. Then*

$$\mathcal{F}(\delta_a^H)(x,s) = e^{\sigma}_{\ominus ix}(a,s)$$

for those $x \in \mathbb{R}$ for which $1 + x\mu(t) \neq 0$, $t \in \mathbb{T}^{\kappa}$.

Proof. We have

$$\begin{aligned}\mathcal{F}(\delta_a^H)(x,s) &= \int_{-\infty}^{\infty} \delta_a(t-a)e^{\sigma}_{\ominus ix}(t,s)\Delta t\\ &= e^{\sigma}_{\ominus ix}(a,s)\end{aligned}$$

for those $x \in \mathbb{R}$ such that $1 + x\mu(t) \neq 0$, $t \in \mathbb{T}^\kappa$. □

Remark 10.1. *1. If $\mathbb{T} = \mathbb{R}$, then we take $s = a$ and we get*

$$\mathcal{F}(\delta_a^H)(x,s) = 1.$$

2. If $a \in \mathbb{T}$ is right-scattered and $s = a$, then

$$\begin{aligned}e^{\sigma}_{\ominus ix}(a,a) &= e^{\log \frac{1}{1+i\mu(a)x}}\\ &= \frac{1}{1+i\mu(a)x}\end{aligned}$$

for those $x \in \mathbb{R}$ for which $1 + i\mu(a)x \neq 0$.

Theorem 10.6. *Let $f : \mathbb{T} \to \mathbb{R}$ be regulated and*

$$F(t) = \int_a^t f(\tau)\Delta\tau, \quad t \in \mathbb{T},$$

for some fixed $a \in \mathbb{T}$. Then

$$\mathcal{F}(F)(x,s) = -\frac{i}{x}\mathcal{F}(f)(x,s)$$

for those $x \in \mathbb{R}$, $x \neq 0$, for which

$$\lim_{t\to\pm\infty} F(t)e_{\ominus ix}(t,s) = 0.$$

Proof. We have

$$\begin{aligned}
\mathcal{F}(F)(x,s) &= \int_{-\infty}^{\infty} F(t)e^{\sigma}_{\ominus ix}(t,s)\Delta t \\
&= \int_{-\infty}^{\infty} F(t)\left(1+\mu(t)(\ominus(ix))(t)\right)e_{\ominus ix}(t,s)\Delta t \\
&= \int_{-\infty}^{\infty} F(t)\frac{1}{1+i\mu(t)x}e_{\ominus ix}(t,s)\Delta t \\
&= -\frac{1}{ix}\int_{-\infty}^{\infty} F(t)\frac{-ix}{1+i\mu(t)x}e_{\ominus ix}(t,s)\Delta t \\
&= \frac{i}{x}\int_{-\infty}^{\infty} F(t)(\ominus ix)(t)e_{\ominus ix}(t,s)\Delta t \\
&= \frac{i}{x}\int_{-\infty}^{\infty} F(t)e^{\Delta}_{\ominus ix}(t,s)\Delta t \\
&= \frac{i}{x}\left(\lim_{t\to\infty} F(t)e_{\ominus ix}(t,s) - \lim_{t\to-\infty} F(t)e_{\ominus ix}(t,s)\right) \\
&\quad -\frac{i}{x}\int_{-\infty}^{\infty} f(t)e^{\sigma}_{\ominus ix}(t,s)\Delta t \\
&= -\frac{i}{x}\mathcal{F}(f)(x,s)
\end{aligned}$$

for those $x \in \mathbb{R}$, $x \neq 0$, for which

$$\lim_{t\to\pm\infty} F(t)e_{\ominus ix}(t,s) = 0.$$

This completes the proof. □

11. Generalized Shifts

Suppose that $t_0 \in \mathbb{T}$, $f \in \mathcal{C}^1_{rds}(\mathbb{T})$, $\sup\mathbb{T} = \infty$ and $\inf\mathbb{T} = -\infty$.

Definition 11.1 (Generalized Shift(Generalized Delay) of a Function). *The solution of the generalized shifting problem*

$$\begin{aligned}
u^{\Delta_t}(t,\sigma(s)) &= -u^{\Delta_s}(t,s), \quad t,s \in \mathbb{T}, \\
u(t,t_0) &= f(t), \quad t \in \mathbb{T},
\end{aligned} \tag{11.1}$$

is denoted by $\widetilde{f}$ *and it is called the generalized shift or generalized delay of* f.

We will investigate the problem (11.1) for existence of solutions. Suppose the following.

(H1) $f \in \mathcal{C}^1_{rd}(\mathbb{T})$ and $|f(t)| \leq A$, $t \in \mathbb{T}$, for some positive constant A.

(H2) The function $g \in \mathcal{C}_{rd}(\mathbb{T} \times \mathbb{T})$, $g \neq 0$ on $\mathbb{T} \times \mathbb{T}$, except finite many points $\{(t^1, s^1), \ldots, (t^k, s^k)\}$ different than $(t, t_0) \in \mathbb{T} \times \mathbb{T}$, satisfies

$$\begin{aligned}\left|\int_{t_0}^{t}\int_{t_0}^{s} |g(t_1, s_1)| \left(|s_1 - t_0| + |t_1 - t_0|\right) \Delta s_1 \Delta t_1\right| &\leq B,\\ \left|\int_{t_0}^{t} |g(t_1, s)| \left(|s - t_0| + |t_1 - t_0|\right) \Delta t_1\right| &\leq B,\\ \left|\int_{t_0}^{s} |g(t, s_1)| \left(|s_1 - t_0| + |t - t_0|\right) \Delta s_1\right| &\leq B,\end{aligned}$$

$s, t \in \mathbb{T}$, where B is a positive constant, $B < \frac{1}{2}$.

(H3) There exists a positive constant ϵ such that

$$\epsilon(1 + 2B) \leq 1.$$

Example 11.1. *There exist a function* $g \in \mathcal{C}^1_{rd}(\mathbb{T}\times\mathbb{T})$, $g \neq 0$ *on* $\mathbb{T}\times\mathbb{T}$, *except finite many points* $\{(t^1, s^1), \ldots, (t^k, s^k)\}$ *different than* $(t, t_0) \in \mathbb{T} \times \mathbb{T}$, *and constants* ϵ, A *and* B *that satisfy* $(H1)$-$(H3)$. *Really, let*

$$A = 1, \quad B = \frac{1}{4}, \quad \epsilon = \frac{1}{200}$$

and

$$g(t, s) = \frac{(\sigma(s) + s)(\sigma(t) + t)}{100(1 + s^2)(1 + (\sigma(s))^2)(1 + t^2)(1 + (\sigma(t))^2)(1 + (s - t_0)^2)(1 + (t - t_0)^2)},$$

$t, s \in \mathbb{T}$, *and*

$$h(t) = \frac{1}{1 + t^2}, \quad t \in \mathbb{T}.$$

Since $\sigma(s) + s \geq 0$, $\sigma(t) + t \geq 0$, $s, t \in \mathbb{T}$, *we have that* $g(t, s) = 0$ *if* $0 \in \mathbb{T}$ *is right-dense and* $s = 0$ *or* $t = 0$. *Then*

$$h^{\Delta}(t) = -\frac{\sigma(t) + t}{(1 + t^2)(1 + (\sigma(t))^2)}, \quad t \in \mathbb{T}.$$

Moreover,

$$\left| \int_{t_0}^{t} \int_{t_0}^{s} |g(t_1, s_1)| \left(|s_1 - t_0| + |t_1 - t_0| \right) \Delta s_1 \Delta t_1 \right|$$

$$\begin{aligned}
&= \frac{1}{100} \\
&\times \left| \int_{t_0}^{t} \int_{t_0}^{s} \frac{(\sigma(s_1) + s_1)(\sigma(t_1) + t_1) \left(|s_1 - t_0| + |t_1 - t_0| \right)}{(1 + s_1^2)(1 + (\sigma(s_1))^2)(1 + t_1^2)(1 + (\sigma(t_1))^2) \left(1 + (s_1 - t_0)^2\right) \left(1 + (t_1 - t_0)^2\right)} \right. \\
&\quad \left. \Delta s_1 \Delta t_1 \right| \\
&= \frac{1}{100} \left| \int_{t_0}^{t} \int_{t_0}^{s} \frac{(\sigma(s_1) + s_1)(\sigma(t_1) + t_1)}{(1 + s_1^2)(1 + (\sigma(s_1))^2)(1 + t_1^2)(1 + (\sigma(t_1))^2)} \right. \\
&\quad \left(\frac{|s_1 - t_0|}{(1 + (s_1 - t_0)^2)(1 + (t_1 - t_0)^2)} \right. \\
&\quad \left. \left. + \frac{|t_1 - t_0|}{(1 + (s_1 - t_0)^2)(1 + (t_1 - t_0)^2)} \right) \Delta s_1 \Delta t_1 \right| \\
&\le \frac{1}{50} \left| \int_{t_0}^{t} \int_{t_0}^{s} \frac{(\sigma(s_1) + s_1)(\sigma(t_1) + t_1)}{(1 + s_1^2)\left(1 + (\sigma(s_1))^2\right)\left(1 + t_1^2\right)\left(1 + (\sigma(t_1))^2\right)} \Delta s_1 \Delta t_1 \right| \\
&= \frac{1}{50} \left| \int_{t_0}^{t} \int_{t_0}^{s} h^{\Delta}(s_1) h^{\Delta}(t_1) \Delta s_1 \Delta t_1 \right| \\
&= \frac{1}{50} \left| \left(\frac{1}{1 + t_0^2} - \frac{1}{1 + s^2} \right) \left(\frac{1}{1 + t_0^2} - \frac{1}{1 + t^2} \right) \right| \\
&\le B, \quad t, s \in \mathbb{T},
\end{aligned}$$

and

$$\left| \int_{t_0}^{t} |g(t_1, s)| \left(|s - t_0| + |t_1 - t_0| \right) \Delta t_1 \right|$$

$$
\begin{aligned}
&= \frac{1}{100}\\
&\times \left|\int_{t_0}^{t} \frac{(\sigma(s)+s)(\sigma(t_1)+t_1)\left(|s-t_0|+|t_1-t_0|\right)}{(1+s^2)(1+(\sigma(s))^2)(1+t_1^2)(1+(\sigma(t_1))^2)\left(1+(s-t_0)^2\right)\left(1+(t_1-t_0)^2\right)}\Delta t_1\right|\\
&= \frac{1}{100}\left|\int_{t_0}^{t} \frac{(\sigma(s)+s)(\sigma(t_1)+t_1)}{(1+s^2)(1+(\sigma(s))^2)(1+t_1^2)(1+(\sigma(t_1))^2)}\left(\frac{|s-t_0|}{\left(1+(s-t_0)^2\right)\left(1+(t_1-t_0)^2\right)}\right.\right.\\
&\left.\left.+\frac{|t_1-t_0|}{\left(1+(s-t_0)^2\right)\left(1+(t_1-t_0)^2\right)}\right)\Delta t_1\right|\\
&\leq \frac{1}{50}\left|\int_{t_0}^{t} \frac{(\sigma(s)+s)(\sigma(t_1)+t_1)}{(1+s^2)\left(1+(\sigma(s))^2\right)\left(1+t_1^2\right)\left(1+(\sigma(t_1))^2\right)}\Delta t_1\right|\\
&\leq \frac{1}{50}\left|\int_{t_0}^{t} \frac{\sigma(t_1)+t_1}{\left(1+t_1^2\right)\left(1+(\sigma(t_1))^2\right)}\Delta t_1\right|\\
&= \frac{1}{50}\left|\int_{t_0}^{t} h^{\Delta}(t_1)\Delta t_1\right|\\
&= \frac{1}{50}\left|\frac{1}{1+t_0^2}-\frac{1}{1+t^2}\right|\\
&\leq B, \quad s,t\in\mathbb{T},
\end{aligned}
$$

and

$$
\left|\int_{t_0}^{s} |g(t,s_1)|\left(|s_1-t_0|+|t-t_0|\right)\Delta s_1\right|
$$

$$
\begin{aligned}
&= \frac{1}{100} \\
&\times \left| \int_{t_0}^{s} \frac{(\sigma(s_1)+s_1)(\sigma(t)+t)\left(|s_1-t_0|+|t-t_0|\right)}{(1+s_1^2)\left(1+(\sigma(s_1))^2\right)(1+t^2)\left(1+(\sigma(t))^2\right)\left(1+(s_1-t_0)^2\right)\left(1+(t-t_0)^2\right)}\Delta s_1 \right| \\
&= \frac{1}{100}\left| \int_{t_0}^{s} \frac{(\sigma(s_1)+s_1)(\sigma(t)+t)}{(1+s_1^2)\left(1+(\sigma(s_1))^2\right)(1+t^2)\left(1+(\sigma(t))^2\right)} \left(\frac{|s_1-t_0|}{\left(1+(s_1-t_0)^2\right)\left(1+(t-t_0)^2\right)} \right.\right. \\
&\quad \left.\left. + \frac{|t-t_0|}{\left(1+(s_1-t_0)^2\right)\left(1+(t-t_0)^2\right)} \right)\Delta s_1 \right| \\
&\le \frac{1}{50}\left| \int_{t_0}^{s} \frac{(\sigma(s_1)+s_1)(\sigma(t)+t)}{(1+s_1^2)\left(1+(\sigma(s_1))^2\right)(1+t^2)\left(1+(\sigma(t))^2\right)}\Delta s_1 \right| \\
&\le \frac{1}{50}\left| \int_{t_0}^{s} \frac{\sigma(s_1)+s_1}{(1+s_1^2)\left(1+(\sigma(s_1))^2\right)}\Delta s_1 \right| \\
&= \frac{1}{50}\left| \int_{t_0}^{s} h^{\Delta}(s_1)\Delta s_1 \right| \\
&= \frac{1}{50}\left| \frac{1}{1+t_0^2} - \frac{1}{1+s^2} \right| \\
&\le B, \quad t, s \in \mathbb{T}.
\end{aligned}
$$

Let $E = C^1_{rd}(\mathbb{T}\times\mathbb{T})$ be endowed with the norm

$$
\|u\| = \max\{\|u\|_\infty, \|u^{\Delta_t}\|_\infty, \|u^{\Delta_s}\|_\infty\}, \tag{11.2}
$$

provided it exists, where $\|u\|_\infty = \sup_{(t,s)\in\mathbb{T}\times\mathbb{T}} |u(t,s)|$. Note that E is a Banach space with respect to the norm (11.2). With $\tilde{\tilde{\tilde{E}}}$ we will denote the set of all equicontinuous families in E with respect to the norm (11.2). Let $\tilde{\tilde{E}} = \overline{\tilde{\tilde{\tilde{E}}}}$ be the closure of $\tilde{\tilde{\tilde{E}}}$, $\tilde{E} = \tilde{\tilde{E}} \cup \{f\}$. For $u \in E$, define the operators

$$
S_0u(t,s) = \int_{t_0}^{s}(u(t,\sigma(s_1)) - u(t_0,\sigma(s_1)))\Delta s_1 - \int_{t_0}^{t}(f(t_1) - u(t_1,s))\Delta t_1,
$$

$$
S_1u(t,s) = \int_{t_0}^{t}\int_{t_0}^{s} g(t_1,s_1)S_0u(t_1,s_1)\Delta s_1\Delta t_1, \quad t,s \in \mathbb{T}.
$$

Lemma 11.1. *Suppose that* $(H1)$ *holds. If* $u \in E$ *satisfies the equation*

$$
S_0u(t,s) = 0, \quad (t,s) \in \mathbb{T}\times\mathbb{T}, \tag{11.3}
$$

then u *is a solution to the IVP* (11.1).

Proof. Let $u \in E$ satisfies the equation (11.3). Then

$$0 = \int_{t_0}^{s} (u(t,\sigma(s_1)) - u(t_0,\sigma(s_1)))\,\Delta s_1 - \int_{t_0}^{t} (f(t_1) - u(t_1,s))\,\Delta t_1, \quad (t,s) \in \mathbb{T}\times\mathbb{T}. \tag{11.4}$$

For $s = t_0$, we get

$$0 = \int_{t_0}^{t} (f(t_1) - u(t_1,t_0))\,\Delta t_1, \quad t \in \mathbb{T}.$$

We differentiate the last equation with respect to t and we find

$$f(t) = u(t,t_0), \quad t \in \mathbb{T}.$$

Now, we differentiate the equation (11.4) with respect to s and we arrive at

$$0 = u(t,\sigma(s)) - u(t_0,\sigma(s)) + \int_{t_0}^{t} u^{\Delta_s}(t_1,s)\Delta t_1, \quad (t,s) \in \mathbb{T}\times\mathbb{T},$$

which we differentiate with respect to t and we find

$$u^{\Delta_t}(t,\sigma(s)) + u^{\Delta_s}(t,s) = 0, \quad (t,s) \in \mathbb{T}\times\mathbb{T}.$$

This completes the proof. □

Lemma 11.2. *Suppose* $(H1)$ *and* $(H2)$*. If* $u \in E$ *satisfies the equation*

$$S_1u(t,s) = 0, \quad (t,s) \in \mathbb{T}\times\mathbb{T}, \tag{11.5}$$

then u *is a solution to the IVP* (11.1).

Proof. Let $u \in E$ satisfies the equation (11.5). Then we differentiate the equation (11.5) with respect to t and s and we get

$$g(t,x)S_0u(t,s) = 0, \quad (t,s) \in \mathbb{T}\times\mathbb{T}.$$

Hence,

$$S_0u(t,s) = 0, \quad (t,s) \in \mathbb{T}\times\mathbb{T}\backslash\{(t^1,s^1),\ldots,(t^k,s^k)\}.$$

Since $S_0u(\cdot,\cdot)$ is a continuous function on $\mathbb{T}\times\mathbb{T}$, we get

$$\begin{aligned} 0 &= \lim_{(t,s)\to(t^1,s^1)} S_0u(t,s) \\ &= S_0u(t^1,s^1) \\ &\vdots \\ &= \lim_{(t,s)\to(t^k,s^k)} S_0u(t,s) \\ &= S_0u(t^k,s^k). \end{aligned}$$

Therefore

$$S_0u(t,s)=0, \quad (t,s)\in\mathbb{T}\times\mathbb{T}.$$

Now, we apply Lemma 11.1 and we get the desired result. This completes the proof. □

Lemma 11.3. *Suppose* $(H1)$ *and* $(H2)$. *For* $u\in E$ *for which* $\|u\|\le A$, *we have*

$$\|S_1u\|\le 2AB.$$

Proof. Let $u\in E$ be arbitrarily chosen so that $\|u\|\le A$. Then

$$\begin{aligned} |S_1u(t,s)| &= \Big|\int_{t_0}^{t}\int_{t_0}^{s} g(t_1,s_1)\Big(\int_{t_0}^{s_1}(u(t_1,\sigma(s_2))-u(t_0,\sigma(s_2)))\,\Delta s_2 \\ &\quad -\int_{t_0}^{t_1}(f(t_2)-u(t_2,s_1))\,\Delta t_2\Big)\Delta s_1\Delta t_1\Big| \\ &\le \Big|\int_{t_0}^{t}\int_{t_0}^{s}|g(t_1,s_1)|\Big(\Big|\int_{t_0}^{s_1}(|u(t_1,\sigma(s_2))|+|u(t_0,\sigma(s_2))|)\,\Delta s_2\Big| \\ &\quad +\Big|\int_{t_0}^{t_1}(|f(t_2)|+|u(t_2,s_1)|)\,\Delta t_2\Big|\Big)\Delta s_1\Delta t_1\Big| \\ &\le 2A\Big|\int_{t_0}^{t}\int_{t_0}^{s}|g(t_1,s_1)|\,(|s_1-t_0|+|t_1-t_0|)\,\Delta s_1\Delta t_1\Big| \\ &\le 2AB, \quad t,s\in\mathbb{T}, \end{aligned}$$

and

$$
\begin{aligned}
\left|\frac{\Delta}{\Delta t}S_1u(t,s)\right| &= \Bigg|\int_{t_0}^{s} g(t,s_1)\bigg(\int_{t_0}^{s_1}(u(t,\sigma(s_2))-u(t_0,\sigma(s_2)))\Delta s_2 \\
&\quad -\int_{t_0}^{t}(f(t_1)-u(t_1,s))\Delta t_1\bigg)\Delta s_1\Bigg| \\
&\le \Bigg|\int_{t_0}^{s}|g(t,s_1)|\bigg(\Bigg|\int_{t_0}^{s_1}(|u(t,\sigma(s_2))|+|u(t_0,\sigma(s-2))|)\Delta s_2\Bigg| \\
&\quad +\Bigg|\int_{t_0}^{t}(|f(t_1)|+|u(t_1,s)|)\Delta t_1\Bigg|\bigg)\Delta s_1\Bigg| \\
&\le 2A\Bigg|\int_{t_0}^{s}|g(t,s_1)|\left(|s_1-t_0|+|t-t_0|\right)\Delta s_1\Bigg| \\
&\le 2AB,\quad t,s\in\mathbb{T},
\end{aligned}
$$

and

$$
\begin{aligned}
\left|\frac{\Delta}{\Delta s}S_1u(t,s)\right| &= \Bigg|\int_{t_0}^{t} g(t_1,s)\bigg(\int_{t_0}^{s}(u(t_1,\sigma(s_1))-u(t_0,\sigma(s_1)))\Delta s_1 \\
&\quad -\int_{t_0}^{t_1}(f(t_2)-u(t_2,s))\Delta t_2\bigg)\Delta t_1\Bigg| \\
&\le \Bigg|\int_{t_0}^{t}|g(t_1,s)|\bigg(\Bigg|\int_{t_0}^{s}(|u(t_1,\sigma(s_1))|+|u(t_0,\sigma(s_1))|)\Delta s_1\Bigg| \\
&\quad +\Bigg|\int_{t_0}^{t_1}(|f(t_2)|+|u(t_2,s)|)\Delta t_2\Bigg|\bigg)\Delta t_1\Bigg| \\
&\le 2A\Bigg|\int_{t_0}^{t}|g(t_1,s)|\left(|s-t_0|+|t_1-t_0|\right)\Delta t_1\Bigg| \\
&\le 2AB,\quad t,s\in\mathbb{T}.
\end{aligned}
$$

Therefore

$$\|S_1u\|\le 2AB.$$

This completes the proof. □

The following theorem will be used to prove existence of solutions of the generalized shifting problem.

Theorem 11.1. *Let E be a Banach space, Y a closed, convex subset of E, U be any open subset of Y with $0\in U$. Consider two operators T and S, where*

$$Tx=\varepsilon x,\ x\in\overline{U},$$

for $\varepsilon>0$ and $S:\overline{U}\to E$ be such that

(i) $I - S : \overline{U} \to Y$ *continuous, compact and*

(ii) $\{x \in \overline{U} : x = \lambda(I - S)x, \quad x \in \partial U\} = \emptyset,$ *for any* $\lambda \in \left(0, \frac{1}{\varepsilon}\right)$.

Then there exists $x^* \in \overline{U}$ *such that*

$$Tx^* + Sx^* = x^*.$$

Proof. We have that the operator $\frac{1}{\varepsilon}(I - S) : \overline{U} \to Y$ is continuous and compact. Suppose that there exist $x_0 \in \partial U$ and $\mu_0 \in (0, 1)$ such that

$$x_0 = \mu_0 \frac{1}{\varepsilon}(I - S)x_0,$$

that is

$$x_0 = \lambda_0 (I - S)x_0$$

where $\lambda_0 = \mu_0 \frac{1}{\varepsilon} \in \left(0, \frac{1}{\varepsilon}\right)$. This contradicts the condition (ii). From Leray-Schauder nonlinear alternative, it follows that there exists $x^* \in \overline{U}$ so that

$$x^* = \frac{1}{\varepsilon}(I - S)x^*$$

or

$$\varepsilon x^* + Sx^* = x^*,$$

or

$$Tx^* + Sx^* = x^*.$$

□

Theorem 11.2. *Suppose* $(H1)$-$(H3)$. *Then the problem* (11.1) *has at least one solution in* $C^1_{rd}(\mathbb{T} \times \mathbb{T})$.

Proof. Let

$$U = \{u \in \widetilde{E} : \|u\| < A\}.$$

For $u \in \overline{U}$, define the operators

$$\begin{aligned} T(u)(t) &= \epsilon u(t), \\ S(u)(t) &= u(t) - \epsilon u(t) - \epsilon S_1(u)(t), \quad t \in \mathbb{T}. \end{aligned}$$

For $u \in \overline{U}$, we have

$$\begin{aligned}\|(I-S)(u)\| &= \|\epsilon u + \epsilon S_1(u)\| \\ &\le \epsilon\|u\| + \epsilon\|S_1(u)\| \\ &\le \epsilon A + 2\epsilon AB.\end{aligned}$$

Thus, $S : \overline{U} \to E$ is continuous and $(I-S)(\overline{U})$ resides in a compact subset of E. Now, suppose that there is a $u \in E$ so that $\|u\| = A$ and

$$u = \lambda(I-S)(u)$$

or

$$u = \lambda\epsilon\,(I + S_1)(u), \tag{11.6}$$

for some $\lambda \in \left(0, \frac{1}{\epsilon}\right)$. Then, using that

$$S_1 u(t_0, s) = S_1(t, t_0) = 0, \quad t, s \in \mathbb{T},$$

we get

$$\begin{aligned}u(t_0, s) &= \lambda\epsilon u(t_0, s), \quad s \in \mathbb{T}, \\ u(t, t_0) &= \lambda\epsilon u(t, t_0), \quad t \in \mathbb{T},\end{aligned}$$

whereupon $\lambda\epsilon = 1$. This is a contradiction. Consequently

$$\{u \in E : u = \lambda_1(I-S)(u),\ \|u\| = A\} = \emptyset$$

for any $\lambda_1 \in \left(0, \frac{1}{\epsilon}\right)$. Then, from Theorem 11.1, it follows that the operator $T + S$ has a fixed point $u^* \in E$. Therefore

$$\begin{aligned}u^*(t) &= T(u^*)(t) + S(u^*)(t) \\ &= \epsilon u^*(t) + u^*(t) \\ &\quad -\epsilon u^*(t) - \epsilon S_1(u^*)(t), \quad t \in \mathbb{T},\end{aligned}$$

whereupon

$$S_1(u^*)(t) = 0, \quad t \in \mathbb{T}.$$

From here, u^* is a solution to the problem (11.1). This completes the proof. □

12. Generalized Convolutions

Suppose that $\sup \mathbb{T} = \infty$ and $\inf \mathbb{T} = -\infty$.

Definition 12.1. *For given functions $f, g : \mathbb{T} \to \mathbb{R}$, the generalized convolution $f\widetilde{\star}g$ is defined by*

$$(f\widetilde{\star}g)(t) = \int_{-\infty}^{\infty} \widetilde{f}(t, \sigma(s))g(s)\Delta s, \quad t \in \mathbb{T}.$$

Theorem 12.1. *Let $f, g : \mathbb{T} \to \mathbb{R}$ be given functions such that*

$$\lim_{u \to \pm\infty} \widetilde{f}(t, u)\widetilde{g}(u, \sigma(s)) = 0.$$

Then the generalized shift of the generalized convolution $f\widetilde{\star}g$ is given by

$$\left(\widetilde{f\widetilde{\star}g}\right)(t, s) = \int_{-\infty}^{\infty} \widetilde{f}(t, \sigma(u))\widetilde{g}(u, s)\Delta u, \quad t, s \in \mathbb{T}.$$

Proof. Let

$$F(t, s) = \int_{-\infty}^{\infty} \widetilde{f}(t, \sigma(u))\widetilde{g}(u, s)\Delta u, \quad t, s \in \mathbb{T}.$$

Then

$$\begin{aligned} F(t, t_0) &= \int_{-\infty}^{\infty} \widetilde{f}(t, \sigma(u))\widetilde{g}(u, t_0)\Delta u \\ &= \int_{-\infty}^{\infty} \widetilde{f}(t, \sigma(u))\widetilde{g}(u)\Delta u \\ &= \left(\widetilde{f}\widetilde{\star}\widetilde{g}\right)(t), \quad t \in \mathbb{T}, \end{aligned}$$

and

$$F^{\Delta_t}(t, \sigma(s)) + F^{\Delta_s}(t, s))\widetilde{g}^{\Delta_s}(u, s)\Delta u =$$

$$\begin{aligned}
&= \int_{-\infty}^{\infty} \widetilde{f}^{\Delta_t}(t,\sigma(u))\widetilde{g}(u,\sigma(s))\Delta u + \int_{-\infty}^{\infty} \widetilde{f}(t,\sigma(u) \\
&= -\int_{-\infty}^{\infty} \widetilde{f}^{\Delta_u}(t,u)\widetilde{g}(u,\sigma(s))\Delta u + \int_{-\infty}^{\infty} \widetilde{f}(t,\sigma(u))\widetilde{g}^{\Delta_s}(u,s)\Delta u \\
&= -\lim_{u\to\infty} \widetilde{f}(t,u)\widetilde{g}(u,\sigma(s)) + \lim_{u\to-\infty} \widetilde{f}(t,u)\widetilde{g}(u,\sigma(s)) \\
&\quad + \int_{-\infty}^{\infty} \widetilde{f}(t,\sigma(u))\widetilde{g}^{\Delta_u}(u,\sigma(s))\Delta u + \int_{-\infty}^{\infty} \widetilde{f}(t,\sigma(u))\widetilde{g}^{\Delta_s}(u,s)\Delta u \\
&= \int_{-\infty}^{\infty} \widetilde{f}(t,\sigma(u))\left(\widetilde{g}^{\Delta_u}(u,\sigma(s)) + \widetilde{g}^{\Delta_s}(u,s)\right)\Delta u \\
&= 0, \quad t,s \in \mathbb{T}.
\end{aligned}$$

This completes the proof. □

Theorem 12.2 (Associativity of the Generalized Convolution)**.** *Let* $f, g, h : \mathbb{T} \to \mathbb{R}$ *be given functions so that*

$$\lim_{u\to\pm\infty} \widetilde{f}(t,u)\widetilde{g}(u,\sigma(s)) = 0.$$

Then

$$(f\widetilde{\star}g)\widetilde{\star}h = f\widetilde{\star}(g\widetilde{\star}h).$$

Proof. Applying Theorem 12.1, we get

$$\begin{aligned}
\left((f\widetilde{\star}g)\widetilde{\star}h\right)(t) &= \int_{-\infty}^{\infty} \left(\widetilde{f\star g}\right)(t,\sigma(u))h(u)\Delta u \\
&= \int_{-\infty}^{\infty} \left(\int_{-\infty}^{\infty} \widetilde{f}(t,\sigma(v))\widetilde{g}(v,\sigma(u))\Delta v\right) h(u)\Delta u \\
&= \int_{-\infty}^{\infty} \widetilde{f}(t,\sigma(v))\left(\int_{-\infty}^{\infty} \widetilde{g}(v,\sigma(u))h(u)\Delta u\right)\Delta v \\
&= \int_{-\infty}^{\infty} \widetilde{f}(t,\sigma(v))(g\widetilde{\star}h)(v)\Delta v \\
&= \left(f\widetilde{\star}(g\widetilde{\star}h)\right)(t), \quad t \in \mathbb{T}.
\end{aligned}$$

This completes the proof. □

Theorem 12.3. *Let $f, g : \mathbb{T} \to \mathbb{R}$ be given functions so that g is infinitely many Δ-differentiable. Then*

$$
\begin{aligned}
(f\widetilde{\star}g)^{\Delta^k}(t) &= -\sum_{l=0}^{k-1}\left(\lim_{u\to\infty} f(\cdot,u)g^{\Delta^{k-1-l}}(u) - \lim_{u\to-\infty} f(\cdot,u)g^{\Delta^{k-1-l}}(u)\right)^{\Delta^l}(t) \\
&\quad + \left(f\widetilde{\star}g^{\Delta^k}\right)(t), \quad t\in\mathbb{T}, \quad k\in\mathbb{N}.
\end{aligned} \tag{12.1}
$$

Proof. We will use the classical mathematical induction.

1. Let $k = 1$. Then

$$
\begin{aligned}
(f\widetilde{\star}g)^{\Delta}(t) &= \int_{-\infty}^{\infty} \widetilde{f}^{\Delta_t}(t,\sigma(u))g(u)\Delta u \\
&= -\int_{-\infty}^{\infty} \widetilde{f}^{\Delta_u}(t,u)g(u)\Delta u \\
&= -\left(\lim_{u\to\infty}\widetilde{f}(t,u)g(u) - \lim_{u\to-\infty}\widetilde{f}(t,u)g(u)\right) + \int_{-\infty}^{\infty}\widetilde{f}(t,\sigma(u))g^{\Delta}(u)\Delta u \\
&= -\left(\lim_{u\to\infty}\widetilde{f}(t,u)g(u) - \lim_{u\to-\infty}\widetilde{f}(t,u)g(u)\right) + \left(f\widetilde{\star}g^{\Delta}\right)(t), \quad t\in\mathbb{T}.
\end{aligned}
$$

2. Assume that (12.1) holds for some $k \in \mathbb{N}$.
3. We will prove that

$$
\begin{aligned}
(f\widetilde{\star}g)^{\Delta^{k+1}}(t) &= -\sum_{l=0}^{k}\left(\lim_{u\to\infty} f(\cdot,u)g^{\Delta^{k-l}}(u) - \lim_{u\to-\infty} f(\cdot,u)g^{\Delta^{k-l}}(u)\right)^{\Delta^l}(t) \\
&\quad + \left(f\widetilde{\star}g^{\Delta^{k+1}}\right)(t), \quad t\in\mathbb{T}.
\end{aligned}
$$

Really, we have

$$\begin{aligned}
&(f\widetilde{\star}g)^{\Delta^{k+1}}(t)\\
&= \left((f\widetilde{\star}g)^{\Delta^k}\right)^{\Delta}(t)\\
&= -\sum_{l=0}^{k-1}\left(\lim_{u\to\infty} f(\cdot,u)g^{\Delta^{k-1-l}}(u) - \lim_{u\to-\infty} f(\cdot,u)g^{\Delta^{k-1-l}}(u)\right)^{\Delta^{l+1}}(t)\\
&\quad + \left(f\widetilde{\star}g^{\Delta^k}\right)^{\Delta}(t)\\
&= -\sum_{l=1}^{k}\left(\lim_{u\to\infty} f(\cdot,u)g^{\Delta^{k-l}}(u) - \lim_{u\to-\infty} f(\cdot,u)g^{\Delta^{k-l}}(u)\right)^{\Delta^{l}}(t)\\
&\quad + \int_{-\infty}^{\infty} f^{\Delta_t}(t,\sigma(u))g^{\Delta^k}(u)\Delta u\\
&= -\sum_{l=1}^{k}\left(\lim_{u\to\infty} f(\cdot,u)g^{\Delta^{k-l}}(u) - \lim_{u\to-\infty} f(\cdot,u)g^{\Delta^{k-l}}(u)\right)^{\Delta^{l}}(t)\\
&\quad - \int_{-\infty}^{\infty} f^{\Delta_u}(t,u)g^{\Delta^k}(u)\Delta u\\
&= -\sum_{l=0}^{k}\left(\lim_{u\to\infty} f(\cdot,u)g^{\Delta^{k-l}}(u) - \lim_{u\to-\infty} f(\cdot,u)g^{\Delta^{k-l}}(u)\right)^{\Delta^{l}}(t)\\
&\quad + \int_{-\infty}^{\infty} f(t,\sigma(u))g^{\Delta^{k+1}}(u)\Delta u\\
&= -\sum_{l=0}^{k}\left(\lim_{u\to\infty} f(\cdot,u)g^{\Delta^{k-l}}(u) - \lim_{u\to-\infty} f(\cdot,u)g^{\Delta^{k-l}}(u)\right)^{\Delta^{l}}(t)\\
&\quad + \left(f\star g^{\Delta^{k+1}}\right)(t), \quad t\in\mathbb{T}.
\end{aligned}$$

This completes the proof. □

Theorem 12.4. *If $\widetilde{f}$ has partial Δ-derivatives of all orders, then*

$$\widetilde{f}^{\Delta_t^k}(t,t) = f^{\Delta^k}(t_0), \quad t\in\mathbb{T}, \quad k\in\mathbb{N}_0.$$

Here $\widetilde{f}^{\Delta_t}$ indicates the Δ-derivative with respect to its first variable.

Proof. We have

$$\widetilde{f}^{\Delta_t^k}(t,t_0) = f^{\Delta^k}(t), \quad t \in \mathbb{T}.$$

Let

$$F(t) = \widetilde{f}^{\Delta_t^k}(t,t), \quad t \in \mathbb{T}.$$

Then

$$F^{\Delta}(t) = \left(\widetilde{f}^{\Delta_t}(t,\sigma(t)) + \widetilde{f}^{\Delta_s}(t,t)\right)^{\Delta_t^k} = 0, \quad t \in \mathbb{T}.$$

Thus,

$$\begin{aligned} F(t) &= F(t_0) \\ &= \widetilde{f}^{\Delta_t^k}(t_0,t_0) \\ &= f^{\Delta^k}(t_0), \quad t \in \mathbb{T}. \end{aligned}$$

This completes the proof. □

Theorem 12.5. *Let $f : \mathbb{T} \to \mathbb{R}$ be a given function so that*

$$\lim_{t \to \pm\infty} \widetilde{f}(t,\sigma(\tau)) e_{\ominus ix}(t,\sigma(\tau)) = 0, \tag{12.2}$$

for those $x \in \mathbb{R}$ for which $1 + ix\mu(t) \neq 0$, $t \in \mathbb{T}$. Define

$$\psi(t) = \int_{-\infty}^{\infty} \widetilde{f}(\tau,t) e_{\ominus ix}(\sigma(\tau),t)\Delta\tau, \quad t \in \mathbb{T}.$$

Then ψ is a constant and

$$\psi(t) = \mathcal{F}(f)(x,t_0), \quad t \in \mathbb{T}.$$

Proof. We have

$$\psi(t) = \int_{-\infty}^{\infty} \frac{\widetilde{f}(\tau,t)}{e_{\ominus ix}(t,\sigma(\tau))}\Delta\tau, \quad t \in \mathbb{T}.$$

Hence,

$$\begin{aligned}
\psi^\Delta(t) &= \int_{-\infty}^{\infty} \frac{\widetilde{f}^{\Delta_t}(\tau,t)e_{\ominus ix}(t,\sigma(\tau)) - (\ominus ix)(t)e_{\ominus ix}(t,\sigma(\tau))\widetilde{f}(\tau,t)}{e_{\ominus ix}(t,\sigma(\tau))e_{\ominus ix}(\sigma(t),\sigma(\tau))}\Delta\tau \\
&= \int_{-\infty}^{\infty} \frac{\widetilde{f}^{\Delta_t}(\tau,t)}{e_{\ominus ix}(\sigma(t),\sigma(\tau))}\Delta\tau + ix\int_{-\infty}^{\infty} \frac{\widetilde{f}(\tau,t)}{(1+ix\mu(t))e_{\ominus ix}(\sigma(t),\sigma(\tau))}\Delta\tau \\
&= -\int_{-\infty}^{\infty} \widetilde{f}^{\Delta_\tau}(\tau,\sigma(t))e_{\ominus ix}(\sigma(\tau),\sigma(t))\Delta\tau + ix\int_{-\infty}^{\infty} \frac{\widetilde{f}(\tau,t)}{e_{\ominus ix}(t,\sigma(\tau))}\Delta\tau \\
&= ix\psi(t) - \int_{-\infty}^{\infty} \widetilde{f}^{\Delta_\tau}(\tau,\sigma(t))e_{\ominus ix}(\sigma(\tau),\sigma(t))\Delta\tau \\
&= ix\psi(t) - \Big(\lim_{\tau\to\infty}\widetilde{f}(\tau,\sigma(t))e_{\ominus ix}(\tau,\sigma(t)) - \lim_{\tau\to-\infty}\widetilde{f}(\tau,\sigma(t))e_{\ominus ix}(\tau,\sigma(t))\Big) \\
&\quad + \int_{-\infty}^{\infty} \widetilde{f}(\tau,\sigma(t))(\ominus ix)(\tau)e_{\ominus ix}(\tau,\sigma(t))\Delta\tau \\
&= ix\psi(\tau) - ix\int_{-\infty}^{\infty} \frac{\widetilde{f}(\tau,\sigma(t))}{1+ix\mu(\tau)}e_{\ominus ix}(\tau,\sigma(t))\Delta\tau \\
&= ix\psi(t) - ix\int_{-\infty}^{\infty} \widetilde{f}(\tau,\sigma(t))e_{\ominus ix}(\sigma(\tau),\sigma(t))\Delta\tau \\
&= ix\psi(t) - ix\psi(\sigma(t)) \\
&= -ix\mu(t)\psi^\Delta(t), \quad t\in\mathbb{T},
\end{aligned}$$

whereupon

$$(1+ix\mu(t))\psi^\Delta(t) = 0, \quad t\in\mathbb{T},$$

and

$$\psi(t) = \psi(t_0), \quad t\in\mathbb{T}.$$

Therefore

$$\begin{aligned}
\psi(t) &= \int_{-\infty}^{\infty} \widetilde{f}(\tau,t_0)e_{\ominus ix}(\sigma(\tau),t_0)\Delta\tau \\
&= \int_{-\infty}^{\infty} f(\tau)e_{\ominus ix}(\sigma(\tau),t_0)\Delta\tau \\
&= \mathcal{F}(f)(x,t_0), \quad t\in\mathbb{T}.
\end{aligned}$$

This completes the proof. □

13. The Generalized Convolution Theorem

Theorem 13.1 (Generalized Convolution Theorem). *Let $f, g : \mathbb{T} \to \mathbb{R}$ be given functions so that* (12.2) *holds. Then*

$$\mathcal{F}(f\widetilde{\star}g)(x,t_0) = \mathcal{F}(f)(x,t_0)\mathcal{F}(g)(x,t_0).$$

Proof. We have

$$\begin{aligned}
\mathcal{F}(f\widetilde{\star}g)(x,t_0) &= \int_{-\infty}^{\infty} (f\widetilde{\star}g)(\tau)e_{\ominus ix}(\sigma(\tau),t_0)\Delta\tau \\
&= \int_{-\infty}^{\infty}\left(\int_{-\infty}^{\infty}\widetilde{f}(\tau,\sigma(s))g(s)\Delta s\right)e_{\ominus ix}(\sigma(s),t_0)e_{\ominus ix}(\sigma(\tau),\sigma(s))\Delta\tau \\
&= \int_{-\infty}^{\infty} g(s)e_{\ominus ix}(\sigma(s),t_0)\left(\int_{-\infty}^{\infty}\widetilde{f}(\tau,\sigma(s))e_{\ominus ix}(\sigma(\tau),\sigma(s))\Delta\tau\right)\Delta s \\
&= \int_{-\infty}^{\infty} g(s)e_{\ominus ix}(\sigma(s),t_0)\psi(\sigma(s))\Delta s \\
&= \mathcal{F}(f)(x,t_0)\int_{-\infty}^{\infty} g(s)e_{\ominus ix}(\sigma(s),t_0)\Delta s \\
&= \mathcal{F}(f)(x,t_0)\mathcal{F}(g)(x,t_0).
\end{aligned}$$

This completes the proof. □

Conclusion

We have shown that the bilateral Laplace transform on time scales converges absolutely and uniformly on $\mathbb{C}_{s,\alpha,\gamma}$, its derivatives are uniformly continuous on $\mathbb{C}_{s,\alpha,\gamma}$, and established an inversion formula for functions that have finitely many poles of finite order. After that, we showed that the bilateral Laplace transform on time scales can be used to define a Fourier transform on time scales which can be used to solve partial dynamic equations. One interesting future application of this bilateral Laplace transform might be in defining analogues for moment-generating functions for probability distributions defined on time scales. Work in that direction for a unilateral transform was initiated in [1, Definition 20], which defined moment-generating functions for time scales with a minimum value of 0. Of particular interest is an investigation of a proper analogue of the Schwartz class of functions for the continuous transform; finding it would allow for a major expansion of the solutions of partial dynamic equations. Another interesting topic for

Chapter 2

The Laplace Transform: Generalizations and Applications

Paulo M. Guzmán[1,2,*]
Juan E. Nápoles Valdés[2,3,†]
Sergio Bermudo[4,‡]
[1]Facultad de Ciencias Agrarias, Universidad Nacional del Nordeste Corrientes, Argentina
[2]Facultad de Ciencias Exactas y Naturales y Agrimensura Universidad Nacional del Nordeste Corrientes, Argentina
[3]Facultad Regional Resistencia , Universidad Tecnol´ogica Nacional Chaco, Argentina
[4]Department of Economics, Quantitative Methods and Economic History Pablo de Olavide University, Sevilla, Spain

Abstract

In this chapter we present a generalization of the Laplace Transform and develop its most important properties based on a generalized derivative and its corresponding integral operator. We study its fundamental properties and use this transform to solve various types of generalized differential equations as well as study the stability of linear systems.

Keywords: generalized derivative, generalized Laplace transform, generalized differential equation

* Corresponding Author's Emails: paulo.guzman@comunidad.unne.edu.ar, guzmanpaulomatias@gmail.com
† Corresponding Author's Email: jnapoles@exa.unne.edu.ar
‡ Corresponding Author's Email: sbernav@upo.es

In: The Laplace Transform and Its Applications in Science
Editor: Victor Martinez-Luaces
ISBN: 979-8-89113-747-9

AMS Subject Classification (2020): Primary 35R11, Secondary 35A20,35C05

Preliminaries

Since its emergence, associated with the French mathematician Pierre-Simon Laplace (1749-1827),and systematically developed by the British physicist Oliver Heaviside (1850-1925), the Laplace Transform has had a high impact on mathematics.

The complete history of this Transform goes back a little further in the past, specifically in 1744, when Leonhard Euler was investigating other types of integrals. Euler, however, did not pursue the theme and abandoned it, this study was continued by Lagrange, who made some modifications to Euler's work and obtained some advances.

This work by Lagrange came to the attention of Laplace 38 years later, in 1782, and he picked it up where Euler left off. But it was not until 3 years later, in 1785, when Laplace had a stroke of genius which forever changed the way we solve differential equations. He obtained new results and was able to show the breadth and strength of such a transform around 1809, when he began to use infinity as an integral condition.

The Laplace transform is a particular integral transform invented (by the) to simplify the solution of many differential equations that describe physical processes. This transform allows us to get explicit solutions for initial value problems, and it is especially useful when the inhomogeneous term is either piecewise discontinuous or impulsive or periodic. It also solves integral equations and some partial differential equations.

One of the mathematical areas that is in constant development is that of Differential Equations (using new operators or defined on different functional spaces), and their solution methods, in particular, due to the multiplicity of applications and its own theoretical development, over time, researchers and productions related to this area have been increasing. Different aspects of this increase and its overlaps with the development of Mathematics itself could be consulted in Nápoles (2007)[1], (2006)[2], (2006)[3] y (2006)[4].

With the appearance of the Fractional Calculus and the use of new local differential operators since the decade of the 60s of the last century, Fleitas (2021)[5], Guzmán (2018)[6], Khalil (2014)[7], Nápoles (2020)[8] and Zhao (2017)[9], theme and successive generalizations were obtained, see for example, Bosch (2021)[10], Fahad (2023)[11], Jarad (2020)[12], Jarad (2018)[13], Khader (2017)[14], Khan (2018)[15], Lugo (2022)[16], Martínez (2022)[17], Molaei (2022)[18], Ozkan (2018) [19], Silva (2018)[20], Vivas (2021)[21], Yavuz (2020)[22] and Younis

(2022)[23]. Modern definitions of derivatives and integrals of non-integer order play a vital role in the theory of fractional calculus. From the original definitions of Riemann-Liouville, to the most current of Atangana-Baleanu, through many others, a body of theory has been formed that is beginning to have applications in various fields. On the other hand, although local differential operators have been known since the 1960s, it was not until 2014 when a complete formalization was achieved with the conformable derivative given by Khalil (2014)[7]. In 2018 we defined a new type of differential operator, called non conformable (see Guzmán (2018)[6], (2020)[24], (2020)[25], Martínez (2020)[26], Nápoles (2020)[27], (2019)[28], (2018)[29]) and we consolidated our ideas with a generalized derivative definition (see Nápoles (2020)[8], (2020)[30], (2020)[31]). In this way, a new area has been formed in the Mathematical Sciences, which we call Generalized Calculus, with many applications and important theoretical results.

As said before, in Nápoles (2020)[8] (see also Fleitas (2021)[5] and Zhao (2017)[9]) a generalized fractional derivative was defined in the following way.

Definition 1 *Given a function $\delta : [0, +\infty) \to \mathbb{R}$, the N-derivative of δ of order α is defined by*

$$N_Y^\alpha \delta(x) = \lim_{\varepsilon \to 0} \frac{\delta(x + \varepsilon Y(x, \alpha)) - \delta(x)}{\varepsilon} \tag{1}$$

for all $x > 0$, $\alpha \in (0, 1)$ being $Y(x, \alpha)$ some absolutely continuous function.

If δ is α–differentiable in some $(0, \alpha)$, and $\lim_{x \to 0^+} N_Y^\alpha \delta(x)$ exists, then we define $N_Y^\alpha \delta(0) = \lim_{x \to 0^+} N_Y^\alpha \delta(x)$. Note that, if δ is differentiable, then $N_Y^\alpha \delta(x) = Y(x, \alpha)\delta'(x)$, where $\delta'(x)$ is the ordinary derivative.

This generalized differential operator is a generalization of many of the known local operators (for example, the conformable and non-conformable derivative given in Khalil (2014)[7] and Guzmán (2018)[6], respectively) and has shown its usefulness in various applications, for example, in Guzmán (2020)[24], (2020)[25], Nápoles (2020)[27], (2020)[30], (2020)[31], Abreu-Blaya (2020)[32], Fleitas(2019) [33], (2019)[34]. One of the most required properties of a derivative operator is the Chain Rule (see Nápoles (2020)[8]), to calculate the derivative of compound functions, which does not exist in the case of classical fractional derivatives

$$N_Y^\alpha (\delta \circ \sigma)(x) = N_Y^\alpha \delta(\sigma(x)) = \delta'(\sigma(x)) N_Y^\alpha \sigma(x). \tag{2}$$

Now, we give the definition of a general fractional integral (see Guzmán (2020)[35]). Throughout the work we will consider that the integral operator kernel X defined below is an absolutely continuous function (for additional details, the interested reader can consult Fleitas (2021) [5] and Zhao (2017)[9]).

Definition 2 *Let I be an interval in $\mathbb{R}$, $a, x \in I$ and $\alpha \in \mathbb{R}$. The integral operator J, is defined for every locally integrable function δ on I as (see Guzmán (2020)[35])*

$$J^{\alpha}_{Y,a}(\delta)(x) = \int_a^x \frac{\delta(r)}{Y(r,\alpha)} dr, x > a. \tag{3}$$

$$J^{\alpha}_{Y,b}(\delta)(x) = \int_x^b \frac{\delta(r)}{Y(r,\alpha)} dr, b > x. \tag{4}$$

Remark 3 *As pointed out in Guzmán (2020)[35], many fractional integral operators can be obtained as particular cases of the previous one, under certain choices of the Y kernel. For example, if $Y(x-r,\alpha) = \Gamma(\alpha)(x-r)^{1-\alpha}$ the right Riemann-Liouville integral is obtained (similarly to the left), further details on Fractional Calculus and fractional integral operators linked to the generalized integral of the previous definition can be found in Khalil (2014)[7], Abdeljawad (2015)[36], Capelas de Oliveira (2014)[37], Cinar (2003)[38], Gorenflo (1997)[39], Katugampola (2014)[40], (2011)[41], Kilbas (1993)[42], Machado (2011)[43].*

The following property is one of the fundamental ones and links the integral operator with the generalized derivative defined above (see also Fleitas (2021)[5] and Zhao (2017)[9]).

Proposition 4 *Let I be an interval in $\mathbb{R}$, $a \in I$, $0 < \alpha \leq 1$ and δ a α-differentiable function on I such that δ' is a locally integrable function on I. Then, we have for all $x \in I$*

$$J^{\alpha}_{Y,a}\big(N^{\alpha}_{Y}(\delta)\big)(x) = \delta(x) - \delta(a). \tag{5}$$

Proposition 5 *Let I be an interval in $\mathbb{R}$, $a \in I$ and $\alpha \in (0,1]$*

$$N^{\alpha}_{Y}\big(J^{\alpha}_{Y,a}(\delta)\big)(x) = \delta(x), \tag{6}$$

for every continuous function δ on I and $a, x \in I$.

Theorem 6 *(Integration by parts) Let $\delta, \sigma : [a,b] \to \mathbb{R}$ be two differentiable functions and $\alpha \in (0,1]$. Then, the following property hold*

$$J^{\alpha}_{Y,a}((\delta)(N^{\alpha}_{Y,a+}\sigma(x))) = [\delta(x)\sigma(x)]_a^b - J^{\alpha}_{Y,a}((\sigma)(N^{\alpha}_{Y,a}\delta(x))). \tag{7}$$

Theorem 7 *If* $\delta : [a,b] \to \mathbb{R}$ *is a continuous function and* $\alpha \in (0,1]$ *then, the following inequality is fulfilled*

$$\left| J^{\alpha}_{Y,a}(\delta)(x) \right| \leq J^{\alpha}_{Y,a} \left| \delta \right| (x). \tag{8}$$

Taking into account the ideas of Martínez (2020)[26] we can define the generalized partial derivatives as follows.

Definition 8 *Given a real valued function* $\delta : \mathbb{R}^n \to \mathbb{R}$ *and* $\overrightarrow{a} = (a_1, , a_n) \in \mathbb{R}^n$ *a point whose ith component is positive, the generalized partial* N*-derivative of* f *of order* α *in the point* $\overrightarrow{a} = (a_1, , a_n)$ *is defined by*

$$N^{\alpha}_{Y_i,x_i}\delta(\overrightarrow{a}) = \lim_{\varepsilon \to 0} \frac{\delta(a_1, .., a_i + \varepsilon Y_i(a_i, \alpha), , a_n) - \delta(a_1, , a_i, ..., a_n))}{\varepsilon}. \tag{9}$$

If it exists, it is denoted $N^{\alpha}_{Y_i,x_i}\delta(\overrightarrow{a})$*, and it is called the ith generalized partial derivative of* δ *of order* $\alpha \in (0,1]$ *at* $\overrightarrow{a}$*.*

In this chapter, we present some of the generalizations of the classical Laplace Transform and its applications to various types of generalized differential equations, such as solving linear differential equations, integral equations, Mass-Spring system, electric circuits, among others.

Main Results

The following definition will be very important for our work since from it the Generalized Laplace Transform is defined.

Definition 9 *Let* $\alpha \in (0,1]$ *and let* γ *be any real number, the generalized exponential order is defined as follows:*

$$E^{N}_{\alpha}(\gamma, x) = e^{(\gamma \mathcal{Y}(x,\alpha))}, \tag{10}$$

with $\mathcal{Y}(x,\alpha) = \int_0^x \frac{dr}{Y(r,\alpha)} = J^{\alpha}_{Y,0}(1)(x)$.

Following the definitions 1, 9 as well as the Chain Rule, we get $N^{\alpha}_{Y}\left\{E^{N}_{\alpha}(\gamma,x)\right\} = \gamma E^{N}_{\alpha}(\gamma, x)$.

Definition 10 *(Vivas (2021)[44]) For $\alpha \in (0,1]$, σ a function and a real number r, the Generalized Laplace Transform is defined as follows*

$$Y(r) = \left(\mathcal{L}_N^{\alpha}\{\sigma(x)\}\right)(r) = J_{Y,0}^{\alpha}\big(E_{\alpha}^{N}(-r,x)\sigma(x)\big)(\infty), \tag{11}$$

and its inverse transform is

$$\sigma(x) = \left(\mathcal{L}_N^{\alpha}\{T(r)\}^{-1}\right)(x) = J_{Y,0}^{\alpha}\Big(Y(x,\alpha)\,E_{\alpha}^{N}(r,x)T(r)\Big)(\infty). \tag{12}$$

Remark 11 *We get the ordinary Laplace Transform in Definition 10 when $Y(x,\alpha) = 1$, and the Conformable Laplace Transform defined in Abdeljawad (2015)[36], if $Y(x,\alpha) = x^{1-\alpha}$ (also see Jarad (2018)[13], Khader (2017)[14], Al-Zhour (2021)[45], Bouaouid (2020)[46], El-Ajou (2020)[47]). We obtain the generalized Laplace transform defined in Jarad (2020)[12], if we consider $Y(x,\alpha) = \frac{1}{\sigma'(x)}$ (see Fahad (2023)[11], Oumarou (2021)[48]).*

Theorem 12 *For any two real functions σ and λ, the Generalized Laplace Transform has the following properties for $\alpha, \beta \in (0,1]$:*

1. $\mathcal{L}_N^{\alpha}\{\alpha\sigma(x) + \beta\lambda(x)\} = \alpha\mathcal{L}_N^{\alpha}\{\sigma(x)\} + \beta\mathcal{L}_N^{\alpha}\{\lambda(x)\}$.
2. $\mathcal{L}_N^{\alpha}\{N_Y^{\alpha}\sigma(x)\} = -\sigma(0) - r\mathcal{L}_N^{\alpha}\{\sigma(x)\}$.
3. $\mathcal{L}_N^{\alpha}\left\{J_{Y,0}^{\alpha}(\sigma(r))(x)\right\} = \frac{1}{r}\mathcal{L}_N^{\alpha}\{\sigma(x)\}$.
4. $\mathcal{L}_N^{\alpha}\{(N_Y^{\alpha})^n\sigma(x)\} = -\sum_{i=1}^{n}(-1)^i r^{n-i}\left((N_Y^{\alpha})^{i-1}\right)\sigma(0) - r^n\mathcal{L}_N^{\alpha}\{\sigma(x)\}$, *where* $(N_Y^{\alpha})^n = \underbrace{N_Y^{\alpha}\circ N_Y^{\alpha}\circ \quad . \quad . \quad . \quad \circ N_Y^{\alpha}}_{n \quad times}$.

Proof. The first equality is verified using properties of the integral.

$$\begin{aligned}
\mathcal{L}_N^{\alpha}\{\alpha\sigma(x) + \beta\lambda(x)\} &= \int_0^{\infty}\frac{E_{\alpha}^{N}(-r,x)(\alpha\sigma(x) + \beta\lambda(x))}{Y(x,\alpha)}dx \\
&= \int_0^{\infty}\frac{\left(E_{\alpha}^{N}(-r,x)\alpha\sigma(x) + E_{\alpha}^{N}(-r,x)\beta\lambda(x)\right)}{Y(x,\alpha)}dx \\
&= \int_0^{\infty}\frac{E_{\alpha}^{N}(-r,x)\alpha\sigma(x)}{Y(x,\alpha)}dx + \int_0^{\infty}\frac{E_{\alpha}^{N}(-r,x)\beta\lambda(x)}{Y(x,\alpha)}dx, \\
&= \alpha\int_0^{\infty}\frac{E_{\alpha}^{N}(-r,x)\sigma(x)}{Y(x,\alpha)}dx + \beta\int_0^{\infty}\frac{E_{\alpha}^{N}(-r,x)\lambda(x)}{Y(x,\alpha)}dx,
\end{aligned}$$

then

$$\mathcal{L}_N^{\alpha}\{\alpha\sigma(x) + \beta\lambda(x)\} = \alpha\mathcal{L}_N^{\alpha}\{\sigma(x)\} + \beta\mathcal{L}_N^{\alpha}\{\lambda(x)\}.$$

By definition 1 and integrating the following by parts, we obtain the second equality.

$$\mathcal{L}_N^{\alpha}\{N_Y^{\alpha}\sigma(x)\} = \int_0^{\infty} \frac{E_{\alpha}^N(-r,x)N_Y^{\alpha}\sigma(x)}{Y(x,\alpha)}dx = \int_0^{\infty} E_{\alpha}^N(-r,x)\sigma'(x)\,dx.$$

We apply the second equality to the function $\lambda(x) = J_{Y,0}^{\alpha}(\sigma(r))(x)$ to obtain the third equality, because $\lambda(0) = 0$ and its N-derivative is $N_Y^{\alpha}\lambda(x) = \sigma(x)$. Finally, iterating the second equality n times, we obtain property $4)$. ■

Since the function $\mathcal{Y}(x,\alpha)$ has the property $\mathcal{Y}'(x,\alpha) > 0$, then we have the following relation between the Generalized Laplace Transform and the classical one.

Theorem 13 *For any real function σ we have*

$$(\mathcal{L}_N^{\alpha}\{\sigma(x)\})(r) = \left(\mathcal{L}\left\{\sigma\left(\mathcal{Y}(x,\alpha)^{-1}\right)\right\}\right)(r) \tag{13}$$

Proof. For the proof we use the change of variables $u = \mathcal{Y}(x,\alpha)$

$$\begin{aligned}(\mathcal{L}_N^{\alpha}\{\sigma(x)\})(r) &= \int_0^{\infty} \frac{E_{\alpha}^N(-r,x)\sigma(x)}{Y(x,\alpha)}dx = \int_0^{\infty} e^{-ru}\sigma\left(\mathcal{Y}(x,\alpha)^{-1}\right)du \\ &= \left(\mathcal{L}\left\{\sigma\left(\mathcal{Y}(x,\alpha)^{-1}\right)\right\}\right)(r).\end{aligned}$$

■

Definition 14 *A function $\delta : [0,\infty) \to \mathcal{R}$ is said to be of $\sigma(x)$ −exponential order if and only if exist non-negative constants M, γ and X such that $|\delta(x)| \leq Me^{\gamma\sigma(x)}$ for $x \geq X$.*

Theorem 15 *If $\delta : [0,\infty) \to \mathcal{R}$ is a piecewise function of $\mathcal{Y}(x,\alpha)$ −exponential order, then the Generalized Laplace Transform exists for $r > \gamma$.*

Proof.

$$\left|\left(J_{Y,0}^{\alpha}E_{\alpha}^N(-r,x)\sigma(x)\right)(\mu)\right| \leq \left(J_{Y,0}^{\alpha}E_{\alpha}^N(-r,x)\left|\sigma(x)\right|\right)(\mu) \leq M\left(J_{Y,0}^{\alpha}e^{-r\mathcal{Y}(x,\alpha)}e^{\gamma\mathcal{Y}(x,\alpha)}\right)(\mu).$$

Applying the change of variables used in Theorem 13, we obtain

$$\begin{aligned}\left|\left(J_{Y,0}^{\alpha}E_{\alpha}^{N}(-r,x)\sigma\left(x\right)\right)\left(\mu\right)\right| &\leq M\int_{0}^{\mu}e^{(\gamma-r)v}dv\\ &= \frac{M}{r-\gamma}\left(1-e^{(\gamma-r)\mu}\right)\xrightarrow{\mu\to+\infty}\frac{M}{r-\gamma}.\end{aligned}$$

■

Theorem 16 *If* $\alpha\in(0,1]$, *then we have*

1. $\mathcal{L}_{N}^{\alpha}\{1\}=\frac{1}{r}$ *for* $r>0$.
2. $\mathcal{L}_{N}^{\alpha}\left\{E_{\alpha}^{N}(\gamma,x)\right\}=\frac{1}{r-\gamma}$ *for* $\gamma<r$.
3. $\mathcal{L}_{N}^{\alpha}\left\{\sigma\left(x\right)E_{\alpha}^{N}(\gamma,x)\right\}=\sigma\left(r-\gamma\right)$.
4. $\mathcal{L}_{N}^{\alpha}\left\{sin\left(\gamma\mathcal{Y}\left(x,\alpha\right)\right)\right\}=\frac{\gamma}{r^{2}+\gamma^{2}}$.
5. $\mathcal{L}_{N}^{\alpha}\left\{cos\left(\gamma\mathcal{Y}\left(x,\alpha\right)\right)\right\}=\frac{r}{r^{2}+\gamma^{2}}$.
6. $\mathcal{L}_{N}^{\alpha}\left\{sinh\left(\gamma\mathcal{Y}\left(x,\alpha\right)\right)\right\}=\frac{\gamma}{r^{2}-\gamma^{2}}$.
7. $\mathcal{L}_{N}^{\alpha}\left\{cosh\left(\gamma\mathcal{Y}\left(x,\alpha\right)\right)\right\}=\frac{r}{r^{2}-\gamma^{2}}$.

Proof.

1. $\mathcal{L}_{N}^{\alpha}\{1\}=\int_{0}^{\infty}\frac{E_{\alpha}^{N}(-r,x)}{Y(x,\alpha)}dx=\int_{0}^{\infty}e^{-ru}du=-\frac{1}{r}\lim_{\mu\to\infty}\left[e^{-r\mu}-1\right]=\frac{1}{r}$.

2. $\mathcal{L}_{N}^{\alpha}\left\{E_{\alpha}^{N}(\gamma,x)\right\}=\int_{0}^{\infty}\frac{E_{\alpha}^{N}(-r,x)E_{\alpha}^{N}(\gamma,x)}{Y(x,\alpha)}dx=\int_{0}^{\infty}e^{-(r-\gamma)u}du=$
$=\frac{1}{r-\gamma}\lim_{\mu\to\infty}\left[1-e^{(\gamma-r)\mu}\right]=\frac{1}{r-\gamma}$.

3. $\mathcal{L}_{N}^{\alpha}\left\{\sigma\left(x\right)E_{\alpha}^{N}(\gamma,x)\right\}=\int_{0}^{\infty}\frac{E_{\alpha}^{N}(-r,x)\sigma(x)E_{\alpha}^{N}(\gamma,x)}{Y(x,\alpha)}dx=$
$=\int_{0}^{\infty}\frac{E_{\alpha}^{N}(-r+\gamma,x)\sigma(x)}{Y(x,\alpha)}du=\sigma\left(r-\gamma\right)$.

4. $\mathcal{L}_{N}^{\alpha}\left\{sin\left(\gamma\mathcal{Y}\left(x,\alpha\right)\right)\right\}=\int_{0}^{\infty}\frac{E_{\alpha}^{N}(-r,x)sin(\gamma\mathcal{Y}(x,\alpha))}{Y(x,\alpha)}dx=$
$=\int_{0}^{\infty}\frac{e^{-rY(x,\alpha)}sin(\gamma\mathcal{Y}(x,\alpha))}{Y(x,\alpha)}dx=\int_{0}^{\infty}e^{-ru}sin\left(\gamma u\right)du$, integrating by parts twice the last integral, we obtain the result.

5. Analogous to the previous item.

6. Knowing that $sinh\,(\alpha) = \frac{e^{\alpha}-e^{-\alpha}}{2}$ we get,

$$\begin{aligned}
&\mathcal{L}_N^{\alpha}\left\{sinh\left(\gamma\mathcal{Y}\left(x,\alpha\right)\right)\right\} = \\
= \;& \frac{1}{2}\int_0^{\infty}\frac{E_{\alpha}^{N}(-r,x)\left(e^{\gamma\mathcal{Y}(x,\alpha)}-e^{-\gamma\mathcal{Y}(x,\alpha)}\right)}{Y(x,\alpha)}dx \\
= \;& \frac{1}{2}\left(\int_0^{\infty}\frac{e^{-(r-\gamma)\mathcal{Y}(x,\alpha)}}{Y(x,\alpha)}du - \int_0^{\infty}\frac{e^{-(r+\gamma)\mathcal{Y}(x,\alpha)}}{Y(x,\alpha)}du\right) \\
= \;& \frac{1}{2}\left(\frac{1}{r-\gamma}-\frac{1}{r+\gamma}\right) \\
= \;& \frac{\gamma}{r^2-\gamma^2}.
\end{aligned}$$

7. We know that $cosh\,(\alpha) = \frac{e^{\alpha}+e^{-\alpha}}{2}$, and proceeding analogously to the previous item, we have

$$\begin{aligned}
&\mathcal{L}_N^{\alpha}\left\{cosh\left(\gamma\mathcal{Y}\left(x,\alpha\right)\right)\right\} = \\
= \;& \frac{1}{2}\int_0^{\infty}\frac{E_{\alpha}^{N}(-r,x)\left(e^{\gamma\mathcal{Y}(x,\alpha)}+e^{-\gamma\mathcal{Y}(x,\alpha)}\right)}{Y(x,\alpha)}dx \\
= \;& \frac{1}{2}\left(\int_0^{\infty}\frac{e^{-(r-\gamma)\mathcal{Y}(x,\alpha)}}{Y(x,\alpha)}du + \int_0^{\infty}\frac{e^{-(r+\gamma)\mathcal{Y}(x,\alpha)}}{Y(x,\alpha)}du\right) \\
= \;& \frac{1}{2}\left(\frac{1}{r-\gamma}+\frac{1}{r+\gamma}\right) \\
= \;& \frac{r}{r^2-\gamma^2}.
\end{aligned}$$

■

To complete the theoretical body the following result is necessary.

Definition 17 *Let δ and σ be two functions which are piecewise continuous at each interval $[0, X]$ and of generalized exponential order. We define the N-convolution of δ and σ by*

$$(\delta * \sigma)_N(x) = \int_0^{x}\delta(\mu)\sigma\left[\mathcal{Y}^{-1}(\mathcal{Y}(x,\alpha)-\mathcal{Y}(\mu,\alpha))\right]\frac{d\mu}{Y(\mu,\alpha)},\, x \le X. \quad (14)$$

The commutativity of the N-convolution is given in the following result.

Lemma 18 *Let δ and σ be two functions which are piecewise continuous at each interval $[0, X]$ and of generalized exponential order. Then*

$$(\delta * \sigma)_N(x) = (\sigma * \delta)_N(x). \tag{15}$$

Proof. From Definition 17, we prove using the change of variables $\mathcal{Y}(u,\alpha) = \mathcal{Y}(x,\alpha) - \mathcal{Y}(\mu,\alpha)$. ■

Below we present the N-Laplace transform of the N-convolution.

Theorem 19 *Let δ and σ be two functions which are piecewise continuous at each interval $[0, X]$ and of generalized exponential order. Then*

$$\mathcal{L}_N^\alpha \{(\delta * \sigma)_N\} = \mathcal{L}_N^\alpha\{\delta\}\mathcal{L}_N^\alpha\{\sigma\}. \tag{16}$$

Proof. Starting from the second member of the previous equality, $\mathcal{L}_N^\alpha\{\delta\}\mathcal{L}_N^\alpha\{\sigma\}$, choosing u such that $\mathcal{Y}(u,\alpha) = \mathcal{Y}(x,\alpha) - \mathcal{Y}(\mu,\alpha)$ we arrive at result, changing the order of integration. ■

Applications to Generalized Differential Equations

It is interesting to study the Cauchy Problem from the Generalized Laplace Transform, as follows

$$N_Y^\alpha (n(x)) = A(x)\, n(x) + \delta(x, n(x)) \tag{17}$$

being n continuous with $n(0) = n_0 + \sigma(n)$, and $0 \le x \le X_0$, A is a sectorial operator which generates a strongly analytic semigroup $(X(x))_{x\ge 0}$ on the Banach Space $(N, \|\cdot\|)$.

If we apply the Generalized Laplace Transform on (17), it gives us

$$\mathcal{L}_N^\alpha \{n(x)\}(r) = (r - A)^{-1}[n_0 + \sigma(n)] + (r - A)^{-1}\mathcal{L}_N^\alpha \{\delta(x, n(x))\}(r), \tag{18}$$

applying the inverse Laplace transform, we have

$$n(x) = X\left(\int_0^x \frac{dr}{Y(r,\alpha)}\right)[n_0 + \sigma(n)] + \int_0^x X\left(\int_0^x \frac{dr}{Y(r,\alpha)} - \int_0^r \frac{du}{Y(u,\alpha)}\right)\delta(r, n(r))\, dr, \tag{19}$$

where $X(x) = \int_0^\infty e^{rx}(r - A(r))^{-1} dr$.

Theorem 20 *Assuming that*

1. *The functions* $\delta(x,\cdot) : N \to N,\quad \delta(\cdot, n) : [0, \mathbb{R}_0] \to N$ *are continuous and there exists a function* $\Delta_m \in L([0,\mathbb{R}_0], \mathbb{R}^+)$ *such that* $\underset{\|n\|\leq m}{Sup} \|\delta(x,n)\| \leq \Delta_m(x)$.
2. *There exists a constant* $P > 0$ *such that* $\|\sigma(y) - \sigma(n)\| < P|y - n|$ *for all* n, y.
3. $(X(x))_{x>0}$ *is compact,*

the Cauchy Problem has at least one solution.

Proof. Let $m \geq \dfrac{\underset{x\in[0,X_0]}{Sup}\left|X\left(\int_0^x \frac{dr}{Y(r,\alpha)}\right)\right|\left[\|n_0\| + \|\sigma(0)\| + \int_0^{X_0} \frac{dr}{Y(r,\alpha)}\|\Delta_m\|_{L^\infty}\right]}{1 - P\underset{x\in[0,X_0]}{Sup}\left|X\left(\int_0^x \frac{dr}{Y(r,\alpha)}\right)\right|}$, let B_m the closed ball centered at the origin of radius m and let $n \in B_m$.

We are going to prove that the operator

$$X\left(\int_0^x \frac{dr}{Y(r,\alpha)}\right)[n_0 + \sigma(n)] + \int_0^x X\left(\int_0^x \frac{dr}{Y(r,\alpha)} - \int_0^r \frac{du}{Y(u,\alpha)}\right)\delta(r, n(r))\, dr$$

has at least one fixed point that is the solution of the Cauchy Problem.

i) The expression $X\left(\int_0^x \frac{dr}{Y(r,\alpha)}\right)[n_0 + \sigma(n)]$ is an contraction operator according with the assumptions 1) and 2).

ii) Let $\Psi(n)(x) = \int_0^x X\left(\int_0^x \frac{dr}{Y(r,\alpha)} - \int_0^r \frac{du}{Y(u,\alpha)}\right)\delta(r, n(r))\, dr$ we will show that $\Psi(n)$ is continuous and compact.

$\Psi(n)$ is continuous: let $(n_j)_{j\in\mathbb{N}}$such that $n_j \to n$ then

$$|\Psi(n_j)(x) - \Psi(n)| \leq$$

$$\underset{x\in[0,X_0]}{Sup}\left|X\left(\int_0^x \frac{dr}{Y(r,\alpha)}\right)\right|\left[\int_0^{X_0} \frac{dr}{Y(r,\alpha)}\|\delta(r,n_j(r)) - \delta(r,n(r))\|_{L^\infty}\right],$$

for the continuity of the function δ we have $\|\delta(r,n_j(r)) - \delta(r,n(r))\|_{L^\infty} \to 0$ as $j \to \infty$ hence $|\Psi(n_j)(x) - \Psi(n)| \to 0$ as $j \to \infty$.

$\Psi(n)$ is compact.

■

Theorem 21 *If we assume that*

1. *The function $\delta(\cdot,n) : [0,X_0] \to N$ is continuous.*
2. *There exists a constant $K > 0$ such that $\|\sigma(y) - \sigma(n)\| < K\|y - n\|$ for all n, y in N,*

then the Cauchy Problem has a unique solution.

Proof. Let

$$\Psi(x) = X\left(\int_0^x \frac{dr}{Y(r,\alpha)}\right)[n_0 + \sigma(n)] +$$
$$+\int_0^x X\left(\int_0^x \frac{dr}{Y(r,\alpha)} - \int_0^r \frac{du}{Y(u,\alpha)}\right)\delta(r,n(r))\,dr$$

and let n, y in N, then

$$\begin{aligned}
&\Psi(n)(x) - \Psi(y)(x) = \\
= \; & X\left(\int_0^x \frac{dr}{Y(r,\alpha)}\right)[\sigma(n) - \sigma(y)] \\
+ \; & \int_0^x X\left(\int_0^x \frac{dr}{Y(r,\alpha)} - \int_0^r \frac{du}{Y(u,\alpha)}\right)[\delta(r,n(r)) - \delta(r,y(r))]\,dr
\end{aligned}$$

taking the norm of the difference and then the supreme for all $x \in [0,X_0]$ we obtain

$$|\Psi(n) - \Psi(y)| \leq K(Y(X_0, \alpha) + 1) \underset{x \in [0, X_0]}{Sup} \left| X\left(\int_0^x \frac{dr}{Y(r, \alpha)} \right) \right| |x - y|$$

By Banach's contraction principle, it is clear that the operator $\Psi(x)$ has a unique fixed point, which is precisely the solution to the Cauchy Fractional Problem. ■

Theorem 22 *Let n and y be solutions for the Cauchy Problem associated to n_0, y_0 respectively. Assuming that the conditions of the Theorem 21 are met, then we have the following estimate*

$$|y - n| \leq \frac{\underset{x \in [0, X_0]}{Sup} \left| X\left(\int_0^x \frac{dr}{Y(r,\alpha)} \right) \right|}{1 - \underset{x \in [0, X_0]}{Sup} \left| X\left(\int_0^x \frac{dr}{Y(r,\alpha)} \right) \right| K\left(1 + \int_0^x \frac{dr}{Y(r,\alpha)} \right)} \tag{20}$$

Remark 23 *In this theorem we refer to the stability of the solution for the Cauchy Problem, that is the dependence of the solution to the initial conditions.*

Proof. For $x \in [0, X_0]$

$$y(x) - n(x) = X\left(\int_0^x \frac{dr}{Y(r, \alpha)} \right) [y_0 - n_0 + \sigma(y) - \sigma(n)] +$$

$$+ \int_0^x X\left(\int_0^x \frac{dr}{Y(r, \alpha)} - \int_0^r \frac{du}{Y(u, \alpha)} \right) [\delta(r, y(r)) - \delta(r, n(r))] dr$$

taking the supreme on both sides we obtain

$$|y - n| \leq$$

$$\leq \underset{x \in [0, X_0]}{Sup} \left| X\left(\int_0^x \frac{dr}{Y(r, \alpha)} \right) \right| \left[\|y_0 - n_0\| + K\left(1 + \int_0^x \frac{dr}{Y(r, \alpha)} \right) |y - n| \right]$$

from this last inequality, we find the estimate. ■

Remark 24 *On other results regarding the stability of the solutions of some generalized differential equations, you can consult Sigarreta [49].*

On the Stability

Let us consider the following system

$$N_Y^{\alpha}\overrightarrow{n}(x) = A\overrightarrow{n}(x) + \overrightarrow{\delta}(x) \tag{21}$$

and its corresponding associated homogeneous system

$$N_Y^{\alpha}\overrightarrow{n}(x) = A\overrightarrow{n}(x)\,. \tag{22}$$

With $\overrightarrow{n} \in \mathbb{R}^n$, $A = [a_{ij}] \in \mathbb{R}^n \times \mathbb{R}^n$, $\overrightarrow{\delta}(x) = [\delta_1(x), \delta_2(x), ..., \delta_n(x)]^T$.

Definition 25 *The solution $\overrightarrow{n}(x)$ of system* (22) *is called stable if, for any initial condition n_0, there exists $\varepsilon > 0$ such that $\|\overrightarrow{n}(x)\| < \varepsilon$ for all $x > 0$. The solution is called asymptotically stable if it is stable and $\|\overrightarrow{n}(x)\| \to 0$ as $x \to \infty$.*

Theorem 26 *The solution of system* (22) *is given by*

$$\overrightarrow{n}(x) = \overrightarrow{n}_0 E_\alpha^N(1, x) \tag{23}$$

whenever the solution is differentiable on $[0, \infty)$ and $E_\alpha^N(1, x)$ is the generalized exponential order of Definition 9.

Proof. It is enough to use the properties of the generalized exponential order function. ■

Theorem 27 *The system* (22) *is is asymptotically stable if and only if the eigenvalues of A have strictly negative real parts.*

Proof. It is enough to take limit in (23). ■

As is known, the eigenvalues are obtained by solving the system $|A - \lambda| = 0$, with A the matrix of system (22).

Now we will see how to obtain the explicit solution of the system (21).

For this, we will apply the generalized Laplace transform $\mathcal{L}_N^\alpha$ on both sides of the system (21), obtain

$$\begin{pmatrix} rN_1(r) - n_1(0) \\ rN_2(r) - n_2(0) \\ .... \\ rN_j(s) - n_j(0) \end{pmatrix} = A \begin{pmatrix} N_1(r) \\ N_2(r) \\ .... \\ N_j(r) \end{pmatrix} + \begin{pmatrix} Y_1(r) \\ Y_2(r) \\ .... \\ Y_j(r) \end{pmatrix} \tag{24}$$

where $N_i(r) = \mathcal{L}_N^{\alpha}[n_i(x)]$, $Y_i(r) = \mathcal{L}_N^{\alpha}[\delta_i(x)]$ with $i = 1, 2, ..., j$.

$$\begin{pmatrix} N_1(r) \\ N_2(r) \\ .... \\ N_j(r) \end{pmatrix} = \begin{pmatrix} \frac{1}{r-a_{11}} \left(n_1(0) + Y_1(r) + \sum_{i=1, i\neq 1}^{j} a_{1i} N_i(r) \right) \\ \frac{1}{r-a_{22}} \left(n_2(0) + Y_2(r) + \sum_{i=1, i\neq 2}^{j} a_{2i} N_i(r) \right) \\ .... \\ \frac{1}{r-a_{jj}} \left(n_j(0) + Y_j(r) + \sum_{i=1, i\neq j}^{j} a_{ji} N_i(r) \right) \end{pmatrix} \tag{25}$$

Solving this system and applying the Inverse Transform we obtain the desired solution.

Remark 28 *The results of this section contain as a particular case, those of Al-Habahbeh (2022)[50], relative to commensurable systems.*

Remark 29 *If we consider $Y(x, \alpha) = x^{1-\alpha}$, that is, the conformable case, the results obtained in Fahad (2023)[11], Khan (2018)[15], Molaei (2022)[18], Ozkan (2018)[19], Silva (2018)[20], Younis (2022)[23] can be generalized with our kernel.*

Conclusion

In this work we have taken up the recently defined Generalized Laplace Transform and have illustrated its usefulness with some applications. Throughout the Chapter, we have shown that several results known from the literature are particular cases of the Generalized Laplace Transform of the Definition 10.

References

[1] Nápoles Valdés, J. E. *Differential equations and contemporaneity*, Revista Brasileira de História da Matemática, 2007, 7(14), 213-232.

[2] Nápoles Valdés, J. E. *Ordinary differential equations as signs of the times*, Revista Eureka, 2006, 21, 39-75.

[3] Nápoles Valdés, J. E. *A century of qualitative theory of differential equations*, Lecturas Matemáticas, 2004, Volumen 25, 59-111.

[4] Nápoles Valdés, J. E. *The historical legacy of ordinary differential equations. (Self) critical considerations*, Boletín de Matemáticas, 1998, 53-79.

[5] Fleitas, A.; Nápoles, J. E.; Rodríguez, J. M.; Sigarreta, J. M. *Note on the generalized conformable derivative*, Revista de la Unin Matemtica Argentina, 2021, Vol. 62, No. 2, Pages 443457. doi.org/10.33044/revuma.1930

[6] Guzmán, P. M.; Langton, G.; Lugo, L. M.; Medina, J., Nápoles Valdés, J. E. *A new definition of a fractional derivative of local type*, J. Math. Anal. 2018, 9:2, 88-98. ISSN 2217-3412.

[7] Khalil, R.; Al Horani, M.; Yousef, A.; Sababheh, M. *A new definition of fractional derivative*. J. Comput. Appl. Math. 2014, 264, 65-70. https://doi.org/10.1016/j.cam.2014.01.002

[8] Nápoles, J. E.; Guzmán, P. M.; Lugo, L. M.; Kashuri, A. *The local non conformable derivative and Mittag Leffler function*, Sigma J Eng & Nat Sci. 2020, 38, 1007-1017.

[9] Zhao, D.; Luo, M. *General conformable fractional derivative and its physical interpretation*, Calcolo, 2017, 54: 903-917, DOI: 10.1007/s10092-017-0213-8.

[10] Bosch, P.; Carmenate, H. J.; Rodríguez, J. M.; Sigarreta, J. M. *On the Generalized Laplace Transform*, Symmetry 2021, 13, 669. https://doi.org/10.3390/sym13040669

[11] Fahad, H. M.; Rehman, M. U.; Fernandez, A. *On Laplace transforms with respect to functions and their applications to fractional differential equations*. Math Meth Appl Sci. 2023; 46(7): 83048323. https://doi.org/10.1002/mma.7772

[12] Jarad, F.; Abdeljawad, T. *Generalized fractional derivatives and Laplace transform*, Discrete and continuous dynamical systems series S, 2020, Volume 13, Number 3, 709-722. doi:10.3934/dcdss.2020039

[13] Jarad, F.; Abdeljawad, T. *A modified Laplace transform for certain generalized fractional operators*, Results in Nonlinear Analysis 1, 2018, No. 2, 88-98.

[14] Khader, A. H. *The conformable Laplace transform of the fractional Chebyshev and Legendre polynnomials*, MSc.Thesis, Zarqa University; 2017.

[15] Khan, N. A.; Razzaq, O. A.; Ayaz, M. *Some properties and applications of conformable fractional Laplace transform (CFLT)*, Journal of Fractional Calculus and Applications, 2018, Vol. 9(1), 72-81.

[16] Lugo, L. M.; Npoles Valds, J. E.; Vivas-Cortez, M. *On the Generalized Laplaces Equation*, Appl. Math. Inf. Sci. 2022, 16, No. 2, 169-176. http://dx.doi.org/10.18576/amis/160204

[17] Martínez, F.; Mohammed, P. O.; Nápoles Valdés, J. E. *Non Conformable Fractional Laplace Transform*, Kragujevac J. of Math., Kragujevac Journal of Mathematics, 2022, Volume 46(3), 341-354. DOI 10.46793/KgJMat2203.341M

[18] Molaei, M.; Dastmalchi Saei, F.; Javidi, M.; Mahmoudi, Y. *Solving a class of ordinary differential equations and fractional differential equations with conformable derivative by fractional Laplace transform*, Turk J Math, 2022, 46: 3025-3044. doi:10.55730/1300-0098.3316

[19] Ozkan, O.; Kurt, A. *The analytical solutions for conformable integral equations and integro-differential equations by conformable Laplace transform*, Opt Quant Electron, 2018, 50:81 https://doi.org/10.1007/s11082-018-1342-2

[20] Silva, F. S.; Moreira, D. M.; Moret, M. A. *Conformable Laplace Transform of Fractional Differential Equations*, Axioms 2018, 7, 55; doi:10.3390/axioms7030055

[21] Vivas-Cortez, M.; Nápoles Valdés, J. E.; Hernández Hernndez, J. E.; Velasco Velasco, J.; Larreal, O. *On Non Conformable Fractional Laplace Transform*, Appl. Math. Inf. Sci. 2021, 15, No. 4, 403-409. http://dx.doi.org/10.18576/amis/150401

[22] Yavuz, M.; Sene, N. *Approximate Solutions of the Model Describing Fluid Flow Using Generalized ρ-Laplace Transform Method and Heat Balance Integral Method*, Axioms 2020, 9, 123, doi:10.3390/axioms9040123

[23] Younis, J.; Ahmed, B.; AlJazzazi, M.; Al Hejaj, R.; Aydi, H. *Existence and Uniqueness Study of the Conformable Laplace Transform*, Journal of Mathematics, 2022, Article ID 4554065, 7 pages doi.org/10.1155/2022/4554065

[24] Guzmán, P. M.; Lugo Motta Bittencurt, L. M.; Nápoles Valdés, J. E. *On the stability of solutions of fractional non conformable differential equations*, Stud. Univ. Babe s-Bolyai Math. 65, 2020, No. 4, 495-502. http://dx.doi.org/10.24193/subbmath.2020.4.02

[25] Guzmán, P. M.; Lugo Motta Bittencurt, L. M.; Nápoles Valdés, J. E. *A note on the qualitative behavior of some nonlinear local improper conformable differential equations*. Journal of Fractional Calculus and Nonlinear Systems, 1(1), 2020 1320. https://doi.org/10.48185/jfcns.v1i1.48

[26] Martínez, F.; Nápoles Valdés, J. E. *Towards a Non-Conformable Fractional Calculus of N-variables*, Journal of Mathematics and Applications, 2020, No 43, 87-98.

[27] Nápoles Valdés, J. E.; Tunc, C. *On the boundedness and oscillation of non-conformable Linard equation*, Journal of Fractional Calculus and Applications, 2020, Vol. 11(2), 92-101.

[28] Nápoles Valdés, J. E.; Rodríguez, J. M.; Sigarreta, J. M. *On Hermite-Hadamard type inequalities for non-conformable integral operators*, Symme-

try 2019, 11, 1108; doi:10.3390/sym11091108.

[29] Nápoles Valdés, J. E.; Guzmán, P. M.; Lugo, L. M. *Some New Results on Nonconformable Fractional Calculus*, Advances in Dynamical Systems and Applications, 2018, ISSN 0973-5321, Volume 13, Number 2, 167175.

[30] Nápoles Valdés, J. E., Quevedo, M. N., Gómez Plata, A. R. *On the asymptotic behavior of a generalized nonlinear equation*, Sigma J Eng & Nat Sci 38 (4), 2020, 2109-2121

[31] Nápoles Valdés, J. E.; Quevedo, M. N. *On the Oscillatory Nature of Some Generalized Emden-Fowler Equation*, Punjab University Journal of Mathematics, 2020, Vol. 52(6), 97-106.

[32] AbreuBlaya, R.; Fleitas, A.; Nápoles Valdés, J. E.; Reyes, R.; Rodríguez, J. M, Sigarreta, J. M. *On the conformable fractional logistic models*, Math Meth Appl Sci. 2020; 1-12. https://doi.org/10.1002/mma.6180

[33] Fleitas, A.; Gómez-Aguilar, J. F.; Nápoles Valdés, J. E.; Rodríguez, J. M.; Sigarreta, J. M. *Analysis of the local Drude model involving the generalized fractional derivative*, Optik, Volume 193, 2019, 163008, ISSN 0030-4026, https://doi.org/10.1016/j.ijleo.2019.163008.

[34] Fleitas, A.; Méndez, J. A.; Nápoles Valdés, J. E.; Sigarreta, J. M. *On the some classical systems of Liénard in general context*, Revista Mexicana de Física, 2019, 65 (6) 618-625.

[35] Guzmán, P. M.; Lugo, L. M.; Nápoles Valdés, J. E.; Vivas, M. *On a New Generalized Integral Operator and Certain Operating Properties*, Axioms 2020, 9, 69. https://doi.org/10.3390/axioms9020069

[36] Abdeljawad, T. *On conformable fractional calculus*, J. Comput. Appl. Math. 2015, 279, 57-66. https://doi.org/10.1016/j.cam.2014.10.016

[37] Capelas de Oliveira, E.; Tenreiro Machado, J. A. *A Review of Definitions for Fractional Derivatives and Integral*, Mathematical Problems in Engineering Volume 2014, Article ID 238459, 6 pages http://dx.doi.org/10.1155/2014/238459

[38] Cinar, I. *On Some Properties of Generalized Riesz Potentials*, Intern. Math. Journal, 2003, Vol. 3, no. 12, 1393-1397.

[39] Gorenflo, R.; Mainardi, F. *Fractional Calculus: Integral and Differential Equations of Fractional Order*, Springer, Wien, 1997, 223-276. https://doi.org/10.48550/arXiv.0805.3823

[40] Katugampola, U.N. *A new approach to generalized fractional derivatives*, Bull. Math. Anal. App. 2014, 6, 1-15. https://doi.org/10.48550/arXiv.1106.0965

[41] Katugampola, U.N. *New approach to a generalized fractional integral*, Appl. Math. Comput. 2011, 218, 860-865. https://doi.org/10.1016/j.amc.2011.03.062

[42] Kilbas, A. A.; Marichev, O. I.; Samko, S. G. *Fractional Integrals and Derivatives, Theory and Applications*. Gordon & Breach, Switzerland, 1993.

[43] Machado, J. T.; Kiryakova, V.; Mainardi, F. *Recent history of fractional calculus*, Commun. Nonlinear Sci. Numer. Simul., 2011, 16, 11401153. https://doi.org/10.1016/j.cnsns.2010.05.027

[44] Vivas-Cortez, M.; Nápoles Valdés, J. E.; Lugo, L. M. *On a Generalized Laplace Transform*, Appl. Math. Inf. Sci. 2021, 15, No. 5, 667-675. http://dx.doi.org/10.18576/amis/150516

[45] Al-Zhour, Z.; Al-Mutairi, N.; Alrawajeh, F.; Alkhasawneh, R. *New theoretical results and applications on conformable fractional Natural transform*, Ain Shams Engineering Journal, Volume 12, Issue 1, 2021, Pages 927-933, ISSN 2090-4479, https://doi.org/10.1016/j.asej.2020.07.006.

[46] Bouaouid, M.; Hilal, K.; Melliani, S. *Existence of mild solutions for conformable fractional differential equations with nonlocal conditions*, Rocky Mountain J. Math. 50, 2020, no. 3, 871-879. doi:10.1216/rmj.2020.50.871

[47] El-Ajou, A.; Oqielat, M.N.; Al-Zhour, Z. et al. *A Class of Linear Non-Homogenous Higher Order Matrix Fractional Differential Equations: Analytical Solutions and New Technique*. Fract Calc Appl Anal 23, 356377 (2020). https://doi.org/10.1515/fca-2020-0017

[48] Oumarou, C. M. S.; Fahad, H. M.; Djida, J.D.; Fernandez, A. *On fractional calculus with analytic kernels with respect to functions*, Technical report, Arxiv.org/abs/2101.03892v1

[49] Sigarreta, J. M.; Bosch, P.; Nápoles, J. E.; Rodríguez, J. M. *Gronwall inequality and fractional differential equations*, submited

[50] Al-Habahbeh, A. *Exact solution for commensurate and incommensurate linear systems of fractional differential equations*, J. Math. Computer Sci., 2022, 28, 123-136. http://dx.doi.org/10.22436/jmcs.028.02.01

[51] Guzmán, P. M.; Nápoles Valdés, J. E. *A note on the oscillatory character of some non conformable generalized Lienard system*, Advanced Mathematical Models & Applications, 2019, Vol.4, No.2, 127-133.

Chapter 3

Laplace Transforms and Probability

Myron Hlynka*

Department of Mathematics and Statistics, University of Windsor, Windsor, Ontario, Canada

Abstract

In this chapter, we consider applications of Laplace Transforms in probability. In particular, we consider the probabilistic interpretation of Laplace Transforms. We see how this interpretation allows for simpler calculation of Laplace Transforms in certain special cases. We consider Laplace transforms of order statistics involving exponential random variables. We also consider use of Laplace transforms in the specific area of queueing theory.

Keywords: Laplace transform, probability, catastrophe, queueing, busy period, exponential random variable

1. Introduction

In this chapter, we consider applications of Laplace transforms (LT) in probability. In particular, we examine a probabilistic interpretation of the Laplace transform, which is not well known, even among specialists in probability. This probabilistic interpretation is especially useful for distributions related to the exponential distribution. We also consider applications of the probabilistic interpretation to proofs in queueing theory. This allows for much shorter derivations than other methods, and gives insights that might otherwise be overlooked.

---

* Corresponding Author's Email: hlynka@uwindsor.ca

In: The Laplace Transform and Its Applications in Science
Editor: Victor Martinez-Luaces
ISBN: 979-8-89113-747-9

2. Laplace Transforms and Generating Functions

We begin by restricting our definition of Laplace transforms to functions with domain $[0, \infty)$

Definition 2.1. *Let $f(x)$ be a function defined on $[0, \infty)$. Then the Laplace transform of $f(x)$ is $L(s) = \int_0^\infty f(x)e^{-sx}dx$.*

If we assume that $f(x)$ is a probability density function (pdf) of a continuous random variable (rv) X with non-negative support (i.e. $f(x) = 0$ for $x < 0$), then we see that $L(s) = E(e^{-sX})$, the expectation of e^{-sX}. We will informally refer to this as the Laplace transform of the random variable X. If $F(x) = P(X \leq x)$ is the cumulative distribution function (cdf) of a continuous random variable X with support on the non-negative real numbers, then define the Laplace Stieljes transform as $L(s) = \int_0^\infty e^{-sx}\, dF(x)$.

In mathematical statistics and in probability theory, random variables are often described by their moment generating functions. The moment generating function of a continuous random variable X, with non-negative support and with pdf $f(x)$, is given by

$$M(t) = \int_0^\infty f(x)e^{xt}dx = E(e^{Xt}) = L(-t). \tag{1}$$

Thus we can move between Laplace transforms and moment generating functions using $L(s) = M(-s)$ or $M(t) = L(-t)$. So $M'(t) = -L'(t)$. It is known that $M'(0) = E(X)$ so

$$L'(0) = -E(X). \tag{2}$$

We also know that $L(0) = M(0) = \int_0^\infty f(x)dx = 1$ for any pdf.

Within the applied probability area of queueing theory, it is most common to use Laplace transforms to express information whereas in other areas of probability, it is more common to use moment generating functions.

3. Probabilistic Interpretation

Laplace transforms of random variables with non-negative support (i.e., the pdf $f(x) = 0$ for $x < 0$) lend themselves very nicely to a probabilistic interpretation. This allows for easier derivation and better understanding of some properties within applied probability.

Theorem 3.1. *Let X is a continuous rv with non-negative support and pdf $f(x)$. If Y is a rv independent of X, with pdf $g(y) = se^{-sy}$, $y \geq 0$, then*

$$P(X < Y) = L(s). \tag{3}$$

Proof.

$$P(X < Y) = \int_0^\infty \int_x^\infty se^{-sy} f(x)\, dy\, dx = \int_0^\infty f(x)e^{-sx}\, dx = L(s). \tag{4}$$

□

The above property is the "probabilistic interpretation of the Laplace transform." i.e. Let X be a random variable with non-negative support. Then the Laplace transform of $f(x)$ is $P(X < Y)$ where Y is an exponential random variable with rate s, (i.e. pdf of Y is $g(y) = se^{-sy}$), and independent of X.

The Laplace transform of the pdf of the random variable X can be thought of as the probability that X wins a race against a random variable Y, where Y is exponentially distributed at rate s. The variable Y is sometimes called a "catastrophe." So the Laplace transform of a random variable X is the probability that X occurs before the catastrophe.

The original development of the probabilistic interpretation of Laplace transforms is due to Van Dantzig (1948)[1], in an obscure article, and the work lay dormant until it was revived and resuscitated by Runnenburg (1965)[2]. A major boost came from the classic queueing theory text by Kleinrock (1975)[3], Queueing Systems, Volume 1, chapter 7 (the catastrophe process, method of collective marks). Other work is due to many others, including Rade [4], Runnenburg [5], Cong ([6],[7]), Hlynka ([8], [9], [10], [11]), Stanford ([9], [11]), Brill([8], [10]). Graduate students of Hlynka and Brill added to the research on the probabilistic interpretation of Laplace transforms ([12], [13], [14], [15], [16], [17], [18], [19], [20]). Among these, Roy (1997)[19] includes a nice description of the method.

One reason that exponential random variables are important is because they are "memoryless." Formally this means the following. If X is exponentially distributed, then $P(X > x + k|X > k) = P(X > x)$ for $x, k > 0$. So if we have waited k time units already, the probability that the remaining wait until the occurrence of an event is the same as if we had just started waiting. This allows us to reset our status every time an event occurs, without keeping track of past history. This allows for major simplifications in derivations.

The following result will be useful for many of our calculations.

Theorem 3.2. *If X and Y are independent with pdf $f(x) = \lambda e^{-\lambda x}$, $x \geq 0$, and pdf $g(y) = \mu e^{-\mu y}$, $y \geq 0$, then $P(X < Y) = \dfrac{\lambda}{\lambda + \mu}$.*

Proof. This is a special case of Theorem 3.1 with $s = \mu$ and $f(x) = \lambda e^{-\lambda x}$. From (4), we have

$$P(X < Y) = \int_0^\infty \lambda e^{-\lambda x} e^{-\mu x}\, dx = \lambda \int_0^\infty e^{-(\lambda+\mu)x}\, dx = \frac{\lambda}{\lambda + \mu}. \qquad \square$$

NOTE: This result generalizes to more than 2 exponential random variables. For example, if X, Y, Z are independent exponential r.v. with rates λ, μ, and τ, then $P(X < min\{Y, Z\}) = \dfrac{\lambda}{\lambda + \mu + \tau}$.

Property 3.3. *The Laplace transform of $f(x) = \lambda e^{-\lambda x}$ is $L(s) = \dfrac{\lambda}{\lambda + s}$.*

Proof. We run a race between an exponential person X with rate λ and a "catastrophe" person Y at rate s, independent of X. The probability that X wins is just the proportion of the rate λ of X relative to the total rate of events $\lambda + s$. So the Laplace transform is $\dfrac{\lambda}{\lambda + s}$. $\square$

Our next result finds the Laplace transform of an Erlang random variable (also called a Gamma random variable), which is defined as the sum of n independent exponential random variables with the same rate.

Property 3.4. *If $X_1, \ldots, X_n$ are independent exponential random variables each with rate parameter λ, then $W = X_1 + \cdots + X_n$ has LT of form $L(s) = \left(\dfrac{\lambda}{\lambda + s}\right)^n$ (This is the LT of an Erlang random variable.)*

Proof. We run a race of $W = X_1 + \cdots + X_n$ against Y, where Y is exponential with rate s, and independent of the X_i. $L(s)$ is the probability that W beats Y. This is like a relay race, with each X_i competing with Y. We first run a race with X_1 competing against Y. X_1 wins with probability $\dfrac{\lambda}{\lambda + s}$. If X_1 wins, then we run a race of X_2 against Y. By the memoryless property, the conditional probability $P(X_1 + X_2 < Y | X_1 < Y) = P(X_2 < Y) = \dfrac{\lambda}{\lambda + s}$. Continuing in this way, we get that the probability that W wins the race is a product of identical probabilities, giving our result. $\square$

4. Laplace Transforms of Order Statistics

If we have a collection of random variables $X_1, X_2, \ldots, X_n$, then the first of these to occur ($min\{X_1, \ldots, X_n\}$) is the first order statistic, the second to occur is the second order statistic, and the last to occur ($max\{X_1, \ldots, X_n\}$) is the n-th order statistic. If the X_i ($i = 1, \ldots, n$) are independent and exponentially distributed, or have a distribution based on exponentials (such as Erlang random variables), then our probabilistic method can be helpful in finding the Laplace transforms of the order statistics. See Horn ([12]) for more complex situations.

Property 4.1. *If X_1 with pdf $f(x) = \lambda e^{-\lambda x}$ (rate parameter λ) is independent of X_2 with pdf $f(x) = \mu e^{-\mu x}$ (rate parameter μ), then $min(X_1, X_2)$ is exponentially distributed with rate parameter $\lambda + \mu$.*

Proof. We can think of a race between three exponential people X_1, X_2, and Y, where Y (the catastrophe) has rate s. The Laplace transform is the probability that either X_1 or X_2 wins the race. Thus

$$L(s) = P(X_1 wins) + P(X_2 wins) = \frac{\lambda}{\lambda + \mu + s} + \frac{\mu}{\lambda + \mu + s} = \frac{(\lambda + \mu)}{(\lambda + \mu) + s}. \tag{5}$$

We recognize this as the L.T. of an exponential r.v. with rate parameter $(\lambda + \mu)$. □

The above result extends to more than two exponential random variables. We next look at the largest order statistic from two random variables. If we were not using the probabilistic method, we would have to first find the pdf of the maximum and then compute the Laplace transform by integration. But our method allows us to avoid finding the pdf and avoid the integration.

Property 4.2. *If X_1, X_2 are independent exponential random variables with rate parameters λ and μ respectively, then the LT of $max(X_1, X_2)$ is*

$$L(s) = \frac{\lambda \mu}{\lambda + \mu + s}\left(\frac{1}{\mu + s} + \frac{1}{\lambda + s}\right). \tag{6}$$

Proof. Again, think of a race between three exponential people X_1, X_2, and Y, where Y (the catastrophe) has rate s. Let $T = max(X_1, X_2)$. Then $L_T(s) = P(T < Y) = P(max(X_1, X_2) < Y)$. The Laplace transform is the probability that both X_1 and X_2 beat Y. If X_1 is first, then we run the remainder of the race between X_2 and Y only. We then need the conditional probability that X_2 beats Y.

Similarly if X_2 is first. So

$$\begin{aligned}L_T(s) &= P(X_1 < X_2 < Y) + P(X_2 < X_1 < Y)\\ &= P(X_1 \text{ is first})P(X_2 < Y|X_1 \text{ first}) + P(X_2 \text{ is first})P(X_1 < Y|X_2 \text{ first}).\\ &= (\frac{\lambda}{\lambda+\mu+s})\frac{\mu}{\mu+s} + (\frac{\mu}{\lambda+\mu+s})\frac{\lambda}{\lambda+s} = \frac{\lambda\mu}{\lambda+\mu+s}(\frac{1}{\mu+s}+\frac{1}{\lambda+s}).\end{aligned}$$

□

Property 4.3. *If $X_1, \ldots, X_n$, are independent exponential random variables with the same rate parameter λ, then the LT of the nth order statistic is given by*

$$L(s) = \prod_{i=1}^{n} \frac{i\lambda}{i\lambda+s}. \quad (7)$$

Proof. Let $T = max\{X_i\}$. Let Y be exponentially distributed with rate s, independent of the X_i. Then $L_T(s) = P(T < Y) = P(max(X_i) < Y)$. The Laplace transform is the probability that all of the X_i beat Y. We begin with all n of the X_i racing against Y. So the total rate of the X_i is $n\lambda$, and the probability that one of them beats Y is $\frac{n\lambda}{n\lambda+s}$. When (if) one of the X_i wins, the memoryless property allows us to restart the race with the remaining $(n-1)$ of the X_i competing against Y. The probability that another X_i wins on the second part of the race is $\frac{(n-1)\lambda}{(n-1)\lambda+s}$. Continuing in the manner, we find
$L(s) = \prod_{i=1}^{n} \frac{i\lambda}{i\lambda+s}$. □

Property 4.4. *If $X_1, \ldots, X_n$, are independent exponential random variables with the same rate parameter λ, then the L.T. of the kth order statistic is given by $L(s) = \prod_{i=1}^{k} \frac{(n+1-i)\lambda}{(n+1-i)\lambda+s}$.*

Proof. Let $T = k$th order statistic. Let Y be exponentially distributed with rate s, independent of the X_i. Then $L_T(s) = P(T < Y)$. The Laplace transform is the probability that k of the X_i beat Y. We begin with all n of the X_i racing against Y. So the total rate of the X_i is $n\lambda$, and the probability that one of them beats Y is $\frac{n\lambda}{n\lambda+s}$. When (if) one of the X_i wins, the memoryless property allows us to restart the race with the remaining $(n-1)$ of the X_i competing against Y. The probability that another X_i wins on the second part of the race is $\dfrac{(n-1)\lambda}{(n-1)\lambda+s}$. Continuing in the manner for k steps, we find $L(s) = \prod_{i=1}^{k} \dfrac{(n+1-i)\lambda}{(n+1-i)\lambda+s}$ □

5. Applications in Queueing Theory

We present two well known results in queueing theory, which is the study of waiting lines. A standard reference in queueing is [3]. An $M/M/1$ queueing system has exponential interarrival times, exponential service times, and one server. The M in $M/M/1$ stands for "memoryless" or "Markov." A busy period B in a queueing system is the random variable for the time that the server is busy with no break (which may involve only one customer or many customers). Stable queueing systems generally alternate between idle periods and busy periods.

Theorem 5.1. *Assume interarrival times are independent exponential at rate λ and are independent of the service times, which are independent exponential at rate μ. The Laplace transform for the busy period B of an $M/M/1$ queueing system is given by*

$$L(s) = \frac{(\lambda+\mu+s) - \sqrt{(\lambda+\mu+s)^2 - 4\lambda\mu}}{2\lambda}. \tag{8}$$

Proof. We run a race between the busy period B and the independent catastrophe Y (with rate s). Every busy period starts with an arrival to an empty system. The next event is either an arrival or a service completion or a catastrophe. The probability that the first event is a service completion is $\dfrac{\mu}{\lambda+\mu+s}$. In this case the busy period ends and B wins the race. However, suppose the first event is an arrival, with probability $\dfrac{\lambda}{\lambda+\mu+s}$. Then there are two customers in the system. Any further arrivals before the system becomes empty will be associated with one of the two "busy periods" linked to the two customers. By the memoryless property, we now need the "busy periods" associated with each of the two customers, to end before the catastrophe. They are independent so the probability is $L(s)^2$. So $L(s)$ must satisfy the functional equation

$$L(s) = \frac{\mu}{\lambda+\mu+s} + \frac{\lambda}{\lambda+\mu+s}(L(s)^2). \tag{9}$$

Hence we get a quadratic in $L(s)$.

$$\lambda(L(s))^2 - (\lambda + \mu + s)L(s) + \mu = 0. \tag{10}$$

The solution for $L(s)$ is

$$L(s) = \frac{(\lambda + \mu + s) - \sqrt{(\lambda + \mu + s)^2 - 4\lambda\mu}}{2\lambda}.$$

The second root has numerator $(\lambda + \mu + s) + \sqrt{(\lambda + \mu + s)^2 - 4\lambda\mu}$ but $(\lambda+\mu+s)^2 - 4\lambda\mu = (\lambda-\mu)^2 + 2(\lambda+\mu)s + s^2 > 0$ for $s \geq 0$. Since $\lambda < \mu$, the numerator of the second root is greater than 2λ so the second root is greater than 1 (and hence cannot be a probability, and is rejected). □

Property 5.2. *Assume an $M/M/1$ queueing system with arrival rate λ and service rate μ, with $\lambda < \mu$. The expected value of the length of a busy period B is*

$$E(B) = \frac{1}{\mu - \lambda} \tag{11}$$

Proof. Take the derivative of the LT in (8) with respect to s. From the functional equation (9),

$$2\lambda L(s)L'(s) - (\lambda + \mu + s)L'(s) - L(s) = 0. \tag{12}$$

But by (2), $L'(0) = -E(B)$, the negative of the expected length of the busy period. Also $L(0) = 1$. Solving gives $E(B) = -L'(0) = \dfrac{1}{\mu - \lambda}$. □

We present one final queueing theory result to conclude this chapter. We define the system time of a customer to be the total time an arriving customer would spend in the system (waiting plus service).

Theorem 5.3. *Consider an $M/M/1$ queueing system with arrival rate λ and service rate μ, where $\lambda < \mu$. Then the system time of a randomly arriving customer is exponentially distributed with rate $\mu - \lambda$.*

Proof. Let W be the system time. Let $L(s)$ be the Laplace transform of W. Then $L(s) = P(W < Y)$ where Y is exponentially distributed with rate s. There are three discrete random variables which have identical distributions. One is the number of customers in the system averaged over all times. A second is the number of customers in the system as seen by an arriving customer. The third is the number of customers in the system as seen by a departing customer. As explained in Adan [21], the first two are equal by the PASTA principle (Poisson Arrivals See Time

Averages) and the latter two are equal by comparing arrival and departure distributions.

We begin our race between W and Y with the arrival of a customer. That customer will see a system with no customers (probability $1 - \lambda/\mu$) or with at least one customer (probability λ/μ). One way that W will win the race against Y occurs if the system is empty and if the service time of the arriving customer (at rate μ) beats Y. This happens with probability $(1 - \frac{\lambda}{\mu})(\frac{\mu}{\mu + s})$. The second way that W will win the race is if the system is non-empty (with probability λ/μ) and the first customer to be serviced, among those in the system, occurs before Y, and then all the remaining customers finish before the catastrophe Y (product of three probabilities). But this places us back into exactly the original queue count distribution when the original customer arrived so the third probability is a repeat of the original one. Thus we have the functional equation

$$\begin{aligned} L(s) &= P(W < Y) \\ &= P(empty)P(\text{service beats catastrophe}) \\ &\quad + P(nonempty)P(\text{service beats catastrophe})P(W < Y) \\ &= (1 - \lambda/\mu)\frac{\mu}{\mu + s} + (\lambda/\mu)(\frac{\mu}{\mu + s})L(s) \end{aligned} \tag{13}$$

Solving for $L(s)$ gives $L(s) = \dfrac{\mu - \lambda}{\mu - \lambda + s}$. So W is exponentially distributed with rate $\mu - \lambda$. □

Conclusion

We have shown that many results in probability can be presented in terms of a probabilistic interpretation of Laplace transforms. In particular, the power of the probabilistic interpretation of the Laplace transform becomes clear when considering order statistics of exponential random variables. Queueing theory often presents results in terms of Laplace transforms, which contain most of the key performance indicators for modeling queueing systems and showing the effects of changes in the parameter values. Theorem 5.1 and Theorem 5.3 are both well known queueing results. However, the proofs presented here, using the probabilistic interpretation, are both new. This concludes our discussion of the probabilistic interpretation of Laplace transforms, and emphasizes the value of Laplace transforms as a valuable tool in probability.

References

[1] Van Dantzig, D. (1949) Sur la methode des fonctions generatrices. *Colloques internationaux du CNRS.* **13:** 29–45.

[2] Runnenburg, J.T. (1965) On the use of collective marks in queueing theory. In *Congestion Theory, eds. W. L. Smith and W. E. Wilkinson*, University of North Carolina Press, Chapel Hill, 399–438.

[3] Kleinrock, L. (1975). *Queueing Systems, Volume 1*, John Wiley & Sons, New York.

[4] Rade, L. (1972) On the use of generating functions and Laplace transforms in applied probability theory. *International Journal of Mathematical Education in Science and Technology.* **3:** 25–33.

[5] Runnenburg, J.T. (1979) Van Dantzig's collective marks revisited. In *Proceedings of the Bicentennial Congress Wiskundig Genootschap, eds. P.C. Baayen, D. van Dulst and J. Oosterhoff*, Mathematisch Centrum, Amsterdam, 309–329.

[6] Cong, T.D. (1994). On the $M^X/G/\infty$ queue with heterogeneous customers in a batch. *Journal of Applied Probability.* **31:** 280–286.

[7] Cong, T.D. (1994) Application of the method of collective marks to some $M/G/1$ vacation models with exhaustive service. *Queueing Systems.* **16:** 67–81.

[8] Hlynka, M., Brill P.H. and Horn, W. (2010). A method for obtaining Laplace transforms of order statistics of Erlang random variables. *Statistics & Probability Letters.* **80:1** 9–18.

[9] Hlynka, M., Stanford, D., Poon, W. and Wang, T., (1994). Observing Queues before Joining, *Operations Research.* **42** 365–371.

[10] Brill, P.H. and Hlynka, M. (2000) An exponential queue with competition for service. *European Journal of Operational Research.* **126:3** 587–602,

[11] Stanford,D.A. and Hlynka, M. (1996) Observing general service queues before joining. *Operations Research Letters* **18:5)** 237–245.

[12] Horn, W.S. (1999) Laplace transforms of order statistics of Erlang random variables. *University of Windsor Electronic Theses and Dissertations.* https://scholar.uwindsor.ca/etd/2571

[13] Huang, Z. (2016) Probabilistic Laplace transform inversion. Major Paper (M.Sc.) University of Windsor Dept. of Mathematics and Statistics. http://web2.uwindsor.ca/math/hlynka/HuangZishan.pdf

[14] Hussein, M. (2011) Time to Ruin for Loss Reserves. Major Paper (M.Sc.) University of Windsor Dept. of Mathematics and Statistics. http://web2.uwindsor.ca/math/hlynka/mubeenmajorpaper.pdf

[15] Jahan, R. (2008) Control of Queues by Laplace Transform. Major Paper (M.Sc.) University of Windsor Dept. of Mathematics and Statistics. http://web2.uwindsor.ca/math/hlynka/rifatjahanmajorpaper.pdf

[16] Liu, C. (2016). The Coupon Collecting Problem via Laplace Transforms. Major Paper (M.Sc.) University of Windsor Dept. of Mathematics and Statistics.

[17] Nantais, J. (2020) Task Interrupted by a Poisson Process. Major Paper (M.Sc.) University of Windsor Dept. of Mathematics and Statistics.

[18] Prabhu, J. and Hlynka, M. (2021) Laplace transforms of improper random variables. *Applied Mathematical Sciences.* **15:12** 563–568 .

[19] Roy, K.A. (1997) Laplace transforms, probabilities and queues. *University of Windsor Electronic Theses and Dissertations.* https://scholar.uwindsor.ca/etd/2572

[20] Yan, B. (2014) Laplace transforms of busy periods in queues. Major Paper (M.Sc.) University of Windsor Dept. of Mathematics and Statistics. http://web2.uwindsor.ca/math/hlynka/yanbingsen.pdf

[21] Adan, I.(2003) Queueing Theory course notes. https://www.win.tue.nl/ iadan/que/

Chapter 4

Laplace Transforms and Summable Series

Russell L. Herman*
Department of Mathematics and Statistics and Department of Physics and Physical Oceanography, University of North Carolina Wilmington, Wilmington, NC, USA

Abstract

We demonstrate how Laplace transforms have been used to sum many convergent series. We then turn to classic examples of asymptotic series, like Euler's and Stirling's series, to make sense out of divergent series. We provide some historical context to the study of infinite series including the emergence of divergent series in physical settings such as rainbow supernumeraries, leading to Airy functions, and the quantum mechanical anharmonic oscillator. We introduce the Borel-Laplace transform and its use in studying asymptotic expansions, trans-series and resurgence.

Keywords: Laplace transforms, summable series, divergent series, asymptotic expansions, Borel transform, Airy function, Feynman parameters, rainbows, anharmonic oscillator

AMS Subject Classification: 40A, 41A, 44A

---

* Corresponding Author's Email: hermanr@uncw.edu

In: The Laplace Transform and Its Applications in Science
Editor: Victor Martinez-Luaces
ISBN: 979-8-89113-747-9

1. Introduction

Certain infinite series have historically captured the imagination of many (Stillwell, 1989). One of the earliest encounters was by Zeno of Elea (490-430 BC) in presenting his famous paradox of Achilles and of the Turtle. It is well known that it leads to a consideration of convergent geometric series,

$$S = 1 + \frac{1}{2} + \frac{1}{2^2} + \frac{1}{2^4} + \dots.$$

Another interesting series was the Basel Problem,

$$S = 1 + \frac{1}{2^2} + \frac{1}{3^2} + \frac{1}{4^2} + \dots,$$

proposed by Pietro Mengoli (1626-1686) in 1650 and solved by Leonhard Euler (1707-1783) in 1734.

However, not all series converge. The harmonic series,

$$S = 1 + \frac{1}{2} + \frac{1}{3} + \frac{1}{4} + \dots.$$

was shown to be divergent in the 14th century by Nicole Oresme (c. 1320 1382) and Mengoli (1650) provided the first published proof (Bellard, 1997). There were serious considerations of the series

$$1 - 1 + 1 - 1 + 1 - \dots$$

in which even the brightest of mathematicians could not agree as to if the series could be summed and if so, meaning should be given to such a series. More recently, the sum

$$1 + 2 + 3 + \cdots = -\frac{1}{12}$$

has populated social media tempting all sorts of conversations.

If these are concerning, then look at the problem Euler grappled with in 1746 (Euler, 1760),

$$\sum_{n=0}^{\infty} (-1)^n n! = 1 - 1 + 2 - 6 + 24 - 120 + \cdots.$$

The partial sums for this series are

$$1, 0, 2, -4, 20, -100, \dots.$$

Clearly, this is a divergent series. Nonetheless, Euler managed to find a meaning in this series.

In this chapter we will see how the Laplace transform can be used to find sums of convergent series using methods that have found a place in quantum field theory. We will also see how divergent series, in particular asymptotic series, appear in physical problems such as diffraction effects in rainbows, and to make sense out of some divergent series, like Euler's series, by way of the Laplace transform and analytic continuation through what is called Borel summation.

We will begin showing how many convergent series can be summed using Laplace transforms. The result will be an integral which might be easier to compute numerically than summing a slowly convergent series. This will extend what the author had mentioned in *An Introduction to Fourier Analysis* (Herman, 2016).

Some of the techniques used are directly related to methods presented in papers in the 1940s in quantum electrodynamics by Feynman and Schwinger. Other interesting convergent series and techniques will be presented including one example of a series used to compute the nth binary digit of π.

Historically, series have played an important role. One of the most interesting is when Airy and Stokes sought to understand interference fringes sometimes seen below the primary rainbow. This led Stokes to a study of divergent series which appear in many physical applications.

What Stokes unveiled had eventually found its way into recent research in areas such as quantum field theory. Many series resulting from perturbation theories, such as the Wentzel-Kramers-Brillouin (WKB) approximation[1], are divergent. In recent decades tools have been developed to understand divergent series and treasures buried within the series. These have been unlocked using Borel transforms, which are essentially Laplace transforms. We will indicate with a couple of examples what Borel summation of divergent series can provide.

2. Series Summation Using Laplace Transforms

2.1. Convergent Series with Simple Terms

In this section we will show how Laplace transforms can be used to sum series. In 1954 Albert D. Wheelon (1929 - 2013) (Wheelon, 1954) described how series can be summed using Laplace transforms and compiled tables which were later published in 1968 as a small book, *Tables of Summable Series and Integrals Involving*

[1] See footnote 29 for more on the WKB approximation.

Bessel Functions (Wheelon, 1968). In his paper he also mentioned that some of his results were also obtainable by Cauchy's method and referred to a 1938 complex variable text by Thomas Murray MacRobert (1884-1962) (MacRobert, 1917). For some additionally challenging sums Wheelon had referred to Richard Feynman's (1918-1988) 1949 paper (Feynman, 1949) on quantum electrodynamics describing how one can use the Convolution Theorem for Laplace transforms to sum series with denominators that involved products. In this part of the chapter we will explore how these are all connected.

The general idea is to start with an infinite series of the form $\sum_n F(n)$, where $F(n)$ looks like the Laplace transform of some known function $f(t)$,

$$F(s) = \mathcal{L}[f(t)] = \int_0^\infty f(t)e^{-st}\,dt. \tag{1}$$

Inserting this integral into the sum and interchanging the summation and integration, we find

$$\begin{aligned}\sum_{n=0}^{\infty} F(n) &= \sum_{n=0}^{\infty}\int_0^\infty f(t)e^{-nt}\,dt \\ &= \int_0^\infty f(t)\sum_{n=0}^{\infty}\left(e^{-t}\right)^n\,dt \\ &= \int_0^\infty f(t)\frac{1}{1-e^{-t}}\,dt. \end{aligned}\tag{2}$$

The last step was obtained using the sum of a geometric series.

We have turned a problem of summing an infinite series into one of integration. Even if the integral cannot be done in closed form, there are numerical methods which may be more successful in obtaining good approximations to the sum provided it exists. However, the challenge now is being able to carry out the final integral. Let us consider a couple of examples.

First we will find the sum of the alternating harmonic series,

$$\sum_{n=1}^{\infty}\frac{(-1)^{n+1}}{n} = 1 - \frac{1}{2} + \frac{1}{3} + \frac{1}{4} - \dots.$$

Using the Laplace transform $\mathcal{L}[1] = \dfrac{1}{s}$, we can write

$$\sum_{n=1}^{\infty}\frac{(-1)^{n+1}}{n} = \sum_{n=1}^{\infty}(-1)^{n+1}\int_0^\infty e^{-nt}\,dt.$$

Next, we interchange the summation and integration and sum the resulting geometric series.

$$\begin{aligned}
\sum_{n=1}^{\infty} \frac{(-1)^{n+1}}{n} &= \sum_{n=1}^{\infty} (-1)^{n+1} \int_0^{\infty} e^{-nt}\, dt \\
&= -\int_0^{\infty} \left[\sum_{n=1}^{\infty} (-e^{-t})^n \right] dt \\
&= \int_0^{\infty} \frac{e^{-t}}{1+e^{-t}}\, dt \\
&= \int_1^2 \frac{du}{u} = \ln 2. \qquad (3)
\end{aligned}$$

So, using the Laplace transform led to an integral, which in this case we could evaluate.

Now, we turn to the famous Basel Problem (Sandifer, 2003, 2004; Stillwell, 1989). Mengoli posed the Basel Problem in 1650[2] in which one had to find the sum of the series

$$\sum_{n=1}^{\infty} \frac{1}{n^2} = 1 + \frac{1}{2^2} + \frac{1}{3^2} + \frac{1}{4^2} + \cdots.$$

Many mathematicians at the time tried to solve the problem. In 1655 John Wallis (1616 - 1703) claimed to know the value to three decimal places. In 1689 Jacob Bernoulli (1655 - 1705) said it was less than 2. Others either could not do it or could only put bounds on the sum. In 1734 Euler (1740b) solved it by writing a factorization of the sine function (Sandifer, 2004).

In fact, this sum is a special case of the Riemann zeta function,

$$\zeta(s) = \sum_{n=1}^{\infty} \frac{1}{n^s}, \qquad \text{Re}(s) > 1. \qquad (4)$$

The Riemann zeta function[3] is important in the study of prime numbers and more recently has seen applications in the study of dynamical systems. The series in this

[2]Mengoli (1650) suggested in his book that this was a problem too hard for him to solve. There are some references, such as (Sandifer, 2003) which puts the date at 1644. Mengoli also studied other series including finding the sum of the alternating harmonic series.

[3]A translation of Riemann's (1859) paper, "Über die Anzahl der Primzahlen unter einer gegebenen Grösse" is in H. M. Edwards (1974), *Riemann's Zeta Function* and, more recently, in (Murphy, 2020). Riemann had shown that the Riemann zeta function can be obtained through contour integral representation,

$$2 \sin \pi s \Pi(s-1)\zeta(s) = i \oint_C \frac{(-x)^{s-1}}{e^x - 1} dx,$$

example is $\zeta(2)$. We also note that $\zeta(1)$ is the harmonic series, which we noted was known to be divergent by the time Euler tackled the problem. Euler (1737, 1740b) began studying $\zeta(2)$ around 1730 and later considered $\zeta(2n)$. He found a functional equation in 1749 (Euler, 1768) later to be proven by Bernhard Riemann (1826-1866) (Ayoub, 1974).

Related to the Riemann zeta function is the Euler, or Dirichlet, eta function, given by

$$\eta(s) = \sum_{n=1}^{\infty} \frac{(-1)^{n+1}}{n^s}, \quad \text{Re}(s) > 0. \tag{6}$$

This is also referred to as the alternating Riemann zeta function. These functions are connected by

$$\eta(s) = \left(1 - 2^{1-s}\right)\zeta(s). \tag{7}$$

This is seen from

$$\begin{aligned} \zeta(s) - \eta(s) &= \sum_{n=1}^{\infty} \frac{1}{n^s} - \sum_{n=1}^{\infty} \frac{(-1)^{n+1}}{n^s} \\ &= \sum_{n=1}^{\infty} \frac{2}{(2n)^s} \\ &= 2^{1-s}\zeta(s). \end{aligned} \tag{8}$$

Solving for $\eta(s)$ gives the result in Equation (7).[4]

---

for a specific contour C. Here $\Pi(s)$ is Gauss' notation for the Gamma function later given by Legendre as $\Gamma(s)$ (Euler, 1738; Davis, 1959; Gronau, 2003; Sandifer, 2007; Murphy, 2020). These are related by

$$\Gamma(s) = \int_0^{\infty} x^{s-1} e^{-x}\, ds = \Pi(s-1), \quad s > 0, \tag{5}$$

generalizing the factorial, $n! = \Gamma(n+1) = \Pi(n)$, for nonnegative integers, n. Additional history is provided in Herman (2025, p.40).

[4]Special values of the eta function are given by

$$\begin{aligned} \eta(0) &= 1 - 1 + 1 - 1 + \cdots = \frac{1}{2}, \\ \eta(-1) &= 1 - 2 + 3 - 4 + \cdots = \frac{1}{4}, \\ \eta(1) &= 1 - \frac{1}{2} + \frac{1}{3} + \frac{1}{4} - \cdots = \ln 2. \end{aligned} \tag{9}$$

We will discuss these divergent sums later. Note that $\zeta(-1) = 1 + 2 + 3 + \cdots$. But, the Riemann zeta function is not defined for $s = -1$. So, we can use $\eta(s)$ to analytically continue $\zeta(s) = \dfrac{\eta(s)}{1 - 2^{1-s}}$. Setting $s = -1$, we obtain $1 + 2 + 3 + \cdots = \dfrac{\eta(-1)}{1 - 2^2} = -\dfrac{1}{12}$. That is assuming $\eta(-1) = \dfrac{1}{4}$. See Nahin (2021).

Using Laplace transforms, we can provide an integral representation of $\zeta(2)$. The first step is to find the correct Laplace transform pair. The sum involves the function $F(n) = \frac{1}{n^2}$. So, we look for a function $f(t)$ whose Laplace transform is $F(s) = \frac{1}{s^2}$. We know that the inverse Laplace transform of $F(s)$ is $f(t) = t$. As before, we replace each term in the series by a Laplace transform integral, exchange the summation and integration, and sum the resulting geometric series:

$$\begin{aligned}\sum_{n=1}^{\infty} \frac{1}{n^2} &= \sum_{n=1}^{\infty} \int_0^{\infty} te^{-nt}\, dt \\ &= \int_0^{\infty} t \sum_{n=1}^{\infty} (e^{-t})^n\, dt \\ &= \int_0^{\infty} \frac{t}{e^t - 1}\, dt. \end{aligned} \tag{10}$$

So, we have that

$$\int_0^{\infty} \frac{t}{e^t - 1}\, dt = \sum_{n=1}^{\infty} \frac{1}{n^2} = \zeta(2).$$

In general, using the Gamma function in Equation (5) and $\mathcal{L}[t^k] = \Gamma(k+1)/s^{k+1}$, we have the integral representations (interchanging $\{k, s\}$ with $\{s, n\}$)

$$\begin{aligned}\sum_{n=1}^{\infty} \frac{(\pm 1)^{n+1}}{n^s} &= \sum_{n=1}^{\infty} (\pm 1)^{n+1} \frac{1}{\Gamma(s)} \int_0^{\infty} t^{s-1} e^{-nt}\, dt \\ &= \frac{1}{\Gamma(s)} \int_0^{\infty} \frac{t^{s-1}}{e^t \mp 1}\, dt. \end{aligned} \tag{11}$$

Integrals of these types occur often in statistical mechanics in the form of Bose-Einstein integrals or Fermi-Dirac integrals. These integrals are defined slightly differently in various sources. In Pathria (2016) the Bose-Einstein integrals are given as[5]

$$G_n(z) = \int_0^{\infty} \frac{t^{n-1}}{z^{-1}e^t - 1}\, dt, \quad 0 \le z < 1, n > 0, \text{ or } z = 1, n > 1,$$

[5] (Dingle, 1973) defines these as

$$\mathcal{B}_p(x) = \frac{1}{p!} \int_0^{\infty} \frac{t^p}{e^{t-x} - 1}\, dt, \quad p > 0,$$

$$\mathcal{F}_p(x) = \frac{1}{p!} \int_0^{\infty} \frac{t^p}{e^{t-x} + 1}\, dt, \quad p > -1,$$

where $z = e^x$.

and the Fermi-Dirac integrals are

$$F_n(z) = \int_0^\infty \frac{t^{n-1}}{z^{-1}e^t + 1}\, dt, \quad 0 \le z < \infty, n > 0.$$

Note that $G_n(1) = \Gamma(n)\zeta(n)$ and these integrals can be related to Hurwitz-Lerch zeta functions and polylogarithm functions (Chaudhry et al., 2010).

In general, the Riemann zeta function must be tabulated through other means. In some special cases, one can use closed form expressions. For example (Edwards, 1974),

$$\zeta(2n) = (-1)^{n+1}\frac{(2\pi)^{2n}}{2(2n)!}B_{2n}, \tag{12}$$

where the B_n's are the Bernoulli numbers. Bernoulli numbers, discovered by Jacob Bernoulli and independently by Seki Takakazu (1642 - 1708), are defined through the Maclaurin series expansion

$$\frac{x}{e^x - 1} = \sum_{n=0}^{\infty} \frac{B_n}{n!}x^n.$$

The first few Riemann zeta functions are

$$\zeta(2) = \frac{\pi^2}{6}, \quad \zeta(4) = \frac{\pi^4}{90}, \quad \zeta(6) = \frac{\pi^6}{945}.$$

However, we do not have closed forms like Equation (12) for other values. For example, $\zeta(3)$ is called Apéry's constant, named after Roger Apéry (1916-1994), who proved in 1979 that $\zeta(3)$ is irrational (Van der Poorten, 1979). Nahin (2021) explores this problem in some detail. Harold Edwards (1936-2020) describes how the Riemann zeta function is computed using the Euler-Maclaurin summation formula (Edwards, 1974). Tingen (2009) used this to study the Mandelbrot and Julia sets associated with the Riemann zeta function.

2.2. Series with More General Terms

We can extend this method of using Laplace transforms to summing series whose terms take special general forms. This was described by Wheelon first in 1954 (Wheelon, 1954, 1955) and later in his table of sums (Wheelon, 1968). He had noted that in 1949 Feynman provided several useful identities for computing series summations which he needed in quantum electrodynamics (Feynman, 1949).

For example, one such identity is

$$\frac{1}{(a+bn)^2} = -\frac{\partial}{\partial a} \int_0^\infty e^{-s(a+bn)}\, ds.$$

This identity can be shown easily by first noting that

$$\int_0^\infty e^{-s(a+bn)}\, ds = \left[\frac{-e^{-s(a+bn)}}{a+bn}\right]_0^\infty = \frac{1}{a+bn}. \tag{13}$$

Now, we differentiate the result with respect to a and the result follows.

We should note, before going further, that these integrals are nothing but Laplace transforms as defined in Equation (1) but with s replaced by $a + bn$ and t replaced by s. In Equation (13), it is the transform $F(a + bn)$ of $f(s) = 1$. The original integral can be written as

$$-\frac{\partial}{\partial a} \int_0^\infty e^{-(a+bn)t}\, dt = \int_0^\infty te^{-(a+bn)t}\, dt,$$

which is the Laplace transform $F(a + bn)$ of $f(t) = t$. So, we see that in order to evaluate the second integral, we need only differentiate under the integral. This is an example of what has become known as Feynman's Trick.[6]

This identity can be generalized further. We note that the Laplace transform of t^k can be written as

$$\mathcal{L}\left[t^k\right](a+bn) = \int_0^\infty t^k e^{-(a+bn)t}\, dt = \frac{\Gamma(k+1)}{(a+bn)^{k+1}}.$$

So, differentiating Equation (13) k times, we arrive at the more general result,

$$\frac{1}{(a+bn)^{k+1}} = \frac{(-1)^k}{k!} \frac{\partial^k}{\partial a^k} \int_0^\infty e^{-s(a+bn)}\, ds.$$

Feynman (1949) also develops methods for handling several other general sums. Wheelon gives more examples of these (Wheelon, 1954, 1955, 1968). For the first

[6]Feynman (1985) remarked in his book, "I had learned to do integrals by various methods shown in a book that my high school physics teacher Mr. Bader had given me. [It] showed how to differentiate parameters under the integral sign its a certain operation. It turns out thats not taught very much in the universities; they dont emphasize it. But I caught on how to use that method, and I used that one damn tool again and again. [If] guys at MIT or Princeton had trouble doing a certain integral, [then] I come along and try differentiating under the integral sign, and often it worked. So I got a great reputation for doing integrals, only because my box of tools was different from everybody elses, and they had tried all their tools on it before giving the problem to me."

expression, we consider the integral result

$$\int_0^1 \frac{du}{[a(1-u)+bu]^2} = \frac{1}{b-a}\int_a^b \frac{dz}{z^2} = \frac{1}{ab}. \tag{14}$$

where we let $z = a(1-u)+bu = a+(b-a)u$.

Again, if we think of $a(1-u)+bu$ as the Laplace transform variable, s, then the integrand reminds us again of the Laplace transform of t. So, we have

$$\frac{1}{[a(1-u)+bu]^2} = \int_0^\infty te^{-t[a(1-u)+bu]}\,dt.$$

As a result, we have the identity

$$\frac{1}{ab} = \int_0^1 du \int_0^\infty te^{-t[a(1-u)+bu]}\,dt. \tag{15}$$

In the next example, we show how this identity can be useful for summing an infinite series. We will evaluate the series

$$S = \sum_{n=0}^{\infty} \frac{1}{(2n+1)(2n+2)}.$$

We define $a = 2n+1$ and $b = 2n+2$ in the formulae (14)-(15) for $\frac{1}{ab}$. Collecting the n-dependent terms, we can sum the series leaving a double integral computation in ut-space. The details are as follows:

$$\begin{aligned}
\sum_{n=0}^{\infty} \frac{1}{(2n+1)(2n+2)} &= \sum_{n=0}^{\infty}\int_0^1 \frac{du}{[(2n+1)(1-u)+(2n+2)u]^2} \\
&= \sum_{n=0}^{\infty}\int_0^1\int_0^\infty te^{-t(2n+1+u)}\,dt\,du \\
&= \int_0^1\int_0^\infty te^{-t(1+u)}\sum_{n=0}^{\infty} e^{-2nt}\,dt\,du \\
&= \int_0^\infty \frac{te^{-t}}{1-e^{-2t}}\int_0^1 e^{-tu}\,du\,dt \\
&= \int_0^\infty \frac{te^{-t}}{1-e^{-2t}}\frac{1-e^{-t}}{t}\,dt \\
&= \int_0^\infty \frac{e^{-t}}{1+e^{-t}}\,dt \\
&= -\ln(1+e^{-t})\Big|_0^\infty = \ln 2.
\end{aligned} \tag{16}$$

If one reads Feynman's paper (Feynman, 1949), he says in the appendix after identity (15), "suggested by some work of Schwinger's involving Gaussian integrals." That is the only mention that Feynman got his idea from Schwinger's work but no further references were provided. We will generalize identity (15) and provide further examples. Many other results can be found in Wheelon's tables (Wheelon, 1968). Also, Glasser (1971) added some slowly converging series which arise from physics.

2.3. Feynman Parameters

The Feynman identities in the last section are part of a more general approach to what are called Feynman and Schwinger parameters. Feynman and Schwinger parameters are mathematical techniques used in quantum field theory to represent integrals over the momentum space in a more convenient form (Feynman, 1949; Schwinger, 1951; Weinberg, 2005). However, these parameters differ in how they are approached and the types of integrals they are used for.

Feynman parameters are used to represent integrals of the form:

$$\int_0^1 dx_1 \int_0^{1-x_1} dx_2 \cdots \int_0^{1-\sum_{i=1}^{n-1} x_i} dx_n \frac{f(x_1, x_2, \ldots, x_n)}{(a_1x_1 + a_2x_2 + \cdots + a_nx_n + b)^p},$$

where a_i, b, p are constants and $f(x_1, x_2, \ldots, x_n)$ is a function of the x_i's. Feynman parameters are particularly useful in the calculation of loop integrals in perturbative quantum field theory.

Schwinger parameters, named after the physicist Julian Schwinger (1918-1994), are used to represent integrals of the form:

$$\int_0^\infty d\alpha_1 \int_0^\infty d\alpha_2 \cdots \int_0^\infty d\alpha_n \exp\left(-\sum_{i=1}^{n} \alpha_i A_i - \alpha_{n+1} B\right),$$

where A_i and B are operators that act on the quantum field, and the integrals are taken over positive values of the parameters α_i. Schwinger parameters are often used in the path integral formulation of quantum field theory, and can be used to express the propagator of a quantum field as a product of simpler propagators.

In the previous section we used Feynman's trick for computing these propagator integrals in quantum field theory. Feynman (1949) writes

$$a^{-1}b^{-1} = \int_0^1 du\, [au + b(1-u)]^{-2}, \tag{17}$$

which is Equation (14). We note that for other powers of a and b, one can derive identities by successive differentiation with respect to these parameters. Feynman also gives the example

$$a^{-2}b^{-1} = \int_0^1 du\, 2u\, [au + b(1-u)]^{-3}. \tag{18}$$

In the first example in Equation (17), the integrand was written as a Laplace transform giving Equation (15). However, following (Kim et al., 2023), we can rewrite the integral in a more suggestive way using the Dirac delta function.[7] Consider

$$\begin{aligned} I &= \int_0^1 \frac{du}{[au + b(1-u)]^2} \\ &= \int_0^1 \int_0^1 dudv\, \delta(1-u-v) \frac{\Gamma(2)}{[au+bv]^2} \\ &= \int_0^1 \int_0^1 dudv\, \delta(1-u-v) \int_0^\infty d\lambda\, \lambda e^{-\lambda(au+bv)}. \end{aligned}$$

Here we introduced $v = 1 - u$ in the denominator and we used the Laplace transform

$$\int_0^\infty d\lambda\, \lambda^{n-1} e^{-s\lambda} = \frac{\Gamma(n)}{s^n}, \quad \text{Re}(s) > 0$$

with $s = au + bv$ and $n = 2$.

We now use the identity

$$\int_0^\infty d\lambda \frac{1}{\lambda} \delta\left(1 - \frac{\alpha}{\lambda}\right) = 1, \quad \alpha > 0. \tag{19}$$

This follows from the relation

$$\delta(f(\lambda)) = \frac{\delta(\lambda - \lambda_0)}{|f'(\lambda_0)|},$$

where λ_0 is a simple zero of $f(\lambda)$. For this example, we have $[\lambda_0 = \alpha,\ f'(\lambda_0) = \alpha^{-1}]$

$$\frac{1}{\lambda}\delta\left(1 - \frac{\alpha}{\lambda}\right) = \frac{1}{\lambda} \frac{\delta(\lambda - \alpha)}{\left|\frac{\alpha}{\lambda^2}\right|_{\lambda=\alpha}} = \delta(\lambda - \alpha).$$

[7]The Dirac delta function is defined by two properties, $\delta(x) = 0,\ x \neq 0$ and

$$\int_{-\infty}^{\infty} \delta(x)\, dx = 1.$$

More properties and historical context can be found in (Herman, 2025, p.13-14) and (Herman, 2016).

Now, in I we let $x = \lambda u$ and $y = \lambda v$ and use the identity (19) to obtain

$$\begin{aligned} I &= \int_0^1 \int_0^1 dudv\, \delta(1-u-v) \int_0^\infty d\lambda\, \lambda e^{-\lambda(au+bv)} \\ &= \int_0^1 \int_0^1 dudv \int_0^\infty d\lambda \frac{1}{\lambda}\delta\left(1 - \frac{x+y}{\lambda}\right) e^{-\lambda(au+bv)} \\ &= \int_0^\infty \int_0^\infty dxdy\, e^{-(ax+by)} \\ &= \int_0^\infty e^{-ax}\, dx \int_0^\infty e^{-by}\, dy \\ &= \frac{1}{ab}. \end{aligned} \tag{20}$$

This can be extended to products of three factors. Working in the other direction, we have

$$\begin{aligned} \frac{1}{abc} &= \int_0^\infty e^{-ax}\, dx \int_0^\infty e^{-by}\, dy \int_0^\infty e^{-cz}\, dz \\ &= \int_0^\infty \int_0^\infty \int_0^\infty dxdydz\, e^{-(ax+by+cz)} \\ &= \int_0^\infty \int_0^\infty \int_0^\infty dxdydz \left[\int_0^\infty d\lambda \frac{1}{\lambda}\delta\left(1 - \frac{x+y+z}{\lambda}\right)\right] e^{-(ax+by+cz)} \\ &= \int_0^1 \int_0^1 \int_0^1 dudvdw\, \delta(1-u-v-w) \int_0^\infty d\lambda\, \lambda^2 e^{-\lambda(au+bv+cw)} \\ &= \int_0^1 \int_0^1 \int_0^1 dudvdw\, \delta(1-u-v-w) \frac{\Gamma(3)}{(au+bv+cw)^3} \\ &= \int_0^1 \int_0^1 dudv\, H(u+v) H(1-u-v) \frac{\Gamma(3)}{[au+bv+c(1-u-v)]^3} \\ &= 2\int_0^1 \int_0^{1-u} \frac{dvdu}{[au+bv+c(1-u-v)]^3}. \end{aligned} \tag{21}$$

Letting $v = p$ and $u + v = q$, this can be written as

$$\frac{1}{abc} = 2\int_0^1 \int_0^q \frac{dpdq}{[a(q-p)+bp+c(1-q)]^3},$$

or, letting $u = 1 - q$ and $v = q - p$, we find that

$$\frac{1}{abc} = 2\int_0^1 \int_0^q \frac{dpdq}{[a(1-q)+b(q-p)+cp]^3} \tag{22}$$

as presented in Wheelon (1968).

As an example of identity (21), we want to find the sum

$$S=\sum_{n=0}^{\infty}\frac{1}{(2n+1)(2n+2)(2n+3)},$$

which is the series number 3.703 in Wheelon's tables (Wheelon, 1968). We let $a=2n+1$, $b=2n+2$, and $c=2n+3$. Then, we use the identity and simplify:

$$\begin{aligned}
S &= \sum_{n=0}^{\infty}\frac{1}{(2n+1)(2n+2)(2n+3)}\\
&= \sum_{n=0}^{\infty}2\int_0^1\int_0^{1-u}\frac{dudv}{[(2n+1)u+(2n+2)v+(2n+3)(1-u-v)]^3}\\
&= \sum_{n=0}^{\infty}2\int_0^1\int_0^{1-u}\frac{dudv}{[2n-2u-v+3]^3}. \qquad (23)
\end{aligned}$$

Next, we insert the Laplace transform integral, interchange the sum with the integrals, and sum the resulting geometric series.

$$\begin{aligned}
S &= \sum_{n=0}^{\infty}\int_0^{\infty}ds\,s^2\int_0^1 du\int_0^{1-u}e^{-s(2n-2u-v+3)}\,dv\\
&= \int_0^{\infty}ds\,s^2\int_0^1 du\int_0^{1-u}dv\,e^{-s(3-2u-v)}\sum_{n=0}^{\infty}e^{-2ns}\\
&= \int_0^{\infty}ds\frac{s^2e^{-3s}}{1-e^{-2s}}\int_0^1 du\int_0^{1-u}dv\,e^{s(2u+v)}. \qquad (24)
\end{aligned}$$

Finally, we carry out the integrations,

$$\begin{aligned}
S &= \frac{1}{2}\int_0^{\infty}ds\frac{e^{-3s}}{1-e^{-2s}}(1-2e^s+e^{2s})\\
&= \frac{1}{2}\int_0^1 dx\frac{x^2}{1-x^2}\left(1-\frac{2}{x}+\frac{1}{x^2}\right)\\
&= \frac{1}{2}\int_0^1 dx\frac{1-x}{1+x}\\
&= \frac{1}{2}\left[2\ln(x+1)-x\right]_0^1=\ln(2)-\frac{1}{2}. \qquad (25)
\end{aligned}$$

2.4. Other Interesting Series

This method of using transforms to sum series is not new or unique. It is well-known that Fourier transforms also can lead to interesting sums. There have been other more recent papers employing Laplace transforms to sum series such as by Efthimiou (1999, 2006) and Lesko and Smith (2003), who was responding to Efthimiou (1999). Efthimiou had answered a question about summing the series $\sum_{n=1}^{\infty} \frac{1}{n^2 + n/2}$. He showed

$$S(a,b) = \sum_{n=1}^{\infty} \frac{1}{(n+a)(n+b)} = \frac{1}{a-b} \int_0^1 \frac{t^b - t^a}{1-t} = \frac{\psi(b+1) - \psi(a+1)}{b-a},$$

where $\psi(z) = \frac{d}{dz} \ln \Gamma(z)$ is the digamma function.

Efthimiou (2006) considered sums involving trigonometric functions. As examples, he found sums of series like

$$\sum_{n=1}^{\infty} \frac{\cos nx}{n}, \quad \sum_{n=1}^{\infty} \frac{\sin nx}{n}, \quad \sum_{n=1}^{\infty} \frac{\cos nx}{n^2}.$$

For example, if we want to sum the series

$$S = \sum_{n=1}^{\infty} \frac{\cos nx}{n^m}$$

for $0 < x < 2\pi$, $m = 1$ or $0 \le x \le 2\pi$ for $m > 1$, we would insert a Laplace transform integral and obtain

$$\begin{aligned} S &= \frac{1}{(m-1)!} \sum_{n=1}^{\infty} \cos nx \int_0^{\infty} e^{-nt} t^{m-1}\, dt \\ &= \frac{1}{(m-1)!} \int_0^{\infty} \left(\sum_{n=1}^{\infty} \cos nx\, e^{-nt} \right) t^{m-1}\, dt. \end{aligned} \tag{26}$$

The key to going further is to compute the sum. Efthimiou derives

$$\begin{aligned} \sum_{n=1}^{\infty} \sin nx\, e^{-nt} &= \frac{e^{-t} \sin x}{1 - 2\cos x\, e^{-t} + e^{-2t}} \\ \sum_{n=1}^{\infty} \cos nx\, e^{-nt} &= \frac{e^{-t}(\cos x - e^{-t})}{1 - 2\cos x\, e^{-t} + e^{-2t}}. \end{aligned} \tag{27}$$

Setting $m = 1$ in Equation (26), he found

$$\sum_{n=1}^{\infty} \frac{\cos nx}{n} = -\ln\left(2\sin\frac{x}{2}\right), \qquad \sum_{n=1}^{\infty} \frac{\sin nx}{n} = \frac{\pi - x}{2}.$$

In the 1990s there was an effort to compute the dth digit of π and other transcendental numbers in various bases without computing all of the preceding digits (Adamchik and Wagon, 1996, 1997). Bailey et al. (1997) used the identity[8]

$$\pi = \sum_{k=0}^{\infty} \frac{1}{16^k}\left(\frac{4}{8k+1} - \frac{2}{8k+4} - \frac{1}{8k+5} - \frac{1}{8k+6}\right) \tag{28}$$

to obtain the binary and hexidecimal digits of π. Fabrice Bellard (1997) introduced an identity which was used to produce the 40 trillionth bit of π,

$$\begin{aligned} \pi = {} & \frac{1}{2^6}\sum_{k=0}^{\infty} \frac{(-1)^k}{2^{10k}}\left(-\frac{2^5}{4k+1} - \frac{1}{4k+3} + \frac{2^8}{10k+1} - \frac{2^6}{10k+3}\right. \\ & \left. -\frac{2^2}{10k+5} - \frac{2^2}{10k+7} + \frac{1}{10k+9}\right). \end{aligned} \tag{29}$$

There are different ways that these formulae were obtained. Bellard pointed out that these are easy to find using the series expansion

$$-\ln(1-x) = \sum_{k=1}^{\infty} \frac{x^k}{k}, \text{ for } |x| < 1.$$

Noting that $\tan^{-1}\frac{1}{x-1} = \operatorname{Im}\left(\ln\left(1 - \frac{1+i}{x}\right)\right)$ and setting $x = 2$, Equation (28) results.

Combining the series expansion for $\tan^{-1}\left(\frac{1}{x-1}\right)$ with the relation

$$\frac{\pi}{4} = 2\tan^{-1}\frac{1}{2} - \tan^{-1}\frac{1}{7},$$

one can obtain Equation (29). This is just a variation of John Machin's (1680 - 1751) 1709 formula

$$\frac{\pi}{4} = 4\tan^{-1}\frac{1}{5} - \tan^{-1}\frac{1}{239}.$$

[8]This formula was discovered by Plouffe in 1995 and has since been referred to as the BBP formula. It was used to compute the nth hexadecimal digit of π.

Michael Hirschhorn (1998) had shown how formula (28) can be obtained using simple integration methods from undergraduate calculus. He used the relation

$$\sum_{k=0}^{\infty} \frac{x^{8k+r}}{8k+r} = \int_0^x \frac{u^{r-1}}{1-u^8}\, du.$$

We obtained (Herman, 2000) a generalization of this relation beginning with Laplace transforms,

$$\begin{aligned}
\sum_{k=0}^{\infty} \frac{1}{n^k} \frac{1}{mk+r} &= \sum_{k=0}^{\infty} \frac{1}{n^k} \int_0^{\infty} e^{-(mk+r)s}\, ds \\
&= \int_0^{\infty} e^{-rs} \sum_{k=0}^{\infty} \left(\frac{e^{-ms}}{n} \right)^k ds \\
&= \int_0^{\infty} \frac{e^{-rs}}{1 - \frac{1}{n} e^{-ms}}\, ds. \qquad (30)
\end{aligned}$$

The convergence of the geometric series is guaranteed for $n > 1 \geq e^{-ms}$, for $s \in [0, \infty)$ and $m \geq 0$.

The last integral can be rewritten in a simpler form by introducing $\beta = n^{-1/m}$ and making the substitution $x = \beta e^{-s}$. The result is the integral expression for the sum of the infinite series as given above in Equation (30).

$$\sum_{k=0}^{\infty} \frac{1}{n^k} \frac{1}{mk+r} = n^{r/m} \int_0^{n^{-1/m}} \frac{x^{r-1}}{1-x^m}\, dx \qquad (31)$$

$$\sum_{k=0}^{\infty} \frac{(-1)^k}{n^k} \frac{1}{mk+r} = n^{r/m} \int_0^{n^{-1/m}} \frac{x^{r-1}}{1+x^m}\, dx \qquad (32)$$

Using $n = 16$, $m = 8$, and $r = 1, 2, 3, \ldots, 8$, we obtain (28). These sums can be used to find other interesting formulae such as shown also in (Lupas, 2002).

2.5. Cauchy Summation by Contour Methods

Wheelon (1954) mentions several other forms of denominators such as second-degree polynomials using Laplace transform pairs. For example, consider the integrals

$$\begin{aligned}
\frac{1}{a} \int_0^{\infty} \sinh ax e^{-xn}\, dx &= \frac{1}{n^2 - a^2} \\
\frac{1}{a} \int_0^{\infty} \sin ax e^{-xn}\, dx &= \frac{1}{n^2 + a^2}. \qquad (33)
\end{aligned}$$

Carrying out the procedure from before, it is relatively easy to compute the sums

$$\begin{aligned}\sum_{n=0}^{\infty} \frac{1}{n^2 - a^2} &= -\frac{1}{a^2}\left[1 + \pi a \cot a\pi\right],\\ \sum_{n=0}^{\infty} \frac{1}{n^2 + a^2} &= \frac{1}{a^2}\left[1 + \pi a \coth a\pi\right].\end{aligned} \tag{34}$$

But, then he says that these results are also obtainable using Cauchy's method [Augustin-Louis Cauchy (1789 - 1857)] and he referenced MacRobert's text (MacRobert, 1917). So, for completeness we discuss Cauchy's summation by contour methods.[9]

Let $f(z)$ be a meromorphic function with a finite set of poles that are not integers. Furthermore, let us assume that

$$|f(z)| < \frac{M}{|z|^k}$$

for $k > 1$ and M a constant. We seek to show that

$$\sum_{n=-\infty}^{\infty} f(n) = -\sum_k \text{Res}\left[\pi \cot \pi z f(z); z_k\right],$$

where the second sum is over the poles of $f(z)$ that are not integers.

We consider the integral

$$I = \oint_{C_N} \pi \cot \pi z f(z)\, dz, \tag{35}$$

where the contour C_N is shown in Figure 1.

There are simple poles of $g(z) = \pi \cot \pi z f(z)$ at $z = 0, \pm 1, \pm 2, \ldots$. The residues are found as

$$\text{Res}[g(z), z = n] = \lim_{z \to n} \frac{(z-n)\pi \cos \pi z}{\sin \pi z} f(z) = f(n).$$

Thus, for the above contour, the Residue theorem gives

$$\oint_{C_N} \pi \cot \pi z f(z)\, dz = 2\pi i \left[\sum_{n=-N}^{N} f(n) + \sum_k \text{Res}\left[\pi \cot \pi z f(z); z_k\right]\right], \tag{36}$$

[9]Cauchy was instrumental in the development of complex analysis starting with his 1814 paper, later published in 1825. Smithies (1997) details Cauchy's works showing his contributions which occupy most of a first course in complex analysis to date.

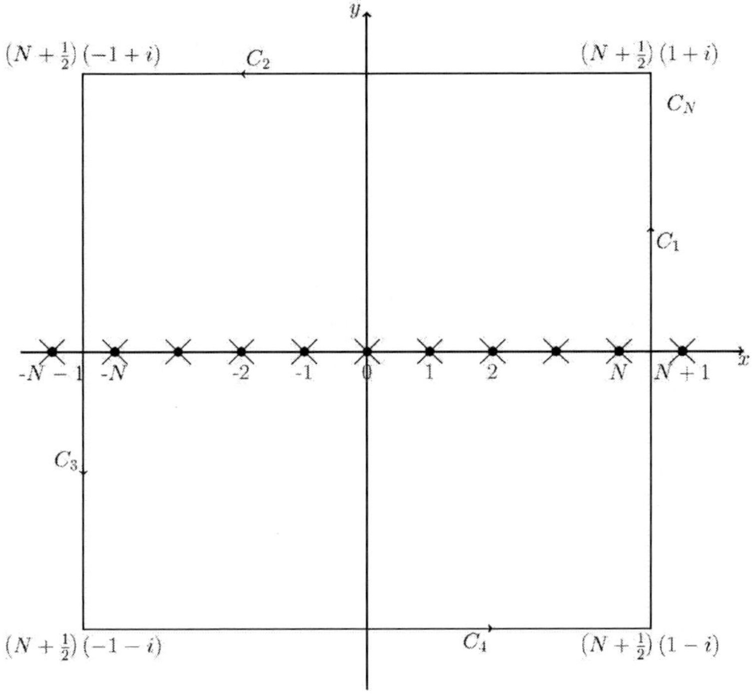

Figure 1. Contour for integration in Equation (35) with simple poles at $z = 0, \pm 1, \pm 2, \ldots$ but additional poles of $f(z)$ not shown. The contour C_N can be broken into four pieces, C_i, i=1,2,3,4.

where the second sum is over the poles of $f(z)$.

Eventually we will let $N \to \infty$. We need to consider the value of the contour integral around the rectangle and equate it to this result. We will show that

$$\lim_{N\to\infty} \oint_{C_N} \pi \cot \pi z f(z)\, dz = 0,$$

leading to the desired result.

The contour C_N can be broken into four pieces, as noted in Figure 1. Along each path we have the inequality

$$\left| \int_C \pi \cot \pi z f(z)\, dz \right| \leq \pi \int_C |\cot \pi z| |f(z)|\, dz.$$

So, we need to get bounds on $|\cot \pi z|$ for each leg of the path. Let $z = x + iy$.

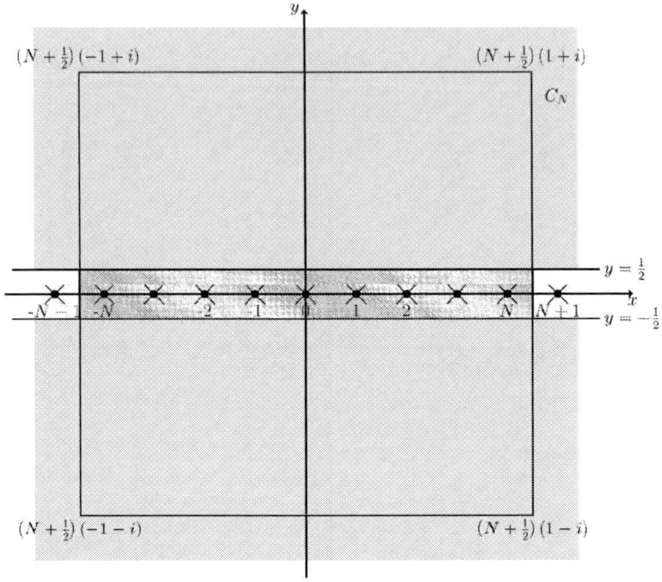

Figure 2. The gray regions, $y > -\frac{1}{2}$ and $y > \frac{1}{2}$, and side edges of the hatched region, $z = \pm\left(N+\frac{1}{2}\right) + iy$, $-\frac{1}{2} \le y \le \frac{1}{2}$, are used to establish that $|\coth \pi z|$ is bounded by a constant independent of N.

Then, we first note that

$$\begin{aligned} |\cot \pi z| &= \left|\frac{e^{i\pi z} + e^{-i\pi z}}{e^{i\pi z} - e^{-i\pi z}}\right| \\ &= \left|\frac{e^{i\pi x - \pi y} + e^{-i\pi x + \pi y}}{e^{i\pi x - \pi y} - e^{-i\pi x + \pi y}}\right| \\ &\le \frac{|e^{i\pi x - \pi y}| + |e^{-i\pi x + \pi y}|}{|e^{i\pi x + \pi y}| - |e^{-i\pi x - \pi y}|} \\ &= \frac{e^{-\pi y} + e^{+\pi y}}{e^{\pi y} - e^{-\pi y}} = \coth \pi y. \end{aligned} \tag{37}$$

We will show that along the contour C_N that $|\cot \pi z| < K$ for some constant K which is independent of N. We will consider different regions as indicated in Figure 2. Namely, we consider the gray regions, $y > -\frac{1}{2}$ and $y > \frac{1}{2}$, and then the section of C_N where $-\frac{1}{2} \le y \le \frac{1}{2}$.

For $y > \frac{1}{2}$ Equation (37) gives $|\coth \pi z| \le \coth \pi/2 \equiv K_1$. Similarly, for $y < -\frac{1}{2}$ we also have $|\cot \pi z| \le K_1$.

In the region $-\frac{1}{2} \leq y \leq \frac{1}{2}$ we consider $z = N + \frac{1}{2} + iy$. Then, we have

$$\begin{aligned} |\cot \pi z| &= \left|\cot \pi \left(N + \frac{1}{2} + iy\right)\right| \\ &= \left|\cot \left(\frac{\pi}{2} + i\pi y\right)\right| \\ &= |\tanh \pi y| \\ &\leq \tanh \frac{\pi}{2} \equiv K_2. \end{aligned} \tag{38}$$

Similarly, for $z = -N - \frac{1}{2} + iy$, $|\cot \pi z| \leq K_2$.

Therefore, we have shown that $|\cot \pi z| \leq K = \max(K_1, K_2)$ independent of N. So, we have the bound

$$\begin{aligned} \left|\oint_{C_N} \pi \cot \pi z f(z)\, dz\right| &\leq \pi \oint_{C_N} |\cot \pi z||f(z)|\, dz \\ &\leq \pi M K \oint_{C_N} \frac{dz}{|z|^k} \\ &\leq \frac{\pi M}{N^k} K \oint_{C_N} dz \\ &= \frac{4\pi M K}{N^k}(2N + 1). \end{aligned} \tag{39}$$

So, we see that as $N \to \infty$, we have

$$\lim_{N \to \infty} \oint_{C_N} \pi \cot \pi z f(z)\, dz = 0.$$

From Equation (36) we have

$$2\pi i \left[\sum_{n=-N}^{N} f(n) + \sum_k \text{Res}\left[\pi \cot \pi z f(z); z_k\right]\right] = 0.$$

This gives the result

$$\sum_{n=-\infty}^{\infty} f(n) = -\sum_k \text{Res}\left[\pi \cot \pi z f(z); z_k\right]. \tag{40}$$

A similar series can be obtained under the same hypotheses:

$$\sum_{n=-\infty}^{\infty} (-1)^n f(n) = -\sum_k \text{Res}\left[\pi \csc \pi z f(z); z_k\right], \tag{41}$$

where the second sum is over the poles of $f(z)$.

As an example, we show that

$$\sum_{n=-\infty}^{\infty} \frac{1}{n^2+a^2} = \frac{\pi}{a} \coth \pi a$$

for $a > 0$.

We identify $f(z) = \dfrac{1}{z^2+a^2}$. This function has simple poles at $z = \pm ia$. The residues of $\pi \cot \pi z f(z)$ are then

$$\begin{aligned} \lim_{z \to \pm ia} (z \mp ia) \frac{\pi \cot \pi z}{z^2+a^2} &= \lim_{z \to \pm ia} (z \mp ia) \frac{\pi \cot \pi z}{(z+ia)(z-ia)} \\ &= \frac{\pi \cot \pi ia}{2ia} \\ &= \frac{\pi}{2a} \coth \pi a. \end{aligned} \tag{42}$$

Adding the two residues gives the result.

3. Airy, Stokes, and the Rainbow

Up to this point we have seen applications of Laplace transforms to the summation of series which mostly are series born of mathematical explorations. But series appear in many applications in the sciences. For example, several interesting physics problems were solved using series expansions. One example is the Kepler problem, named after Johannes Kepler (1571 - 1630) (Colwell, 1993). In that case if one is interested in predicting the position of a planet as it travels around the sun. The solution is obtained by solving the two-body problem. Historically this resulted in a series which led to other interesting series for the expansion coefficients of the original series, which were later called Bessel functions, named for Friedrich Wilhelm Bessel (1784 - 1846).[10]

Another interesting problem came from the study of rainbows such as in Figure 3. We will review the history of this problem and what is found using geometric optics (Boyer, 1959; Corradi, 2016; Jackson, 1999; Khare and Nussenzveig, 1974; Nussenzveig, 1977). However, there are features of the rainbow which cannot be

[10]Bessel was director of the Königsberg Observatory most of his life and made significant contributions to determining stellar distances. While studying the dynamics of gravitational systems, he developed what are now called Bessel functions.

Figure 3. Typical rainbow showing the lower primary rainbow and the upper secondary rainbow. In between is Alexander's dark band. Other interesting features are the color order, polarizations, and other minute details. Note: There are oscillations below the primary bow which aren not often seen. [Arc-en-ciel, Countryside, Fields, `https://www.publicdomainpictures.net/pictures/160000/velka/double-arc-en-ciel-paysage.jpg`.]

described by just reflection and refraction. One needs to include wave interference, a topic that really was not understood until the early seventeenth century. One of the first to come close to providing the mathematical tools for predicting these effects was George Biddle Airy (1801-1892) in an 1838 paper (Airy, 1838). He derived a slowly convergent series approximation of the now termed the Airy integral, giving the Airy function,

$$\mathrm{Ai}(x) = \frac{1}{\pi} \int_0^\infty \cos\left(\frac{t^3}{3} + xt\right)\, dt. \tag{43}$$

It has been shown that the Airy function satisfies the differential equation

$$\frac{d^2y}{dx^2} - xy = 0. \tag{44}$$

Since this is a second order differential equation, there is a second linearly independent solution, $y(x) = \mathrm{Bi}(x)$, the Bairy function, which is given by (DLMF)

$$\mathrm{Bi}(x) = \frac{1}{\pi} \int_0^\infty \left[e^{-\frac{t^3}{3}+xt} + \sin\left(\frac{t^3}{3} + xt\right)\right] dt. \tag{45}$$

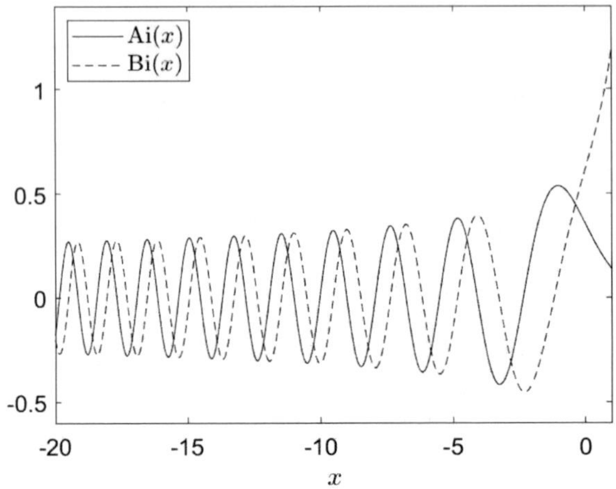

Figure 4. Plots of the Airy functions Ai(x) and Bi(x).

Plots of these functions are shown in Figure 4. These functions play an interesting role in recent work on trans-series and resurgence in which studies of divergent series are a centerpiece and discussed later in the chapter.

How are Airy functions connected to rainbows? We have to go back to the early studies of rainbows. Rainbows fascinated people for centuries. Some of the earliest studies were done in several cultures as people began to understand optical processes. Aristotle (384-322 BCE) was the first to rationally explain that rainbows occurred from the reflection of light from clouds. Alexander of Aphrodisias (3rd-2nd century BCE) noted a dark band between the primary and secondary band, now called Alexander's band as can be seen in Figure 3 (Boyer, 1959; Jackson, 1999; Nussenzveig, 1977).

Modern optics may have begun with Hasan Ibn al-Haytham (c. 965-1050), who was one of the first to understand reflection. Later, Kaml al-Dn al-Fris (1267-1319) attributed rainbows to raindrops. He was a student of Qutb Al-Din al-Shirāzī (12361311), who was a student of Nasir al-Din al-Tusi (1201-1274) a creator of trigonometry. al-Shirāzī advised al-Fris to read Ibn al-Haytham's Book of Optics and Farisi wrote a revision of the book.

Some western scholars were influenced by Persian and Arabic science. Early Europeans understood rainbows from experiments and developing geometric optics. Roger Bacon (1214-1294) observed that the primary rainbow was at 42°. Theodoric of Freiberg (1250-1310) independently claimed that rainbows were due to raindrops and not clouds. Later studies of geometric optics saw contributions from Willebrord Snellius (1580-1626) on refraction; René Descartes (1596-1650) rediscovered in-

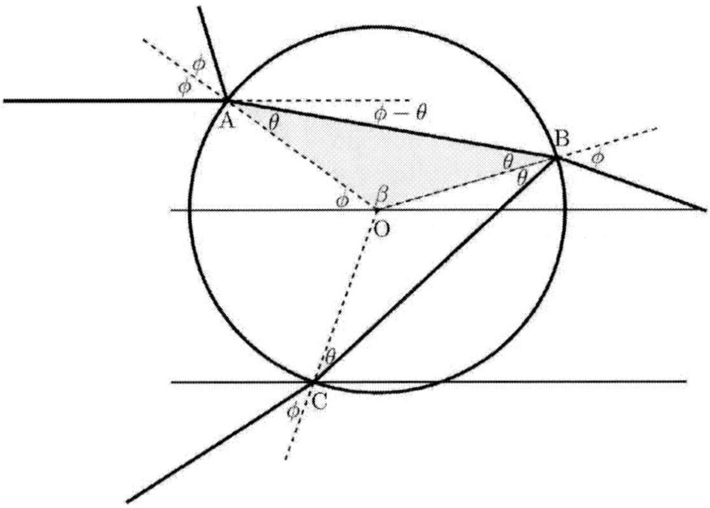

Figure 5. Reflections of an incident ray through a raindrop. A light ray enters the raindrop at point A. It is partially reflected and the rest refracts to B. It is then reflected except for some that leaves the droplet. We follow the ray to point C and follow that portion which leaves the droplet to contribute to the primary rainbow. Further internal scattering takes place to add to a secondary rainbow.

ternal reflections (Descartes, 1637, 1998); and, Pierre de Fermat (1607-1665), was challenged by Descartes to write on the theory of refraction and Snell's Law.[11] Isaac Newton (1642-1727) gave a theory of color and wrote a book on optics (Newton, 1740, 1952).

So, before the 1800's there was some understanding of rainbows. Then, things began to change. In 1687 Christiaan Huygens (1629-1695) proposed a theory of light which was rejected in favour of Newton's corpuscular theory of light. But, he put forward the idea that points on a wavefront act like point sources generating new wavefronts. In 1801 Thomas Young (1773-1829) demonstrated that light was a wave through interference experiments thus supporting Huygen's principles. In 1821 Augustin-Jean Fresnel (1788-1827) adopted Huygens's principle to give an explanation of the propagation and diffraction effects of light. Before we see how wave theory figures into the picture, we will look closer at the geometric optics of the rainbow.

The geometric theory of rainbows relied solely on reflection and refraction. As

[11]There are interesting studies on Descartes work in optics, priority and Fermat's intent to provide a better theory (Indorato and Nastasi, 1989; Malet, 1990).

depicted in Figure 5, a light ray is horizontally incident at point A with an angle of incidence, ϕ. Part of the ray is reflected and part is refracted into the droplet with angle of refraction θ. At point B this ray again is either refracted or reflected.

One can use the laws of optics to eventually determine where the light ray will emerge. [We refer to the above references and Adam (2002a,b, 2017).] The law of Reflection states that the incident angle equals the reflected angle. The Law of Refraction states that the angle of refraction, θ, and angle of incidence are related through Snell's Law, $\sin\phi = n\sin\theta$, where it is assumed that the water droplet is in air and has an index of refraction n.

We can use geometry to determine the scattering angle. At point A, the ray is bent by $\phi - \theta$. At point B the ray reflects and turns by $\beta = \pi - 2\theta$. Then, at point C the ray is bent by another angle $\phi - \theta$. In total, the ray has been bent from its original path by

$$2(\phi - \theta) + \beta = 2\phi - 4\theta + \pi.$$

In principle the internal light ray can undergo k reflections. This would give a scattering angle of

$$\Phi = |2(\phi - \theta) + k\beta| = |2\phi - 2(k+1)\theta + k\pi|. \tag{46}$$

In Figure 6 we show the paths of multiple rays which go through one internal reflection. There are locations where multiple rays meet causing caustics. This focusing of light leads to singularities in geometric optics, or caustics (Casselman, 2009, 2015). Also, there is a minimal angle at which the rays emerge being the source of high intensity.

It is the angle of minimum deviation which is of interest for each color in the rainbow. This can be determined by finding the extrema of Φ in Equation (46) and the angle of minimum deviation, θ_c, after k internal reflections is given by

$$\sin\theta_c = \sqrt{\frac{(k+1)^2 - n^2}{(k+1)^2 - 1}}, \quad k = 1, 2. \tag{47}$$

In Figure 7 we plot the scattering angles for red (660 nm) and violet (410 nm) showing the extrema. The upper curves are for the primary rainbow and the lower curves are for the secondary rainbow. Notice that the primary rays emerge between 40° (violet) and 42° (red).

Geometric optics was fine for describing some of the main features of the rainbow, but there were other observations which could not be explained. It was noticed that there were green, pink, and purple patterns, or fringes, below the primary rainbow.

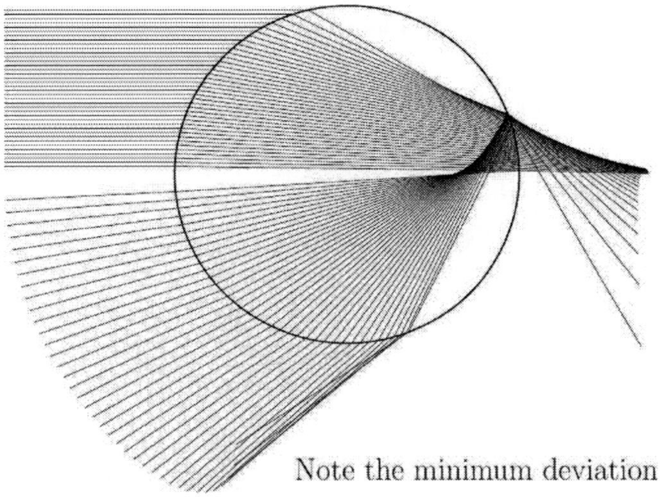

Figure 6. There are multiple incident rays at the top. Using geometric optics, we can see where the internally reflected ray emerges at the bottom. There are locations where these rays meet causing caustics, or regions of high intensity. Also, we see that there is a minimal angle at which the rays emerge.

Thomas Young (1804) suggested these supernumeraries were due to wave interference. Richard Potter (1799-1886) in 1835 also attempted to incorporate wave interference into rainbow studies not knowing of Young's results at the beginning of his studies. Potter (1838) located the caustic and used wave interference to estimate the diameter of a raindrop to produce supernumeraries at about 1/72 inch. This was similar to the 1/76 inch reported in 1804 by Young (1804).

Airy (1838) used Young's wave interference to derive the locations of the fringes. The idea was to superimpose the waves along the rainbow caustic. He determined the equation of the caustic along a cubic path and this led to the Airy integral (43). In order to compute the maxima, he had to approximate the integral. In a second paper, Airy (1849) obtained a converging power series, due to communication with Augustus De Morgan (1806-1871). Stokes (1850) gave this representation as

$$\begin{aligned} u(x) &= A\left\{1+\frac{9x^3}{2\cdot 3}+\frac{9^2x^6}{2\cdot 3\cdot 5\cdot 6}+\frac{9^3x^9}{2\cdot 3\cdot 5\cdot 6\cdot 8\cdot 9}+\cdots\right\} \\ &\quad +B\left\{x+\frac{9x^4}{3\cdot 4}+\frac{9^2x^7}{3\cdot 4\cdot 6\cdot 7}+\frac{9^3x^{10}}{3\cdot 4\cdot 6\cdot 7\cdot 9\cdot 10}+\cdots\right\}. \end{aligned} \tag{48}$$

As noted by Stokes (1857), such a series can be obtained as a series solution of an

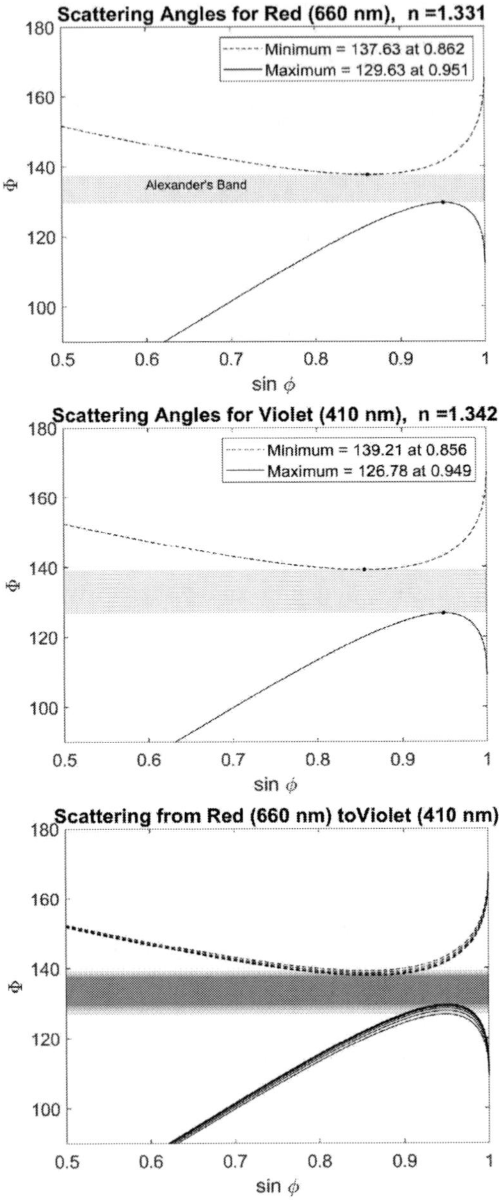

Figure 7. A plot of the scattering angles vs. different incident angles for red (660 nm) and violet (410 nm) rays which have slightly different indices of refraction. The extrema are marked for both the primary and secondary rainbows. In the third plot is shown an overlay for scattering angles for several colors from red to violet highlighting the location of Alexander's band.

Airy differential equation (44) of the form[12]

$$\frac{d^2u}{dx^2} - 9xu = 0.$$

One assumes $u(x) = \sum_{m=0}^{\infty} a_m x^m$. Inserting the series into the differential equation and collecting like powers of x, eventually one finds that

$$2a_2 + \sum_{k=1}^{\infty} \left[(k+1)(k+2)a_{k+2} - 9a_{k-1}\right] x^k = 0.$$

Therefore, $a_2 = 0$ and one has the equation for the expansion coefficients,

$$(k+1)(k+2)a_{k+2} = 9a_{k-1}, \quad k = 1, 2, \ldots.$$

Solving for the a_k's, one gets two linearly independent solutions:

$$\begin{aligned} u_1(x) &= 1 + \sum_{k=1}^{\infty} \frac{9^k}{2 \cdot 3 \cdot 5 \cdot 6 \cdots (3k-1) \cdot (3k)} x^{3k}, \\ u_2(x) &= x + \sum_{k=1}^{\infty} \frac{9^k}{3 \cdot 4 \cdot 6 \cdot 7 \cdots (3k) \cdot (3k+1)} x^{3k+1}, \end{aligned}$$

which can be seen in Equation (48).

The goal was to determine the zeros of Airy's integral,[13]

$$W(m) = \int_0^{\infty} \cos \frac{\pi}{2}(w^3 - mw)\, dw.$$

[12]The Airy function, $\mathrm{Ai}(z)$, is a solution of $y''(z) = zy(z)$. Making the transformation $z = \alpha x$ and $y(z) = u(x)$, we find that $\alpha^3 = 9$. So, we have that $u(x) = \mathrm{Ai}\left(3^{2/3}x\right)$.

In 1909 John William Nicholson (1881-1955) showed in (Nicholson, 1909) that the Airy function is related to modified Bessel functions of order $\frac{1}{3}$. Several of these relations are shown at the NIST Digital Library of Mathematical Functions (DLMF), in Section 9.6 such as

$$\mathrm{Ai}(z) = \frac{1}{\pi}\sqrt{\frac{z}{3}}\mathrm{K}_{\frac{1}{3}}\left(\frac{2}{3}z^{3/2}\right).$$

[13]Stokes (1850) makes a few variable changes relating $W(m)$ to

$$U(n) = \int_0^{\infty} \cos(x^3 - nx)\, dx.$$

Here, $W(m) = \left(\frac{2}{\pi}\right)^{\frac{1}{3}} U(n)$, where $n = \left(\frac{\pi}{2}\right)^{2/3} m$ and $U'' + \frac{n}{3}U = 0$.

Here m is the perpendicular distance between the caustic and point of observation. Airy developed a table for $m = -4$ to $m = 4$ in steps of 0.2. Airy found a converging series ascending in m.

It turned out the computation to find the locations of the supernumeraries was tedious and the convergence slow. In order to compute the series for large values of m, Airy needed ten-figure logarithms and was only able to compute two roots (corresponding to dark bands). However, in 1841 William H. Miller (1801 - 1880) (Miller, 1848) measured 30 dark bands! So, there had to be a better approach to computing the Airy integral. In 1850 George Gabriel Stokes (1819-1903) took up the question (Stokes, 1850).

Stokes (1850) wanted to determine the roots for large m. As he noted, using ascending series was common for physical problems. So, he converted the problem to studying the differential equation for a function $y(x)$ and seek the leading order terms for large x. Then, he sought corrections in the form of series and related the arbitrary coefficients in this solution to those in the convergent series (48). Stokes (1857) later gave the result

$$u(x) = Cx^{-1/4}e^{-2x^{3/2}}\left\{1 - \frac{1\cdot 5}{1\cdot 144x^{\frac{3}{2}}} + \frac{1\cdot 5\cdot 7\cdot 11}{1\cdot 2\cdot 144^2x^3} - \frac{1\cdot 5\cdot 7\cdot 11\cdot 13\cdot 17}{1\cdot 2\cdot 3\cdot 144^3x^{\frac{9}{2}}} + \cdots\right\}$$
$$+Dx^{-1/4}e^{2x^{3/2}}\left\{1 + \frac{1\cdot 5}{1\cdot 144x^{\frac{3}{2}}} + \frac{1\cdot 5\cdot 7\cdot 11}{1\cdot 2\cdot 144^2x^3} + \frac{1\cdot 5\cdot 7\cdot 11\cdot 13\cdot 17}{1\cdot 2\cdot \cdot 3144^3x^{\frac{9}{2}}} + \cdots\right\}.$$

In support of his methods Stokes (1850) gave two other examples based on the functions

$$\begin{aligned} u(x) &= \frac{2}{\pi}\int_0^{\frac{\pi}{2}} \cos(x\cos\theta)\,d\theta, \\ v(x) &= \frac{2}{\pi}\int_0^{x}\int_0^{\frac{\pi}{2}} x'\cos(x'\cos\theta)\,d\theta dx' \\ &= \int_0^x x'u(x')\,dx'. \end{aligned} \tag{49}$$

Stokes said that $v(x)$ occurs for diffraction with a circular aperture and that $u(x)$ occurs in many physical applications. The goal was to find approximations useful for computing $u(x)$ and $v(x)$ for large values of x and to extract the roots of these functions.

Stokes did not refer to $u(x)$ as the zeroth order Bessel function,[14] $J_0(x)$, nor did he

[14] Bessel functions were first encountered in the work of Daniel Bernoulli on heavy chains (1738). Euler studied more general Bessel functions in 1781 and in vibrating membranes in 1764. These functions also ap-

indicate that[15] $v(x) = xJ_1(x)$. Stokes wrote down the differential equations that each satisfied and proceeded from there just as he had done for the Airy function.[16] In each case he arrived at the first 12 nonzero roots of the two Bessel functions.

What is surprising about these series is that they are not convergent series but divergent series. For large z, one can write (Cosnahan, 2021; Mariño, 2021; Silverstone et al., 1985),[17]

$$\mathrm{Ai}(z) \sim \frac{e^{-\frac{2}{3}z^{\frac{3}{2}}}}{2\sqrt{\pi}\, z^{\frac{1}{4}}} \left[\sum_{n=0}^{\infty} \frac{(-1)^n \Gamma(n+\frac{5}{6})\Gamma(n+\frac{1}{6}) \left(\frac{3}{4}\right)^n}{2\pi n! z^{3n/2}} \right], \qquad |\arg(z)| < \pi,$$

and

$$\mathrm{Bi}(z) \sim \frac{e^{\frac{2}{3}z^{\frac{3}{2}}}}{\sqrt{\pi}\, z^{\frac{1}{4}}} \left[\sum_{n=0}^{\infty} \frac{\Gamma(n+\frac{5}{6})\Gamma(n+\frac{1}{6}) \left(\frac{3}{4}\right)^n}{2\pi n! z^{3n/2}} \right], \qquad |\arg(z)| < \frac{\pi}{3}.$$

We see that the coefficients involve the ratio of two Gamma functions to one. So, this series is actually factorially divergent, meaning that the series coefficients have factorial growth as in Equation (58). But, Stokes was able to use the divergent series in his examples to obtain useful results. Furthermore, Stokes' explorations of the supernumeraries in 1850 led to different exponential behavior on the dark side of the rainbow as opposed to the bright side. This puzzled him for years until his 1857 publication (Stokes, 1857, 1868).[18]

Stokes' work has played an important role in understanding divergent series through the work of Robert Dingle (Dingle, 1958a,b,c, 1959a,b, 1973; Dingle and Wilkes, 1959), Michael Berry and Christopher Howls (Berry, 1988, 1989, 1991; Berry and Howls, 1990, 1991), and Jean Écalle (Écalle, 1981). This has led to the theory of trans-series and resurgence (Aniceto et al., 2019; Dunne, 2016; Mariño, 2014, 2021; Marucho, 2004; Mas, 2019). We will discuss other divergent series in

---

peared in the Kepler Problem, describing planetary motion (Gray and Mathews, 1895; Watson, 1922; Dutka, 1995). Joseph-Louis Lagrange (1736-1813) derived a series solution in the study of the Kepler problem. In 1816, Friedrich Wilhelm Bessel (1784-1846) gave an integral representation for the expansion coefficients and in 1824 he presented a thorough study of these functions, which became known as Bessel functions. The earliest treatises devoted to Bessel functions were by Gray and Mathews (1895) and Watson (1922).

[15] $J_n(x)$ is a solution of $x^2y''(x) + xy'(x) + (x^2 - n^2)y(x) = 0$. $J_0(x)$ and $J_1(x)$ are related by the identity $[xJ_1(x)]' = xJ_0(x)$.

[16] According to Watson (1922), the J notation for Bessel functions was first used in 1843 by Peter A, Hansen (1795-1874) and later in 1857 by Oskar Schlomilch (1823 - 1901). The current use of $J_n(x)$ is in the text of Gray and Mathews (1895). So, it may not be surprising that Stokes did not use it in his paper.

[17] Also, see 10.4.59 in Abramowitz and Stegun, *Handbook of Mathematical Functions* (Abramowitz and Stegun, 1965) and 9.7 in (DLMF).

[18] There are other approaches to supernumeraries than Airy's theory such as Mie theory, Debye series, or a modification of Young's interference theory (Lee, 1998; Laven, 2017; Adam, 2017) indicating the range of applicability of Airy's theory.

the next section and see how Laplace transforms figure into making sense of the series. Then, we will return to the behavior of the series found by Stokes for the Airy function.

4. Divergent Series

4.1. Revisiting Euler

We started the chapter showing how some infinite series can be summed using Laplace transforms. When series can be summed, they are often said to be convergent. However, divergent series can also provide something useful as we saw with Stokes' series representation of the Airy integral. In this section we will define special types of divergent series known as an asymptotic series and how Laplace transforms are used to get useful information from them through what is referred to as Borel resummation using a Laplace transform.

The study of divergent series goes back a few centuries (Bromwich, 1908; Hardy, 1949; Kline, 1983; Bagni, 2005, 2009). One of the first discussions is reportedly from 1703 by Guido Grandi (1671-1742) (Tucciarone, 1973) who considered the series

$$1-1+1-1+\cdots.$$

Setting $x=1$ in the geometric sum,

$$1-x+x^2-x^3+\cdots=\frac{1}{1+x},$$

he concluded that

$$1-1+1-1+\cdots=\frac{1}{2}.$$

Tucciarone (1973) said that Gottfried Leibniz (1646-1716) later argued for a similar result in a letter to Christian Wolff (1679 - 1754). He claimed that the sum should be one half since it is the average of the partial sums.

Not everyone was satisfied with this result. Over the years many lent their reasoning for, or against, this result by regrouping the terms, adding zeros, or using different series expansions. Euler tried to define what it meant to "sum" a divergent series. Noting that

$$\frac{1}{1+x}=1-x+x^2-x^3+\cdots$$

converged for $|x|<1$, but as x approached 1 we have in the limit

$$\lim_{x\to 1}\frac{1}{1+x}=1-1+1-1+\cdots,$$

he agreed that this loose definition of sum would result in a value of $\frac{1}{2}$. So, with caution, Euler adopted the idea to take as the sum of the divergent series $\sum a_n$ the limit

$$\lim_{x\to 1} \sum a_n x^n,$$

if it exists.

Leonhard Euler (1760) wrote on divergent series in 1760 mentioning the Grandi series and considered Wallis hypergeometric series with alternating signs

$$\sum_{n=0}^{\infty} (-1)^n n! = 1 - 1 + 2 - 6 + 24 - 120 + \cdots. \tag{50}$$

Euler found the sum of this series was about $.596347362123$. Euler studied the series[19]

$$\begin{aligned} s(x) &= \sum_{n=0}^{\infty} (-1)^n n! x^{n+1} \\ &= x - x^2 + 2x^3 - 6x^4 + 24x^5 - 120x^6 + \cdots, \end{aligned} \tag{51}$$

which he obtained Wallis' series by setting $x = 1$ in Equation (51).

Euler (1760) approached the series (51) in several ways such as continued fractions and difference methods. He also considered the differential equation for $s(x)$,

$$x^2 \frac{ds(x)}{dx} = x - s(x).$$

Using the integrating factor, $\mu(x) = e^{-\frac{1}{x}}$, one finds the solution

$$s(x) = \int_0^x \frac{1}{\xi} e^{\frac{1}{x} - \frac{1}{\xi}} \, d\xi.$$

Making a change of variables $\xi = \dfrac{x}{1+tx}$, we obtain

$$s(x) = \int_0^{\infty} \frac{xe^{-t}}{1+xt} \, dt.$$

[19]Euler's explorations can be seen in `http://eulerarchive.maa.org/docs/originals/E247.pdf` on pages 220 and 230 of *De seriebus divergentibus [On divergent series]* where he uses continued fractions to get this value (Euler, 1760). Euler (1740a) also investigated the harmonic series, which Kowlenko claims to be the first regularization of divergent series in his discussion of Euler's work on series (Kowalenko, 2011). Additional historical perspectives are given by Barbeau (1979) and Kline (1983).

Therefore, Euler assigned the sum of the series

$$s(1) = 1 - 1 + 2 - 6 + 24 - 120 + \cdots = \int_0^\infty \frac{e^{-t}}{1+t}\, dt. \tag{52}$$

We can apply the Laplace transform to sum the Euler series (51) making use of the transform

$$n! = \int_0^\infty t^n e^{-t}\, dt = \mathcal{L}\left[t^n\right]_{s=1}.$$

Then, we have

$$\begin{aligned} \sum_{n=0}^{\infty} (-1)^n n! x^{n+1} &= \sum_{n=0}^{\infty} (-1)^n \left(\int_0^\infty t^n e^{-t}\, dt \right) x^{n+1} \\ &= x \int_0^\infty \left(\sum_{n=0}^{\infty} (-xt)^n \right) e^{-t}\, dt \\ &= \int_0^\infty \frac{x e^{-t}}{1+xt}\, dt. \end{aligned} \tag{53}$$

Therefore, we found the same solution as we had by solving the differential equation.

We will show that the solution can also be written in terms of the Exponential integral,[20]

$$\mathrm{E}_1(z) = \int_z^\infty \frac{e^{-t}}{t}\, dt.$$

In the solution (53) we make a change of variables, $w = 1 + xt$. Then, we obtain

$$\int_0^\infty \frac{x e^{-t}}{1+xt}\, dt = \int_1^\infty \frac{e^{(1-w)/x}}{w}\, dw.$$

[20] If we define $\phi(z) = \sum_{n=0}^{\infty} (-1)^n n! z^{-n-1}$, then we find that $\phi(z)$ satisfies a differential equation,

$$\phi'(z) - \phi(z) = -\frac{1}{z}. \tag{54}$$

One can find the general solution of Equation (54) as

$$\phi(z) = e^z \Gamma(0; z) + C e^z,$$

where $\Gamma(x; z)$ is the incomplete Gamma function, Chapter 8 in the DLMF:

$$\Gamma(x; z) = \int_z^\infty t^{x-1} e^{-t}\, dt = z^x \mathrm{E}_{1-x}(z).$$

So,

$$\Gamma(0, x) = -\mathrm{Ei}(-x) = \mathrm{E}_1(x), \quad x > 0.$$

Now, letting $w = xz$, we find

$$\begin{aligned} \int_0^\infty \frac{xe^{-t}}{1+xt}\,dt. &= e^{\frac{1}{x}} \int_{\frac{1}{x}}^\infty \frac{e^{-z}}{z}\,dz \\ &= e^{\frac{1}{x}} \mathrm{E}_1\left(\frac{1}{x}\right). \end{aligned} \tag{55}$$

The Euler series (51) is an example of a factorially divergent series. We managed to relate it to something we could call a sum by using a Laplace transform. This would be clearer if instead we were to replace x with s^{-1} as

$$\begin{aligned} \sum_{n=0}^\infty (-1)^n \frac{n!}{s^{n+1}} &= \sum_{n=0}^\infty (-1)^n \int_0^\infty t^n e^{-st}\,dt \\ &= \int_0^\infty e^{-st} \left(\sum_{n=0}^\infty (-t)^n \right) dt. \end{aligned} \tag{56}$$

We have related a divergent series to the Laplace transform of a convergent series. This process, which we will use later, is called a Borel resummation. But first, we want to explore a couple of other asymptotic series.

Historically, mathematicians before the nineteenth century had no problem working with divergent series using them to obtain approximate solutions to problems. However, as mathematics was becoming more rigorous, many began to question this practice and called for more definitive definitions. Niels Abel (1802-1829) is quoted as saying, “Divergent series are the invention of the devil, and it is shameful to base on them any demonstration whatsoever” (Maor). Nonetheless, Freeman Dyson (1923-2020) wrote a two-page paper in 1952 explaining why quantum electrodynamics (QED) should be expected to be divergent (Dyson, 1952).

In the nineteenth century new ideas emerged as to what it meant to sum a series. Augustin-Louis Cauchy (1789-1857) introduced in *Cours d’Analyse* (1821) the idea of convergence of series by convergence of the partial sums, as we now teach undergraduates. Abel read this work and added his own ideas as did many in the early 1800s. There were different notions of summability that arose, such as Abel summability, which was similar to Euler’s definition. A series $\sum a_n$ is Abel summable if $\sum a_n z^n$ has a radius of convergence at least 1 and $\lim_{z\to 1} \sum a_n z^n$ exists. Moore (1932) discusses these and other approaches to the definitions of summability.

There was a period when the Abel-Cauchy notions that divergent series should be avoided had a grip on the mathematics community and divergent series were frowned upon until new developments and texts were written in the late nineteenth to early twentieth centuries (Bromwich, 1908; Hardy, 1949). But there were some early discussions, such as from Augustus de Morgan (de Morgan, 1844) and interesting takes like Thomas Bayes (1763) note on the Stirling's series, which we discuss later in the chapter.

Divergent series began to appear in solutions of differential equations, in physical applications, and in attempts to evaluate integrals using series expansions. We will look at a few examples of these and how Laplace transforms come into the story.

4.2. Asymptotic Series from Integrals

We are interested in asymptotic series. Such series can arise from solutions of differential equations or asymptotic expansions of integrals (Bender and Orszag, 1999; Bleistein and Handelsman, 1975; Copson, 2004). Just like the divergent series developed to study supernumeraries, there are many applications in physics which lead to asymptotic series. Thomas John I'Anson Bromwich (1875 - 1929) recalls some of the history Chapter XI in his first edition of *An Introduction to the Theory of Infinite Series* Bromwich (1908) relating to the computation of planetary positions.[21] In quantum mechanics it is well known that the WKB method [See footnote 29.] leads to factorially divergent series (Bender and Wu, 1969; Mariño, 2021).

So, how useful are divergent series? Bromwich (Bromwich, 1908) says about some of the early users of asymptotic series, " In numerical work, however, they naturally used only series which are now called asymptotic; in such series the terms begin to decrease, and reach a minimum, afterwards increasing. If we take the sum to a stage at which the terms are sufficiently small, we may hope to obtain an approximation with a degree of accuracy represented by the last term retained and it can be proved that this is the case with many series which are convenient for numerical calculations."

So, what does this mean. If we have a series representation of a function, then we write the series approximation using the first N terms as

$$f(x) = \sum_{n=0}^{N-1} a_n (x - x_0)^n + R_N(x), \tag{57}$$

[21]Bromwich also provided quotes from Abel and Cauchy indicating their take on asymptotic series. He noted that "Cauchy established the asymptotic property of Stirling's series" and that it was later applied by Stokes 1850; 1857; 1868. He also discussed other applications such as Stirling series and evaluating Fresnel integrals from optics and the theory of deep water waves. The historical section was moved to Chapter XII in the second edition (Bromwich and MacRobert, 1926) revised with the help of Thomas Murray MacRobert (1884-1962).

where $R_N(x)$ is the remainder. Such a series is said to be convergent if $|R_N(x)| \to 0$ as $N \to \infty$, with x fixed. On the other hand, we say that it is asymptotic if $|R_N(x)| \ll (x - x_0)^N$ as $x \to x_0$, for N fixed. In Chapter 8 of Whittaker and Watson (1920) they review Poincaré asymptotics.

We will look at several examples of asymptotic series. Some examples occur from studying the behavior of integrals for large parameters and others occur from developing series solutions of differential equations. We will also be interested in factorially divergent series which are asymptotic. As we vary N, we see that the partial sums,

$$f_N(x) = \sum_{n=0}^{N} a_n x^n$$

appear to approach the true value and then begin to diverge. In practice, one looks for an optimal truncation where N truncates the series in some optimal fashion. Then, we use the approximation where the series is truncated at the optimal N.

For a factorially divergent series, the coefficients grow factorially as

$$a_n \sim A^{-n} n!, \quad n \gg 1. \tag{58}$$

We locate the place for truncation by minimizing N for

$$\left| a_N z^N \right| = cN! \left| \frac{z}{N} \right|^N .$$

The Stirling approximation can be used to determine the optimal truncation to be at a value of $N = |A/z|$ (Mariño, 2021). Then, we can get a handle on the error made to find that the error is exponentially small leading some to investigate asymptotics beyond all orders (Segur et al., 1991).[22]

For the first example, we consider the integral

$$\int_0^\infty \frac{e^{-t/\epsilon}}{1+t}\, dt$$

and evaluate it for small ϵ. The main contribution to the integral occurs for small t. So, we expand the denominator in the integrand and integrate term by term,

$$\begin{aligned} \int_0^\infty \frac{e^{-t/\epsilon}}{1+t}\, dt &= \int_0^\infty e^{-t/\epsilon} \sum_{n=0}^{\infty} (-t)^n \, dt \\ &\sim \epsilon - 1!\epsilon^2 + 2!\epsilon^3 - 3!\epsilon^4 + \cdots . \end{aligned} \tag{59}$$

[22]Other terms that are common are superasymptotics and hyperasymptotics. (Berry and Howls, 1990; Boyd, 1999).

We see that the resulting series is factorially divergent. [This series should look familiar to the reader after reading earlier sections.] It is also asymptotic as seen in Figure 8. In the figure are shown the Nth partial sums of the series in Equation (59) along with the "sum"

$$\int_0^\infty \frac{e^{-t/\epsilon}}{1+t}\,dt = e^{\frac{1}{\epsilon}} E_1\left(\frac{1}{\epsilon}\right).$$

This indicates that the optimal truncation is just before the least term at roughly $N = \frac{1}{\epsilon}$.

Another typical example stems from evaluating the error function for small argument. The error function is given by

$$\text{Erf}\,(z) = \frac{2}{\sqrt{\pi}} \int_0^z e^{-t^2}\,dt.$$

We note that the integrand, e^{-t^2}, is analytic; i.e., it has the power series representation

$$e^{-t^2} = \sum_{n=0}^{\infty} \frac{(-t^2)^n}{n!},$$

which is convergent for $|t| < \infty$. Therefore, integrating term by term, we naturally obtain a series representation for the error function,[23]

$$\begin{aligned} \text{Erf}\,(z) &= \frac{2}{\sqrt{\pi}} \sum_{n=0}^{\infty} \frac{(-1)^n z^{2n+1}}{(2n+1)n!} \\ &= \frac{2}{\sqrt{\pi}} \left(z - \frac{1}{3}z^3 + \frac{1}{10}z^5 - \frac{1}{42}z^7 + \cdots \right). \end{aligned} \tag{60}$$

However, the number of terms need to obtain a desired accuracy is strongly dependent on the argument, z. Let us say that we would like to compute the sum to an accuracy of 10^{-5}. A quick check in Maple gives the results in Table 1.

So, we need another representation for large arguments. For this we will use the complementary error function.

$$\text{Erfc}\,(z) = 1 - \text{Erf}\,(z) = \frac{2}{\sqrt{\pi}} \int_z^\infty e^{-t^2}\,dt.$$

[23] See Jeffreys (1927) for a short history of this notation.

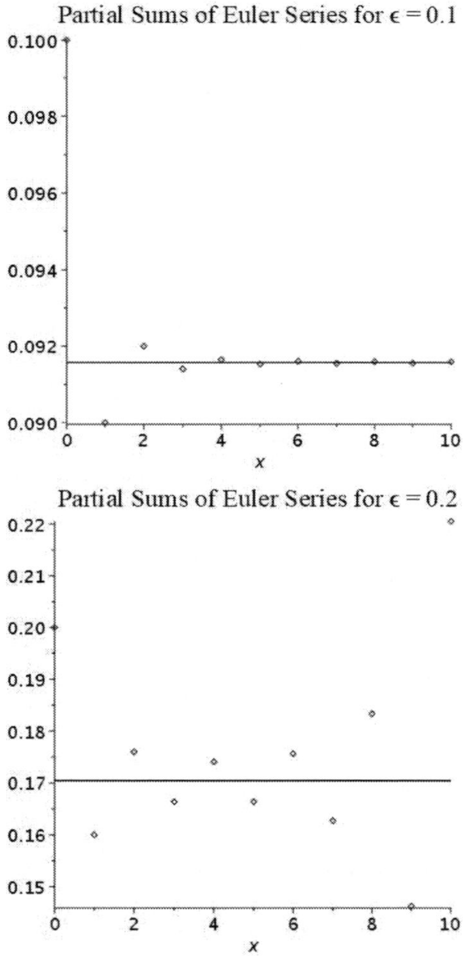

Figure 8. Partial sums for the series (59) for $\epsilon = 0.1, 0.2$ demonstrating that the series is asymptotic. In each case the partial sums seem to converge before eventually diverging. This leads to the idea of an optimal truncation point near $N \approx \dfrac{1}{\epsilon} = 10, 5$, respectively.

Table 1. The number of terms needed in the series (60) for an accuracy of about 10^{-5} for a given z in Maple.

Terms	Up to $z =$
7	1
16	2
30	3
73	5

If we integrate by parts, we obtain a series in powers of z^{-1},

$$\begin{aligned} \int_z^\infty e^{-t^2}\,dt &= \frac{e^{-z^2}}{2z} - \int_z^\infty \frac{e^{-t^2}}{2t^2}\,dt \\ &= \frac{e^{-z^2}}{2z}\left(1 - \frac{1}{2z^2} + \frac{3(1)}{(2z^2)^2} - \cdots\right). \end{aligned} \tag{61}$$

This, in turn, leads to the desired series expansion for the complementary error function,

$$\text{Erfc}\,(z) = \frac{e^{-z^2}}{z\sqrt{\pi}}\left(1 - \frac{1}{2z^2} + \frac{3(1)}{(2z^2)^2} - \frac{5(3)(1)}{(2z^2)^3} + \mathcal{O}(z^{-8})\right).$$

We note that at $z = 2.5$, we only need three terms for an accuracy of 10^{-5} and only need two terms when $z > 3$. However, this expansion is divergent!

We further note that this series for the complementary error function was encountered in the work of Heaviside and described by E. T. Whittaker in Moore's 1921 book (Moore, 1971). In (Herman, 2025, p.44), *Laplace Transforms and Green's Functions*, we derived the function

$$f(t) = 1 - e^{t/a}\text{erfc}\left(\sqrt{\frac{t}{a}}\right) \tag{62}$$

as the solution to one of Heaviside's problems. We obtained a series expansion for small t,

$$\begin{aligned} f(t) &\sim 2\sqrt{\frac{t}{\pi a}}\sum_{k=0}^{\infty}\frac{2^k}{(2k+1)!!}\left(\frac{t}{a}\right)^k + 1 - e^{t/a} \\ &= 2\sqrt{\frac{t}{\pi a}}\left[1 + \frac{2}{3}\frac{t}{a} + \frac{4\,t^2}{15\,a^2} + \frac{8\,t^3}{105\,a^3} + \frac{16\,t^4}{945\,a^4}\right] + V_0\left[1 - e^{t/a}\right], \end{aligned} \tag{63}$$

and a series expansion for large t,

$$\begin{aligned} f(t) &\sim 1-\left(\frac{a}{\pi t}\right)^{1/2}\sum_{k=0}^{\infty}(-1)^k\frac{(2k)!}{2^{2k}k!}\left(\frac{a}{t}\right)^k \\ &= 1-\left(\frac{a}{\pi t}\right)^{1/2}\left(1-\frac{a}{2t}+1\cdot 3\left(\frac{a}{2t}\right)^2-\cdots\right). \end{aligned} \tag{64}$$

We can look at the convergence properties of these series by comparing the Nth partial sums for the small t expansion, Equation (63), and large t expansion, Equation (64), to the exact solution, Equation (62) with $a = 1$. We plot these in Figure 9 with the Nth partial sums ($N = 10, 20, 40$). We see that the regions in which each expansion is applicable depends on N and generally where one series fails the other succeeds. For odd N, the figures are different in that the diverging curves head in opposite directions.

We notice that in the series (64), the coefficients are factorially divergent. From the Legendre Duplication Formula, Section 5.5 number 5 in (DLMF),

$$\Gamma(z)\Gamma(z+\frac{1}{2}) = 2^{1-2z}\sqrt{\pi}\Gamma(2z), \tag{65}$$

we see that the $(2k)!$ factor is actually two factorials.

$$\frac{(2k)!}{k!} = \frac{\Gamma(2k+1)}{k!} = \frac{2k}{2^{1-2k}\sqrt{\pi}k!}\Gamma(k)\Gamma(k+\frac{1}{2}) = \frac{2^{2k}}{\sqrt{\pi}}\Gamma(k+\frac{1}{2}).$$

Another famous asymptotic series comes from the Stirling approximation (Bender and Orszag, 1999; Bleistein and Handelsman, 1975; Dominici, 2006). There is a history tied to questions in probability theory as outlined by Dutka (1991) and by others.[24] In 1733 Abraham de Moivre (1667-1754) gave an approximation to the factorial function,

$$n! \approx Bn^{n+\frac{1}{2}}e^{-n}.$$

[24] Deming (1933) noted that Pearson (1924) commented on a publication of de Moivre (1730) concerning the origins of the normal curve. de Moivre communicated with Stirling who wrote (Stirling, 1730). In 1733 de Moivre added two supplements to his book. The second supplement, "Approximatio ad Summam Terminorum Binomii (a+ b)n in Seriem expansi," contained "the first use of the approximation for large factorials " according to Deming. de Moivre (1756) explored the problem further and noted the slow convergence of the series. Ranjan Roy (1948 - 2020), one of the author's early mentors, provides a more extensive discussion comparing what both de Moivre and Stirling computed, how they were motivated by a problem in probability, and that the first appearance of the Stirling approximation appeared in a letter from Euler to Goldbach in 1744 (Roy, 2011). Roy also describes the further work on the Stirling series by Gauss, Cauchy, and others. Also note Dominici (2006) adds further to the history. It is interesting that the Stirling series appeared in the English translation (Lacroix, 1816) of (Lacroix, 1797).

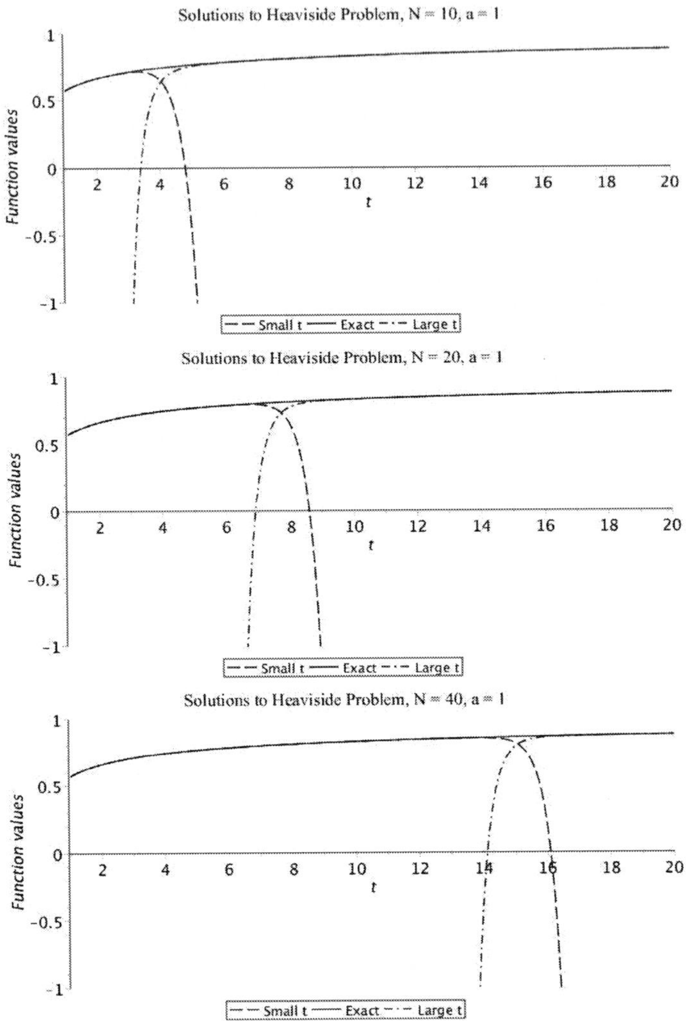

Figure 9. Plots of the solutions of Heaviside's problem with $a = 1$. Here we compare the exact solution in Equation (62) with the Nth partial sums ($N = 10, 20, 40$) for the small and large t expansions in Equations (63) and (64). Note the regions in which each expansion is applicable depends on N and generally where one series fails the other succeeds.

He had estimated the constant

$$\ln B = 1 - \frac{1}{12} + \frac{1}{360} - \frac{1}{1260} + \frac{1}{1680} \approx 2.5074.$$

Not long after, James Stirling (1692-1770) determined the constant to be $\sqrt{2\pi}$.

In 1763 a letter (Bayes, 1763) was presented for the late Reverend Mr. Thomas Bayes (1701-1761) pointing out that

$$\log z! - \left(z + \frac{1}{2}\right) \log z - \log c$$

is lessened by

$$z - \frac{1}{12z} + \frac{1}{360z^3} - \frac{1}{1260z^5} + \frac{1}{1680z^7} - \frac{1}{1188z^9} + \dots$$

provided "z is large, and you take in only a proper number of the first terms of the foregoing series: but the whole series can never properly express any quantity at all ...." Therefore, Bayes knew that the Stirling series was divergent.

One method for deriving the Stirling formula[25] is Laplace's Method (Bender and Orszag, 1999; Bleistein and Handelsman, 1975; Copson, 2004). Assuming $g(t)$ and $f(t)$ are real, continuous functions, then this method is used to find the asymptotic behavior of the integral[26]

$$I(x) = \int_a^b g(t) e^{xf(t)}\, dt \tag{66}$$

as $x \longrightarrow \infty$.

For the Stirling approximation, we use the integral representation of the Gamma function,

$$N! = \Gamma(N+1) = \int_0^\infty t^N e^{-t}\, dt, \tag{67}$$

which was known to Euler.[27] In order to apply Laplace's method to the Gamma

[25] Whittaker and Watson (1920) develop the Stirling approximation using Binet's expression for $\ln \Gamma(z)$.

[26] Related to Laplace's Method is Watson's Lemma. Let $f(t)$ be continuous on $0 \leq t \leq b$ and have the asymptotic expansion

$$f \sim t^\alpha \sum_{n=0}^{\infty} a_n t^{n\beta} \quad \text{as } t \to 0^+.$$

for $\alpha > -1$, and $\beta > 0$ (Bender and Orszag, 1999; Bleistein and Handelsman, 1975; Paris and Kaminski, 2001). Then, the Laplace transform has the asymptotic expansion

$$\mathcal{L}[f(t)] \sim \sum_{n=0}^{N} a_n \frac{\Gamma(\alpha + \beta n + 1)}{s^{\alpha+\beta n+1}} \quad \text{as } s \to \infty.$$

[27] Recall the Laplace transform $L[t^N] = \dfrac{\Gamma(N+1)}{s^{N+1}}$.

function integral, we first rewrite it as

$$N! = \int_0^\infty e^{-t+N\ln t}\, dt.$$

Letting $t = Ns$, we have

$$N! = N^{N+1} \int_0^\infty e^{N(\ln s - s)}\, ds. \tag{68}$$

This is now in the proper form to apply Laplace's method where $f(s) = \ln s - s$, $g(s) = 1$, and $x = N$. We also note that at the endpoint, $s = 0$, the integrand is well defined in the original integral and we'll only need contributions to $f(s)$ near $s = 1$.

Let us focus on the integral

$$I(N) = \int_0^\infty e^{Nf(s)}\, ds$$

for large N. If $f(s)$ has a maximum at a point $s = s_0$, then the main contribution to the integral is near s_0. So, we can replace $f(s)$ in the exponential by

$$f(s) \approx f(s_0) + \frac{1}{2} f''(s_0)(s - s_0)^2.$$

Then, we capture the main contribution by integrating over a neighborhood of s_0, given by $s \in [s_0 - \epsilon, s_0 + \epsilon]$ for small ϵ. This gives

$$\begin{aligned} I(N) &\approx e^{Nf(s_0)} \int_{s_0-\epsilon}^{s_0+\epsilon} e^{\frac{N}{2} f''(s_0)(s-s_0)^2}\, ds \\ &= \frac{e^{Nf(s_0)}}{\sqrt{-Nf''(s_0)}} \int_{-\epsilon\sqrt{-Nf''(s_0)}}^{\epsilon\sqrt{-Nf''(s_0)}} e^{-y^2/2}\, dy \\ &\approx \frac{e^{Nf(s_0)}}{\sqrt{-Nf''(s_0)}} \int_{-\infty}^{\infty} e^{-y^2/2}\, dy \\ &= \frac{\sqrt{2\pi} e^{Nf(s_0)}}{\sqrt{-Nf''(s_0)}}. \end{aligned} \tag{69}$$

Here we let $y^2 = -Nf''(s_0)(s - s_0)^2$.

For the Gamma function integral (67), we defined $f(s) = \ln s - s$ in Equation (68). This function has a maximum value at $s_0 = 1$ as seen in Figure 10. Since

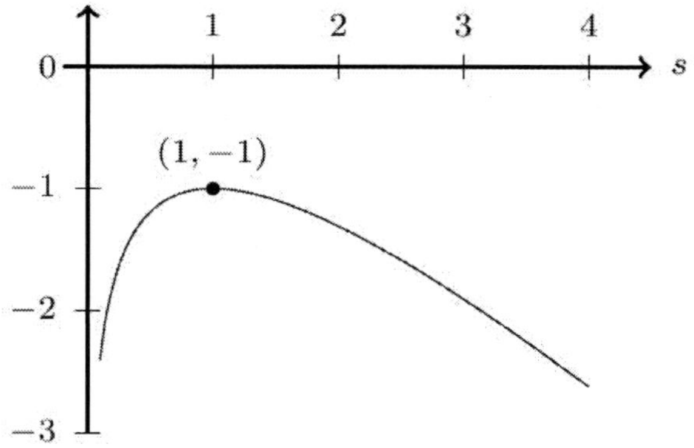

Figure 10. Plot of $f(s) = \ln s - s$ indicating the maximum as $s = 1$.

$f'(s_0) = 0$ and $f''(s_0) = -1$, we obtain the large N approximation

$$N! \approx N^{N+1} \frac{\sqrt{2\pi} e^{Nf(s_0)}}{\sqrt{-Nf''(s_0)}} = \sqrt{2\pi} e^{-N} N^{N+\frac{1}{2}}.$$

Taking the logarithm of this result, yields the Stirling approximation,

$$\log N! \approx \left(N + \frac{1}{2}\right) \log N - N + \log \sqrt{2\pi}.$$

This is the leading order approximation. If we desire more terms, then we add higher order terms in the Taylor series expansion of $f(s)$ (Bender and Orszag, 1999). The general formula can be written in terms of the Riemann zeta functions using the Euler-Maclaurin summation formula[28] and Equation (12).

$$\begin{aligned} \log z! - \left(z + \frac{1}{2}\right) \log z - \log \sqrt{2\pi} + z &= \frac{1}{2\pi^2 z} \sum_{n=0}^{\infty} (-1)^n \frac{(2n)!}{(2\pi z)^n} \zeta(2n+2) \\ &= \frac{1}{12z} - \frac{1}{360z^3} + \frac{1}{1260z^5} - \frac{1}{1680z^7} \\ &\quad + \frac{1}{1188z^9} - \frac{691}{360360z^{11}} + \cdots. \end{aligned} \tag{70}$$

[28] The Euler-Maclaurin summation formula relates the integral of a function to a sum over discrete values of the function,

$$\sum_{i=m}^{n} f(i) = \int_m^n f(x)\,dx + \frac{f(m) + f(n)}{2} + \sum_{k=1}^{\lfloor \frac{p}{2} \rfloor} \frac{B_{2k}}{(2k)!} (f^{(2k-1)}(n) - f^{(2k-1)}(m)) + R_p,$$

where the error R_p can be estimated. A detailed analysis can be found in Edwards (1974) and Kline (1983).

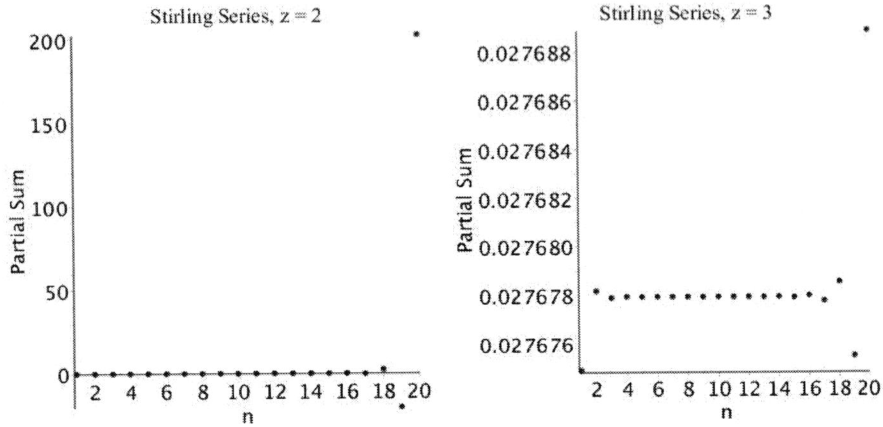

Figure 11. The partial sums for $z = 2$ and $z = 3$ in the Stirling series (70).

To see that this is an asymptotic series, we look at the partial sums at fixed values of z. In Figure 11 we show plots for $z = 2$ and $z = 3$. We see that eventually the partial sums begin to grow.

We can also observe the behavior of the individual terms of the Stirling series (70). In Figure 12 we see that for $z = 2$ and $z = 3$ that the terms begin to decrease and then at a point the factorial divergence takes over. There is a point of optimal truncation at which to terminate the partial sums to obtain the best approximation. Weniger (2010) and Kowalenko (2020) describe methods to compute values of divergent series like the expansion for the exponential integral or Stirling's series.

4.3. Return to the Airy Function

In Section 3. we saw that Stokes derived divergent series expansions in order to describe several physical processes in nature such as the supernumeraries in rainbows. How did he arrive at his results? Essentially, he sought the leading order behavior for large x by studying the differential equation associated with the functions defined by integrals. Then, he sought series expansions for the corrections to this behavior. Some would say that Stokes used what we would now call the method of steepest descent and WKB techniques.[29] This is not surprising as George Green

[29]The WKB Approximation, named after Wentzel-Kramers-Brillouin, is a method to approximate solutions to linear differential equations with spatially varying coefficients. According to Schlissel (1977), this method was discussed by several researchers such as George Green (1793-1841), Joseph Liouville (1809-1882), Lord Rayleigh (1842-1909), Richard Gans (1880-1954), Harold Jeffreys (1891-1989), only to have

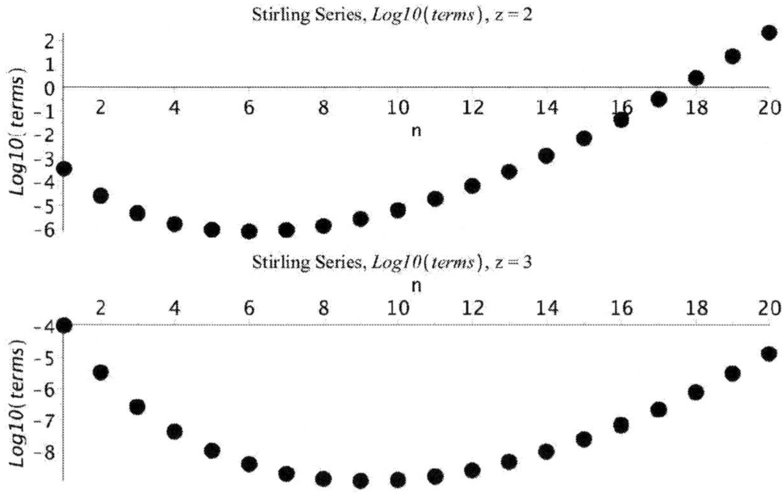

Figure 12. The behavior of the logarithms of the magnitudes of the terms in the Stirling series (70) for $z = 2$ and $z = 3$. Note that the optimal cutoffs are roughly at $n = 6, 10$, respectively.

(1793 - 1841) is often cited (Berry, 1988) as one of the originators of the WKB approximation and Stokes was aware of Green's work through William Thomson (1824-1907). These connections are discussed in the author's Chapter 1, *Laplace Transforms and Green's Functions* in (Herman, 2025).

As a quick review, we indicate how these methods work (Cosnahan, 2021). We begin with the complex version for the Airy function,

$$\text{Ai}(z) = \frac{1}{2\pi} \int_{-\infty}^{\infty} ds\, e^{i(s^3/3+zs)}, \quad z \in \mathbb{C}. \tag{71}$$

The Airy function satisfies the differential equation

$$y''(x) = xy(x). \tag{72}$$

We begin with a WKB-like analysis. As we are interested in large x behaviors, we

the names of Gregor Wentzel (1898-1978), Henrich A. Kramers (1884-1952), and Leon Brillouin (1889-1969) stick as the "WKB" approximation (or, method). This is probably because of they studied applications in quantum mechanics. Sometimes Jeffreys is added giving the WKBJ or JWKB approximation. It has also been called the Liouville-Green method, especially when focusing on asymptotics of solutions of second order differential equations. As noted by Cannell (2001), Green anticipated the use of these methods in his work on water waves in a canal (Green, 1838).

make a change in variables, $z = \epsilon^{\frac{2}{3}} x$ and $y(x) = w(z)$ for $\epsilon \ll 1$. Then,

$$\epsilon^2 w''(z) = z w(z).$$

Assuming that $w(z) = e^{S(z)/\epsilon}$. Then, we find $S(z)$ satisfies the differential equation

$$\epsilon S'' + (S')^2 = z.$$

To leading order (ϵ^0), we have $S'(z) \approx \pm\sqrt{z}$. Therefore, $S(z) \sim \pm\frac{2}{3} z^{\frac{3}{2}}$ as $z \to \infty$.

In order to obtain further approximations, we can develop $S(x)$ in an expansion in powers of ϵ. Then, we assume

$$w(z) = e^{\frac{1}{\epsilon}\sum_{n=0}^{\infty} \epsilon^n S_n(z)}.$$

So, to order ϵ we have

$$\begin{aligned} \epsilon (S_0 + \epsilon S_1)'' + (S_0' + \epsilon S_1')^2 &= z \\ (S_0')^2 + \epsilon\left[S_0'' + 2 S_0' S_1'\right] + O(\epsilon^2) &= z. \end{aligned} \tag{73}$$

We already know $S_0'(z) \sim \pm\sqrt{z}$. So, to order ϵ,

$$S_1' = -\frac{S_0''}{2 S_0'} = -\frac{1}{4z}.$$

Integrating, we find that $S_1 \sim -\frac{1}{4}\ln z$ and we can write the leading order behavior,

$$w(z) \sim e^{\frac{1}{\epsilon}\left(\pm\frac{2}{3} z^{\frac{3}{2}} - \frac{\epsilon}{4}\ln z\right)}.$$

Therefore, we have for the leading order behavior for $y(x)$ as $x \to \infty$,

$$y(x) \sim e^{\frac{1}{\epsilon}\left(\pm\frac{2}{3}(\epsilon^{2/3} x)^{\frac{3}{2}} - \frac{\epsilon}{4}\ln \epsilon^{2/3} x\right)} = c x^{-1/4} e^{\pm\frac{2}{3} x^{\frac{3}{2}}},$$

for some multiplicative constant c.

At this point we could determine the next S_n's or we could let

$$y(x) = x^{-1/4} e^{\pm\frac{2}{3} x^{\frac{3}{2}}} v(x) \tag{74}$$

and solve the differential equation satisfied by $v(x)$ using a power series method.[30] The end result is that we find the asymptotic behavior of the Airy functions (Dunne, 2019),

$$\left.\begin{array}{c} 2\mathrm{Ai}(x) \\ \mathrm{Bi}(x) \end{array}\right\} = \frac{e^{\mp\frac{2}{3}x^{3/2}}}{2\pi^{3/2}x^{1/4}} \sum_{n=0}^{\infty} (\mp 1)^n \frac{\Gamma\left(n+\frac{1}{6}\right)\Gamma\left(n+\frac{5}{6}\right)}{n!(\frac{4}{3}x^{3/2})^n}.$$

Another common approach is to use the Method of Steepest Descent (Bender and Orszag, 1999). This is useful for integrals of the form

$$I(x) = \int_C h(t)e^{x\rho(t)}\, dt \tag{75}$$

as $x \to \infty$. Here C is a contour in the complex-t domain, and $h(t)$, $\rho(t)$ are analytic functions. The idea is to deform the contour to a new one for which $\mathrm{Im}\, \rho(t) = $ const. Along this new contour one can apply the Laplace Method,[31] which we described for handling the integral in Equation (66).

Let us first consider integrating along the real axis. We will demonstrate this for the complex form of the Airy integral in Equation (71). Let $s = -i\zeta$, then

$$\mathrm{Ai}(z) = \frac{1}{2\pi i} \int_{-\infty}^{\infty} d\zeta\, e^{(\zeta z - \zeta^3/3)}, \quad z \in \mathbb{C}.$$

Defining $\varphi(\zeta) = \zeta z - \zeta^3/3$, the critical points are found when $\varphi'(\zeta) = 0$, or $\zeta_0 = \pm\sqrt{z}$. Assuming that for large x the main contribution come from points near the critical points, we can deform the contour through a stationary point. For example, near $\zeta_0 = \sqrt{x}$, let $z = \sqrt{x} + \xi$ for ξ small. Then,

$$\varphi(z) = \varphi(\zeta_0 + \xi) \approx \varphi(\zeta_0) + \frac{1}{2}\varphi''(\zeta_0)\xi^2 + \frac{1}{6}\varphi'''(\zeta_0)\xi^3.$$

[30] Inserting the expression in Equation (74) in the Airy differential equation (72), we find that $v(x)$ satisfies (Bender and Orszag, 1999; Cosnahan, 2021; Mariño, 2021)

$$x^2 v''(x) \pm \left(2x^{5/2} \mp \frac{1}{2}x\right) v'(x) + \frac{5}{16}v(x) = 0.$$

Assuming that $v(x) \sim \sum_{n\geq 0} a_n^{\pm} x^{\alpha x}$ as $x \to \infty$, we find that $\alpha = -\frac{3}{2}$ and

$$a_n^{(\pm)} = \frac{1}{2\pi} \frac{\Gamma\left(n+\frac{1}{6}\right)\Gamma\left(n+\frac{5}{6}\right)}{\Gamma(n+1)} \left(\pm\frac{3}{4}\right)^n.$$

[31] A similar method known as the Method of Stationary Phase is useful for integrals of the form

$$I(x) = \int_a^b h(t)e^{ixf(t)}\, dt. \tag{76}$$

Here $h(t)$ and $f(t)$ are real. This method might be useful for contours for which Re $\rho(t) = $ const.

Inserting this into the Airy integral, we have

$$\begin{aligned}\mathrm{Ai}(z) &= \frac{1}{2\pi i}\int_{-\infty}^{\infty} d\zeta\, e^{(\zeta z-\zeta^3/3)}\\ &\sim \frac{e^{-\frac{2}{3}x^{3/2}}}{2\pi i}\int_{-\infty}^{\infty} e^{\sqrt{x}\xi^2-\frac{1}{3}\xi^3}\,d\xi, \quad (x\to\infty)\\ &= \frac{e^{-\frac{2}{3}x^{3/2}}}{2\pi x^{1/4}}\int_{-\infty}^{\infty} e^{-u^2}e^{-iu^3x^{-3/4}/3}\,d\xi,\end{aligned}$$

where the last integral came from the substitution $\xi = iux^{-1/4}$.

We can expand the second exponential factor in powers of $x^{3/4}$ and compute the resulting Gaussian integrals. After a little work (Cosnahan, 2021), we find

$$\mathrm{Ai}(x) \sim \frac{e^{-\frac{2}{3}x^{\frac{3}{2}}}}{2\sqrt{\pi}\,x^{\frac{1}{4}}}\left[\sum_{n=0}^{\infty}\left(-\frac{3}{4}\right)^n \frac{\Gamma(n+\frac{5}{6})\Gamma(n+\frac{1}{6})}{2\pi\Gamma(n+1)}x^{-\frac{3n}{2}}\right].$$

In summary, Airy derived an integral representation to describe supernumeraries but was only able to determine two of the bands due to the highly oscillatory nature of the integrand. In 1841 Miller (1848) demonstrated that there were many more, which inspired Stokes (1850, 1857, 1868) to look closer at the Airy function. He performed an asymptotic expansion of the Airy function and was able to produce more of the bands.

But Stokes was perplexed that there were different types of solutions on the bright side of the rainbow as opposed to the dark side. On the dark side there was a decaying exponential behavior, while on the bright side there was oscillatory behavior which essentially is the sum of two oscillatory exponentials. So, in his later paper he explored this behavior in the complex plane. This was understood by following the saddle points in the integral.

The Airy integral has the stationary points $s = \pm(-z)^{1/2}$. For $z \in \mathbb{R}$ and $\mathrm{Re}(z) \gg 0$, one stationary point contributes, and the integral is exponentially small. For $z \in \mathbb{R}$ and $\mathrm{Re}(z) \ll 0$, both stationary points contribute, leading to oscillatory behavior.

Stokes studied the Airy function in the complex domain to understand the change from exponential to oscillatory behavior. At some place between the positive real and negative real axes a second exponential has to be emerge. This happens near what are now termed Stokes' lines as depicted in Figure 13. In Figure 14 are shown the real and imaginary parts of $\mathrm{Ai}(z)$, clipped to vertical interval $[-1.0, 1.0]$.

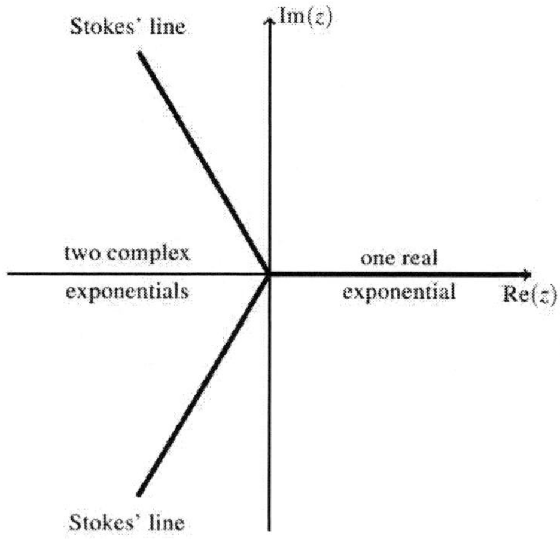

Figure 13. Stokes studied the Airy function in the complex domain. At a place between the positive real and negative real axes a second exponential emerges near what are now referred to as Stokes lines.

Stokes computed the Airy function on either side of the line at $\arg(z) = \frac{2\pi}{3}$ and compared the results from the convergent series noting that the subdominant (exponentially small) series is needed to ensure accuracy at his selected point on the imaginary axis. This type of behavior for other asymptotic series has been the basis of recent work on transseries and resurgence such as seen in the discussions by Aniceto et al. (2019), Dunne (2016), and Mariño (2014).

Berry and Howls have described this behavior in a number of papers (Berry, 1989, 1991; Berry and Howls, 1990, 1991). For example, Berry (1991) shows how the truncated Airy series behaves for complex arguments as one progresses from the positive real axis towards the negative real axis. He writes the first term in the expansion as

$$A_1(z) = \frac{1}{2z^{\frac{1}{4}}\sqrt{\pi}} e^{\frac{F}{2}},$$

where he refers to $F = -\frac{4}{3}z^{\frac{3}{2}}$ as the singulant. Then, he adds more terms,

$$A_2(z) = \frac{1}{2z^{\frac{1}{4}}\sqrt{\pi}} e^{\frac{F}{2}} \sum_{n=0}^{N} \frac{\Gamma(n+\frac{5}{6})\Gamma(n+\frac{1}{6})}{2\pi\Gamma(n+1)F^n}.$$

In Figure 15 we show, as Berry had, plots of the Airy function, $A_1(z)$, and $A_2(z)$ in the complex plane for $z = 1.7171e^{i\theta}$, $\theta \in [0, \pi]$. The three curves begin on the

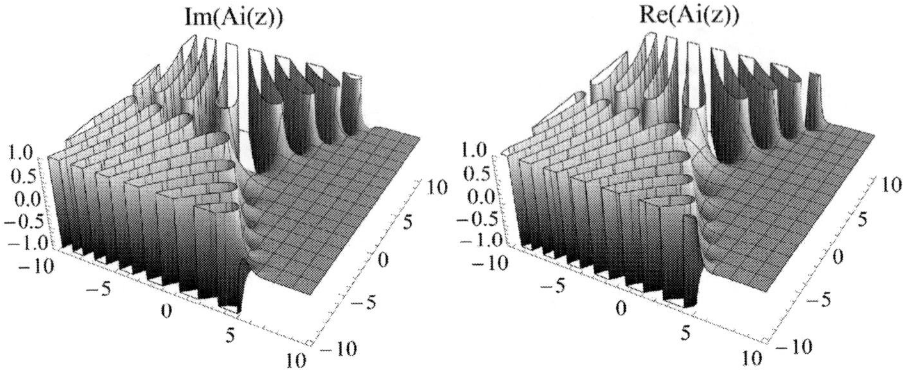

Figure 14. Vertically truncated plots of the real and imaginary parts of $\mathrm{Ai}(z)$ over the domain $\{z|z = x + iy$ and $\max(|x|, |y|) \leq 10\}$.

real axis, but as z approaches the negative real axis, the curves depart with $A_2(z)$ computed for $N = 3$.

$A_2(z)$ seems to significantly diverge from the Airy function once its passes the Stokes' line in the z-plane as $\theta = \frac{2\pi}{3}$. As Stokes noted, one needs to keep the subdominant terms in the D series.

In order to see this, Berry turns on the subdominant series at $\theta = \frac{2\pi}{3}$. So, he considers[32]

$$\begin{aligned} A_3(z) &= \frac{1}{2z^{\frac{1}{4}}\sqrt{\pi}}\left[e^{\frac{F}{2}}\sum_{n=0}^{N}\frac{\Gamma(n+\frac{5}{6})\Gamma(n+\frac{1}{6})}{2\pi\Gamma(n+1)F^n}\right. \\ &\left. +ie^{-\frac{F}{2}}\sum_{n=0}^{N}(-1)^n\frac{\Gamma(n+\frac{5}{6})\Gamma(n+\frac{1}{6})}{2\pi\Gamma(n+1)F^n}H\left(\theta-\frac{2\pi}{3}\right)\right], \end{aligned}$$

where $H(x)$ is the Heaviside step function,

$$H(t) = \begin{cases} 1, & t > 0, \\ 0, & t < 0. \end{cases}$$

A plot of the Airy function and $A_3(z)$ are shown in Figure 16. One can see that the addition of the previously exponentially small terms does a good job for θ-values bringing z past the Stokes' line.

[32]Berry (1991) calls this function $A_4(z)$.

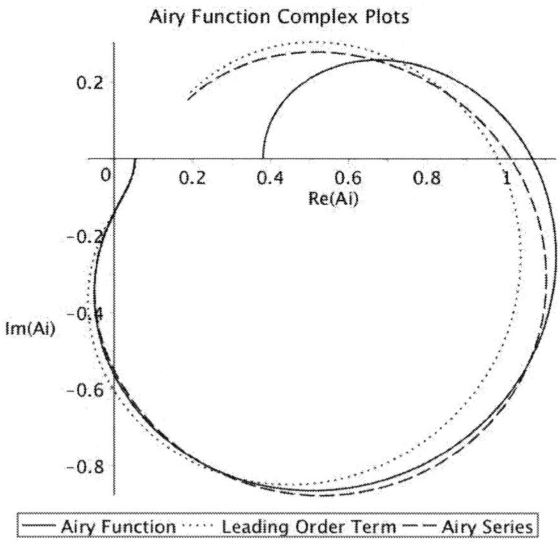

Figure 15. Plots of the Airy function, the leading order term, $A_1(z)$, and the series $A_2(z)$ for $N = 3$ in the complex plane for $z = 1.7171e^{i\theta}$, $\theta \in [0, \pi]$.

Stokes noted a discontinuity in the series behavior as one crosses the Stokes' line. However, Berry goes on to show more detail in what he call error function smoothing (Berry, 1989, 1991). This involves doing an asymptotic analysis on the remainder terms that were dropped in the truncation process. These terms are factorially divergent and can be put through a resummation process. This leads to information about nearby saddle point contributions and lead to deeper ideas involving resurgence. The resummation of divergent series involves the Laplace transform. On the other hand, there have been other methods for smoothing the Stokes phenomenon, such as using Mellin-Barnes integrals (Paris, 1992; Paris and Kaminski, 2001).

5. Borel Transform and Asymptotic Series

Up to this point we have seen several asymptotic series. These series are divergent, but it has long been known that if one keeps an optimal number of terms, the truncated series provides a useful approximation to some function. The function having the series as its expansion can sometimes be found by relating it to a convergent expansion using a Borel transform.

We recall the Euler series (51). We rearranged the series using the Laplace trans-

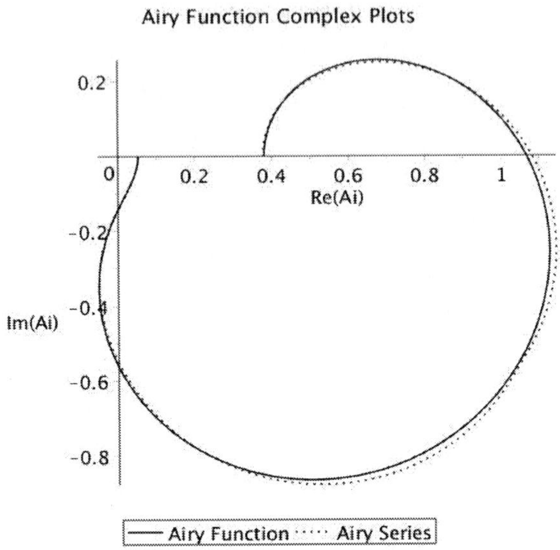

Figure 16. A plot of the Airy function and $A_3(z)$ in the complex plane for $z = 1.7171e^{i\theta}$, $\theta \in [0, \pi]$. The addition of exponentially small terms gives a better representation of the Airy function past the Stokes' line.

form. Replacing x with z^{-1}, we have

$$\begin{aligned} \sum_{n=0}^{\infty}(-1)^n \frac{n!}{z^{n+1}} &= \sum_{n=0}^{\infty}(-1)^n \int_0^{\infty} s^n e^{-sz}\, ds \\ &= \int_0^{\infty} e^{-sz}\left(\sum_{n=0}^{\infty}(-1)^n s^n\right) ds. \end{aligned} \tag{77}$$

In this way we related the asymptotic series, $\sum a_n z^{-n-1}$, to a convergent series, $\sum \frac{a_n}{n!} s^n$. In this case, $a_n = (-1)^n n!$. This mapping is called a Borel transform.

More generally, we consider a factorially divergent series $f(z) = \sum a_n z^{-n-1}$. We then divide out the factorial and use the Laplace transform to connect the divergent series to a convergent one:

$$\begin{aligned}
\sum a_n z^{-n-1} &= \sum \frac{a_n}{n!}\frac{n!}{z^{n+1}} \\
&= \sum \frac{a_n}{n!}\int_0^\infty s^n e^{-sz}\,ds \\
&= \int_0^\infty e^{-sz}\left(\sum \frac{a_n}{n!}s^n\right)ds \\
&\equiv \int_0^\infty e^{-sz}\mathcal{B}[f](s)\,ds. \qquad (78)
\end{aligned}$$

Here we have defined the Borel transform as

$$\mathcal{B}(f)[s] = \sum \frac{a_n}{n!}s^n$$

and the inverse Borel transform as

$$S[f](z) = \int_0^\infty e^{-zs}\mathcal{B}(f)[s]\,ds.$$

We see that the inverse Borel transform is nothing other than a Laplace transform. This is also referred to as Borel summation, which is a summation method for divergent series and was introduced by Émile Borel (1871-1956) in (Borel, 1899). Kowalenko (2020) calls this a regularization of the divergent series. It is a way to remove infinity from the divergent series in order to make it summable.[33]

The application of the Borel resummation process is captured in Figure 17. At this point we should note that there are different definitions used in the literature. The derivation based on the treatment of Euler's series would lead a mathematician to choosing the schematic in Figure 17. In another version, a physicist might start with $f(z) = \sum a_n g^n$, where g is some coupling constant. Then, the Borel transform would be

$$\mathcal{B}(f)[s] = \sum \frac{a_n}{n!}s^n.$$

The relation is subtler than just letting $g = z^{-1}$, but either approach works.

The main idea is that the Borel transform leads to a convergent series. This will happen if the original series is Gervey-1, or, loosely, the original series is factorially divergent for which (Mariño, 2021)

$$a_n \sim A^{-n} n!, \quad n \gg 1.$$

[33]Kowalenko (2020) also notes that Borel summation may not lead to efficient forms of computation. Kowalenko and coworkers proposed a numerical technique, Mellin-Barnes regularization, which produced twenty decimal place accuracy.

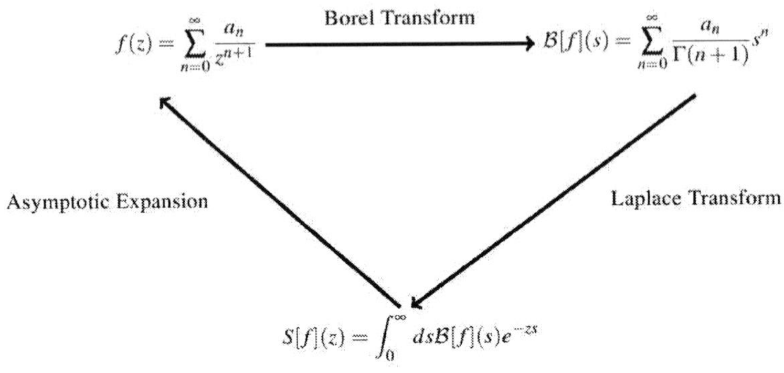

Figure 17. The Borel transform and Borel resummation Scheme.

In order to demonstrate how Borel resummation works, we will apply it to two examples. First we apply it to the Airy series and then we will show how it works with the series coming out of the study of the quantum anharmonic oscillator.

5.1. The Airy Function

Consider the Airy function asymptotic series written as

$$\mathrm{Ai}(x) \sim \frac{e^{-\frac{2}{3}x^{\frac{3}{2}}}}{2\sqrt{\pi}\,x^{\frac{1}{4}}} \left(\frac{4}{3}x^{\frac{3}{2}}\right) \left[\sum_{n=0}^{\infty} (-1)^n \frac{\Gamma(n+\frac{5}{6})\Gamma(n+\frac{1}{6})}{2\pi\Gamma(n+1)} \left(\frac{3}{4}x^{-\frac{3}{2}}\right)^{n+1}\right].$$

Letting $z = \frac{4}{3}x^{\frac{3}{2}}$, we can write the bracketed series as[34]

$$Z(z) = \sum_{n=0}^{\infty} (-1)^n \frac{\Gamma(n+\frac{5}{6})\Gamma(n+\frac{1}{6})}{2\pi\Gamma(n+1)} z^{-n-1}.$$

Then, the Borel transform is given by

$$\mathcal{B}[Z](s) = \sum_{n=0}^{\infty} (-1)^n \frac{\Gamma(n+\frac{5}{6})\Gamma(n+\frac{1}{6})}{2\pi\Gamma(n+1)\Gamma(n+1)} s^n.$$

---

[34]We could change the index using $n = k+1$. Then,

$$Z(z) = \sum_{n=0}^{\infty} (-1)^n \frac{\Gamma(n+\frac{5}{6})\Gamma(n+\frac{1}{6})}{2\pi\Gamma(n+1)} z^{-n} = \sum_{k=-1}^{\infty} (-1)^{k+1} \frac{\Gamma(k+\frac{11}{6})\Gamma(k+\frac{7}{6})}{2\pi\Gamma(k+2)} z^{-k-1}.$$

This series can be compared to the series for a hypergeometric function [See Erdélyi et al. (1953), Chapter 15 of DLMF, or Gasper and Rahman (2004)],

$$ {}_2F_1(a, b; c; z) = \frac{\Gamma(c)}{\Gamma(a)\Gamma(b)} \sum_{n=0}^{\infty} \frac{\Gamma(a+n)\Gamma(b+n)}{\Gamma(c+n)n!} z^n. \tag{79} $$

In particular, we have that

$$ \begin{aligned} {}_2F_1\left(\frac{1}{6}, \frac{5}{6}; 1; z\right) &= \frac{\Gamma(1)}{\Gamma\left(\frac{1}{6}\right)\Gamma\left(\frac{5}{6}\right)} \sum_{n=0}^{\infty} \frac{\Gamma\left(n+\frac{1}{6}\right)\Gamma\left(n+\frac{5}{6}\right)}{\Gamma(n+1)n!} z^n \\ &= \frac{1}{2\pi} \sum_{n=0}^{\infty} \frac{\Gamma\left(n+\frac{1}{6}\right)\Gamma\left(n+\frac{5}{6}\right)}{\Gamma(n+1)n!} z^n, \end{aligned} \tag{80} $$

where we used Euler's reflection formula,

$$ \Gamma(x)\Gamma(1-x) = \frac{\pi}{\sin \pi x}, $$

to obtain

$$ \frac{\Gamma(1)}{\Gamma\left(\frac{1}{6}\right)\Gamma\left(\frac{5}{6}\right)} = \frac{\sin\frac{\pi}{6}}{\pi} = \frac{1}{2\pi}. $$

So, we can apply the Laplace transform to recover $Z(z)$,

$$ Z(z) = 2\pi \int_0^{\infty} ds\, e^{-sz} {}_2F_1\left(\frac{1}{6}, \frac{5}{6}; 1; -s\right). $$

Therefore, we have

$$ \mathrm{Ai}(x) \sim \frac{4x^{\frac{5}{4}}\sqrt{\pi}e^{-\frac{2}{3}x^{\frac{3}{2}}}}{3} \int_0^{\infty} ds\, e^{-\frac{4}{3}x^{\frac{3}{2}}s} {}_2F_1\left(\frac{1}{6}, \frac{5}{6}; 1; -s\right). $$

Similarly,

$$ \mathrm{Bi}(x) \sim \frac{4x^{\frac{5}{4}}\sqrt{\pi}e^{\frac{2}{3}x^{\frac{3}{2}}}}{3} \int_0^{\infty} ds\, e^{-\frac{4}{3}x^{\frac{3}{2}}s} {}_2F_1\left(\frac{1}{6}, \frac{5}{6}; 1; s\right). $$

Typically, one would then develop the connection formulae between the Airy and Bairy functions using analytic properties of hypergeometric functions using identities in Erdélyi et al. (1953) or DLMF. We will demonstrate this with the example in the next section.

5.2. Quartic Partition Function

Problems in quantum mechanics are notorious for producing divergent expansions. For a typical problem, we can begin with the Schrödinger equation or the partition function, given by

$$Z(\hbar) = \frac{1}{2\pi} \int_\Gamma dz\, e^{-\frac{1}{\hbar}V(z)}, \tag{81}$$

where $V(z)$ is the potential and $\hbar$ is Planck's constant divided by 2π. The contour Γ is determined from a saddle point analysis.

As an example, consider the rescaled Schödinger equation for a quartic potential,

$$-\frac{1}{2}\psi''(x) + \frac{1}{2}x^2\psi(x) + gx^4\psi(x) = E\psi(x), \tag{82}$$

where g is a coupling constant. For small coupling, one can assume that the energies are close to those of the harmonic oscillator. The numerical results confirm this. How does one analytically obtain approximations to the energies? The usual methods for doing so are either Rayleigh-Schrödinger perturbation theory or a WKB analysis (Berera and Del Debbio, 2021; Merzbacher, 1998; Schwinger, 2003). Bender and Wu (1969, 1973) had found for the ground state energy that corrections to the ground state energy in powers of the coupling parameter are given as[35]

$$E_0 = \frac{1}{2} + \frac{3}{4}g - \frac{21}{8}g^2 + \frac{333}{16}g^3 - \frac{30885}{128}g^4 + \frac{916731}{256}g^5 + \mathcal{O}(g^6). \tag{83}$$

Writing

$$E_0 = \frac{1}{2} + \sum_{n=1}^{\infty} A_n g^n,$$

they found that for large n,[36]

$$\begin{aligned} A_n &\sim (-1)^{n+1}\sqrt{\frac{6}{\pi^3}}3^n\Gamma(n+\frac{1}{2}) \\ &= -\sqrt{\frac{6}{\pi^2}}\left(-\frac{3}{2}\right)^n (2n-1)!!. \end{aligned} \tag{84}$$

---

[35] Actually, they solved the problem

$$\left[-\frac{d^2}{dx^2} + \frac{1}{4}x^2 + \frac{1}{4}\lambda x^4\right]\psi(x) = E\psi(x).$$

Letting $x = \sqrt{2}q$, we obtain Equation (82) and get the same energies.

[36] Recall that $\Gamma(n+\frac{1}{2}) = \frac{(2n)!}{4^n n!}\sqrt{\pi} = \frac{(2n-1)!!}{2^n}\sqrt{\pi}$.

Thus, the series is divergent with factorial growth.

We will consider the quartic potential, following Aniceto et al. (2019),

$$V(z) = \frac{1}{2}z^2 - \frac{\lambda}{24}z^4. \tag{85}$$

The partition function can be shown to satisfy a differential equation. First, we rescale the integration variable $z = \sqrt{\hbar}z'$ so that the quartic partition function takes the form

$$Z(x) = \frac{\sqrt{\hbar}}{2\pi}\int_{-\infty}^{\infty} dz\, e^{-\frac{1}{2}z^2+\frac{x}{24}z^4}. \tag{86}$$

The integral can actually be shown to be given by

$$\sqrt{\frac{3}{x}}e^{-\frac{3}{4x}}K_{\frac{1}{4}}\left(-\frac{3}{4x}\right).$$

This can be used to show that $Z(x)$ satisfies the differential equation

$$16x^2Z''(x) + (32x - 24)Z'(x) + 3Z(x) = 0. \tag{87}$$

Moreover, if one lets $x = \dfrac{b}{v}$ and $Z(x) = x^a u(x) = v^{-a}y(v)$, then one obtains the differential equation

$$2bv^2y''(v) + \left[-4abv + 3v^2\right]y'(v) + 2\left[(a^2 + a + \frac{3}{16})b - \frac{3av}{2}\right]y(v) = 0. \tag{88}$$

Eliminating the constant term in the second pair of brackets leads to a Kummer differential equation.[37] This is accomplished by setting $a = -\dfrac{1}{4}, -\dfrac{3}{4}$. For example, for $a = -\dfrac{3}{4}$, Equation (88) becomes

$$vy''(v) + \left[\frac{3}{2} + \frac{3}{2b}v\right]y'(v) + \frac{9}{8b}y(v) = 0. \tag{89}$$

---

[37] Kummer's equation is given by

$$xy''(x) + (\beta - x)\,y'(x) - \alpha y(x) = 0.$$

There are a number of notations for the solutions, which are confluent hypergeometric functions such as $U(\alpha, \beta, x)$, $M(\alpha, \beta, x)$, and ${}_1F_1\,(\alpha; \beta; x)$.

Setting $b = -\dfrac{3}{2}$, we obtain

$$vy''(v) + \left(\frac{3}{2} - v\right) y'(v) - \frac{3}{4} y(v) = 0.$$

This is Kummer's equation and the solutions are Kummer functions or confluent hypergeometric functions, ${}_1F_1\left(\frac{3}{4}, \frac{3}{2}; v\right)$. [See Chapter 13 in the DLMF.] Had we selected $a = -\frac{1}{4}$, then b is the same and the solutions would be ${}_1F_1\left(\frac{1}{4}, \frac{1}{2}; v\right)$. These functions are related to the modified Bessel functions as expected [Section 10.39 in DLMF].

We will now seek series solutions of Equation (87) of the form

$$Z(x) = e^{-A/x}\Phi(x) = e^{-A/x}\sum_{n=0}^{\infty} c_n x^n$$

and show that they result in factorially divergent series. Inserting this into the differential equation, we obtain

$$\sum_{n=0}^{\infty} 8[2A+3]Ac_n x^{n-2} - \sum_{n=0}^{\infty} 8[4A+3]nc_n x^{n-1} + \sum_{n=0}^{\infty} [16n^2+16n+3]c_n x^n = 0.$$

The leading order term comes from $n = 0$ in the first term, giving

$$8[2A+3]Ac_0 = 0.$$

Therefore, we have for $c_0 \neq 0$ that either $A = 0$ or $A = -3/2$.

Setting $A = 0$, we obtain

$$-\sum_{n=0}^{\infty} -24nc_n x^{n-1} + \sum_{n=0}^{\infty} [16n^2 + 16n + 3]c_n x^n = 0.$$

Reindexing, we obtain that the coefficients satisfy the recursion relation

$$(16k^2 + 16k + 3)c_k = 24(k+1)c_{k+1}c_0, \quad k = 0, 1, 2 \ldots. \tag{90}$$

The solution is found as

$$c_k = \frac{(2k - \frac{1}{2})!}{\sqrt{\pi}6^k k!} c_0 = \left(\frac{1}{6}\right)^k \frac{\Gamma\left(2k + \frac{1}{2}\right)}{\sqrt{\pi}\Gamma(k+1)} c_0. \tag{91}$$

Setting $A = -3/2$, we obtain

$$-\sum_{n=0}^{\infty} 24nc_n x^{n-1} + \sum_{n=0}^{\infty} [16n^2 + 16n + 3]c_n x^n = 0.$$

Reindexing, we obtain that the coefficients satisfy the recursion relation

$$(16k^2 + 16k + 3)c_k = -24(k+1)c_{k+1}, \quad k = 0, 1, 2 \ldots. \tag{92}$$

The solution is found as

$$c_k = (-1)^k \frac{(2k - \frac{1}{2})!}{\sqrt{\pi}6^k k!} c_0 = (-1)^k \left(\frac{1}{6}\right)^k \frac{\Gamma\left(2k + \frac{1}{2}\right)}{\sqrt{\pi}\Gamma(k+1)} c_0. \tag{93}$$

Therefore, we have two solutions for $\Phi(x)$ just as we had for the Airy equation.

$$\Phi_0(x) = c_0 \sum_{n=0}^{\infty} \frac{\Gamma\left(2n + \frac{1}{2}\right)}{\sqrt{\pi}\Gamma(n+1)} \left(\frac{x}{6}\right)^n, \tag{94}$$

$$\Phi_1(x) = d_0 \sum_{n=0}^{\infty} \frac{\Gamma\left(2n + \frac{1}{2}\right)}{\sqrt{\pi}\Gamma(n+1)} \left(-\frac{x}{6}\right)^n. \tag{95}$$

We can determine c_0 by evaluating $Z(0)$ using (86), we have

$$\begin{aligned} Z(0) &= \frac{\sqrt{\hbar}}{2\pi} \int_{-\infty}^{\infty} dz\, e^{-\frac{1}{2}x^2} \\ &= \sqrt{\frac{\hbar}{2\pi}} = c_0. \end{aligned} \tag{96}$$

In order to determine d_0 in terms of c_0, we would need to consider the asymptotics of modified Bessel functions. Using the results of Paris (2017), we can confirm the relationship $d_0 = -i/\sqrt{2}c_0$ which is stated in (Aniceto et al., 2019).

As in other examples, the series solutions in Equations (94)-(95) are divergent series. We can see this by again using the Legendre Duplication Formula,

$$\Gamma(z)\Gamma(z + \frac{1}{2}) = 2^{1-2z}\sqrt{\pi}\Gamma(2z). \tag{97}$$

Letting $z = n + \frac{1}{4}$, we have

$$\begin{aligned}\Gamma\left(2n+\frac{1}{2}\right) &= \Gamma\left(2(n+\frac{1}{4})\right)\\ &= \frac{\Gamma\left(n+\frac{1}{4}\right)\Gamma\left(n+\frac{3}{4}\right)}{2^{1/2-2n}\sqrt{\pi}}\\ &= \frac{4^n}{\sqrt{2\pi}}\Gamma\left(n+\frac{1}{4}\right)\Gamma\left(n+\frac{3}{4}\right). \qquad (98)\end{aligned}$$

Therefore, the coefficients in the series for $\Phi_0(x)$ can be written as

$$\left(\frac{1}{6}\right)^n \frac{\Gamma\left(2n+\frac{1}{2}\right)}{\sqrt{\pi}\Gamma(n+1)} = \frac{1}{\pi\sqrt{2}}\left(\frac{2}{3}\right)^n \frac{\Gamma\left(n+\frac{1}{4}\right)\Gamma\left(n+\frac{3}{4}\right)}{\Gamma(n+1)}.$$

In the limit as $n \to \infty$ these terms behave like $n!\left(\frac{2}{3}\right)^n$ and so the series for $\Phi_0(x)$ and $\Phi_1(x)$ diverge. In order to get information out of these divergent series, we apply Borel summation.

In this case, we will use the Borel transformation of $f(x) = \sum_{n=0}^{\infty} a_n x^n$ as given in physics literature (Dunne and Ünsal, 2014; Dunne, 2016) [38] by

$$\mathcal{B}[f](s) = \sum_{n=0}^{\infty} \frac{a_n}{n!} s^n.$$

Applying the Borel transform, we find the series

$$\begin{aligned}\mathcal{B}[\Phi_0](s) &= \frac{c_0}{\pi\sqrt{2}}\sum_{n=0}^{\infty}\frac{\Gamma\left(n+\frac{1}{4}\right)\Gamma\left(n+\frac{3}{4}\right)}{\Gamma(n+1)n!}\left(\frac{2}{3}s\right)^n, \qquad (100)\\ \mathcal{B}[\Phi_1](s) &= \frac{d_0}{\pi\sqrt{2}}\sum_{n=0}^{\infty}\frac{\Gamma\left(n+\frac{1}{4}\right)\Gamma\left(n+\frac{3}{4}\right)}{\Gamma(n+1)n!}\left(-\frac{2}{3}s\right)^n. \qquad (101)\end{aligned}$$

[38] In other literature the Borel transform is defined differently. For example, one might find (Aniceto et al., 2019)

$$\hat{f}(s) = \sum_{n=1}^{\infty} \frac{a_n}{(n-1)!} s^{n-1}.$$

The relation between these series is given by

$$\hat{f}(s) = \frac{d\mathcal{B}[f](s)}{ds}. \qquad (99)$$

As with the Ariy function, these series can be related to hypergeometric functions. In particular, we have that

$$
\begin{aligned}
{}_2F_1\left(\frac{1}{4},\frac{3}{4};1;z\right) &= \frac{\Gamma(1)}{\Gamma\left(\frac{1}{4}\right)\Gamma\left(\frac{3}{4}\right)}\sum_{n=0}^{\infty}\frac{\Gamma\left(n+\frac{1}{4}\right)\Gamma\left(n+\frac{3}{4}\right)}{\Gamma(n+1)n!}z^n \\
&= \frac{1}{\sqrt{2}\pi}\sum_{n=0}^{\infty}\frac{\Gamma\left(n+\frac{1}{4}\right)\Gamma\left(n+\frac{3}{4}\right)}{\Gamma(n+1)n!}z^n. \qquad (102)
\end{aligned}
$$

Therefore, the Borel transformations of Φ_0 and Φ_0 are

$$
\begin{aligned}
\mathcal{B}[\Phi_0](s) &= c_0\,{}_2F_1\left(\frac{1}{4},\frac{3}{4};1;\frac{2}{3}s\right), \qquad (103)\\
\mathcal{B}[\Phi_1](s) &= d_0\,{}_2F_1\left(\frac{1}{4},\frac{3}{4};1;-\frac{2}{3}s\right). \qquad (104)
\end{aligned}
$$

From these results, we will compute Borel resummations, or appropriate Laplace transforms, of the hypergeometric functions in (103)-(104). These functions have singularities for $s = \pm\frac{3}{2}$, since ${}_2F_1(a,b;c;z)$ can have a singularity at $z=1$.

When there are singularities in the Borel plane, then the Borel summation (Laplace transform) is ill-defined for $\arg(z)=\theta$. Then, one can deform the contour above, or below, the rays $e^{i\theta}R^+$. So, one can perform what is called a lateral Borel resummation (Mariño, 2021) defined by

$$
S_\theta[f](z) = \int_0^{\infty e^{i\theta}} e^{-zs}\mathcal{B}(f)[s]\,ds.
$$

For this problem, there are two such rays, $\theta = 0,\pi$. We first consider the function in Equation (103) with $\theta = 0$. Taking the contour slightly above the positive real axis and then below it, we can determine the jump across the branch cut. We denote these integrals by

$$
\begin{aligned}
S_0\Phi_0^+ &= c_0\int_0^\infty e^{-z\zeta}\,{}_2F_1\left(\frac{1}{4},\frac{3}{4};1;\frac{2}{3}\zeta+i\epsilon\right)d\zeta \qquad (105)\\
S_0\Phi_0^- &= c_0\int_0^\infty e^{-z\zeta}\,{}_2F_1\left(\frac{1}{4},\frac{3}{4};1;\frac{2}{3}\zeta-i\epsilon\right)d\zeta. \qquad (106)
\end{aligned}
$$

To evaluate the difference (jump) in these functions, leading to a connection formula, we will use an identity 15.2.3 in the DLMF

$$
\begin{aligned}
&{}_2F_1(a,b;c;x+i0) - {}_2F_1(a,b;c;x-i0)\\
&= \frac{2\pi i}{\Gamma(a)\Gamma(b)}(x-1)^{c-a-b}\,{}_2F_1(c-a,c-b;c-a-b+1;1-x) \qquad (107)
\end{aligned}
$$

Applying the identity to the integrands in Equation (106), we have

$$ {}_2F_1\left(\frac{1}{4},\frac{3}{4};1;\frac{2}{3}\zeta+i\epsilon\right) - {}_2F_1\left(\frac{1}{4},\frac{3}{4};1;\frac{2}{3}\zeta-i\epsilon\right) = \sqrt{2}i\,{}_2F_1\left(\frac{3}{4},\frac{1}{4};1;1-\frac{2}{3}\zeta\right). \tag{108} $$

Therefore,

$$ \begin{aligned} S_0\Phi_0^+ - S_0\Phi_0^- &= \sqrt{2}ic_0\int_0^\infty e^{-z\zeta}\,{}_2F_1\left(\frac{3}{4},\frac{1}{4};1;1-\frac{2}{3}\zeta\right)d\zeta \\ &= \sqrt{2}ic_0\int_{3/2}^{-\infty} e^{-z(s+3/2)}\,{}_2F_1\left(\frac{3}{4},\frac{1}{4};1;-\frac{2}{3}s\right)ds \\ &= \sqrt{2}ic_0e^{-3z/2}\int_0^{-\infty} e^{-zs}\,{}_2F_1\left(\frac{3}{4},\frac{1}{4};1;-\frac{2}{3}s\right)ds \\ &= \sqrt{2}ie^{-3z/2}\frac{c_0}{d_0}S_\pi\Phi_1 \\ &= -2e^{-3z/2}S_\pi\Phi_1 \end{aligned} \tag{109} $$

Following the same process, we have jumps across the $\theta = \pi$ ray connecting to the $\theta = 0$ function.

$$ S_\pi\Phi_1^+ - S_\pi\Phi_1^- = \sqrt{2}ie^{-3z/2}\frac{d_0}{c_0}S_0\Phi_0 = e^{-3z/2}S_0\Phi_0. $$

So, we have shown that one can find functions through Borel resummation whose asymptotic series in different regions of the complex plane lead to the given asymptotic expansions. Encoded in these functions is some information about the physical problem.

Conclusion

Laplace transforms have a number of applications, especially to solving differential equations. These naturally lead to applications in areas of physics, electrical engineering and signal processing. However, as a transform they can be used in other applications in mathematics. In this chapter we explored how Laplace transforms are useful in summing or resumming infinite series. Along the way we have pointed out some interesting history behind a few well known series expansions.

We began by looking at their use in summing convergent series. We started with an infinite series of the form $\sum_{n=0}^{\infty} F(n)$, where

$$ F(s) = \int_0^\infty f(t)e^{-st}\,dt $$

for some function $f(t)$. Inserting this Laplace transform integral into the sum and interchanging the summation and integration, we find that

$$\sum_{n=0}^{\infty} F(n) = \int_0^{\infty} \frac{1}{1-e^{-t}} f(t)\, dt.$$

This process is useful if we can either find a closed form for the integral or efficiently compute it as compared to the original series, which might converge slowly.

The study of infinite series has a long history and we delved into some of that, highlighting some interesting series and people in the process. In particular, we saw how using some of Feynman's tricks led to other versions families of series which could be summed.

An interesting application of series in history was when people tried to apply wave interference to understand some of the intricacies of the rainbow, This led to infinite series as one of many examples where infinite series arises in physical investigations. In many cases, divergent series arise and in the last several decades our understanding of divergent series went from the work of the devil, as Abel would put it, to interesting expansions full of more detail than expected when placed in the complex domain. This leads to interesting notions of resurgence and has been applied to many fields in order to extract nonperturbative solutions from perturbative ones.

Laplace transforms come into play with divergent series in the resummation process. Namely, for a factorially divergent series $f(z) = \sum a_n z^n$ one does a Borel transform, which is a formal inverse Laplace transform, to arrive at a new series $\mathcal{B}[f](s) = \sum \frac{a_n}{n!} s^n$, which can be summed. One then applies a Laplace transform,

$$S[f](z) = \int_0^{\infty} \mathcal{B}[f](s) e^{-sz}\, ds,$$

to produce a function which has as it asymptotic expansion in certain regions of the complex plane the original divergent series. Of course, the details are a little more complicated, but one can explore the literature if one wants to learn more (Aniceto et al., 2019; Bender and Heissenberg, 2017; Costin and Dunne, 2017; Dorigoni, 2019; Dunne and Ünsal, 2014; Dunne, 2016; Mariño, 2021).

Over the last several decades people have exploited these simple ideas to not only get a handle on the interesting and deep mathematics behind these processes, but also to learn new physics buried in the asymptotic tails of divergent series.

References

Abramowitz, M. and Stegun, I. A. Handbook of Mathematical Functions with Formulas. *Graphs, and Mathematical Tables*, 1965.

Adam, J. A. Like a Bridge over Colored Water: A Mathematical Review of *The Rainbow Bridge: Rainbows in Art, Myth, and Science*. *Not. Am. Math. Soc.*, 49 (11):1360–1371, 2002a. Review of book by R. Lee, Jr. and A. Fraser.

Adam, J. A. The Mathematical Physics of Rainbows and Glories. *Physics Reports*, 356(4-5):229–365, 2002b.

Adam, J. A. *Rays, Waves, and Scattering: Topics in Classical Mathematical Physics*, volume 56. Princeton University Press, 2017.

Adamchik, V. and Wagon, S. π: A 2000-year Search Changes Direction. *Mathematica in Education and Research*, 5(1):11–19, 1996.

Adamchik, V. and Wagon, S. A Simple Formula for π. *Am. Math. Mon.*, 104(9): 852–855, 1997.

Airy, G. B. On the Intensity of Light in the Neighbourhood of a Caustic. *Trans. Camb. Philos. Soc.*, 6:379402, 1838.

Airy, G. B. Supplement to a Paper on the Intensity of Light in the Neighbourhood of a Caustic. *Trans. Camb. Philos. Soc.*, 8:595–599, 1849.

Aniceto, I., Başar, G., and Schiappa, R. A Primer on Resurgent Transseries and their Asymptotics. *Phys. Rep.*, 809:1–135, 2019.

Ayoub, R. Euler and the Zeta Function. *The American Mathematical Monthly*, 81 (10):1067–1086, 1974.

Bagni, G. T. Mathematics Education and Historical References: Guido Grandi's Infinite Series. *Normat-Nordisk Matematisk Tidsskrift*, 53(4):173–185, 2005.

Bagni, G. T. Infinite Series from History to Mathematics Education. *Colección Digital Eudoxus*, 1(5), 2009.

Bailey, D., Borwein, P., and Plouffe, S. On the Rapid Computation of Various Polylogarithmic Constants. *Math. Comput.*, 66(218):903–913, 1997.

Barbeau, E. J. Euler Subdues a Very Obstreperous Series. *Am. Math. Mon.*, 86(5): 356–372, 1979.

Bayes, T. A letter from the late Reverend Mr. Thomas Bayes, F.R.S. to John Canton, M.A. and F.R.S. *Phil. Trans. Roy. Soc. of Lond.*, 53:269–271, 1763.

Bellard, F. A New Formula to Compute the nth Binary Digit of π. *Formerly available online*, 1997.

Bender, C. M. and Heissenberg, C. Convergent and Divergent Series in Physics. *preprint, arXiv:1703.05164*, 2017. Lecture notes of the 22nd "Saalburg" Summer School (2016).

Bender, C. M. and Orszag, S. A. *Advanced Mathematical Methods for Scientists and Engineers I: Asymptotic Methods and Perturbation Theory*. Springer Science & Business Media, 1999.

Bender, C. M. and Wu, T. T. Anharmonic Oscillator. *Phys. Rev.*, 184(5):1231–

1260, 1969.

Bender, C. M. and Wu, T. T. Anharmonic Oscillator. II. A Study of Perturbation Theory in Large Order. *Phys. Rev. D*, 7(6):1620–1636, 1973.

Berera, A. and Del Debbio, L. *Quantum Mechanics*. Cambridge University Press, 2021.

Berry, M. V. Stokes' Phenomenon; Smoothing a Victorian Discontinuity. *Publications Mathématiques de l'IHÉS*, 68:211–221, 1988.

Berry, M. V. Uniform Asymptotic Smoothing of Stokes Discontinuities. *Proc. Roy. Soc. Lond. A. Math. and Phys. Sci.*, 422(1862):7–21, 1989.

Berry, M. V. Asymptotics, Superasymptotics, Hyperasymptotics... In Segur, H., Tanveer, S., and Levine, H., editors, *Asymptotics beyond All Orders*, pages 1–14. Springer, 1991.

Berry, M. V. and Howls, C. J. Hyperasymptotics. *Proc. Roy. Soc. Lond. A. Math. and Phys. Sci.*, 430(1880):653–668, 1990.

Berry, M. V. and Howls, C. J. Hyperasymptotics for Integrals with Saddles. *Proc. Roy. Soc. Lond. A. Math. and Phys. Sci.*, 434(1892):657–675, 1991.

Bleistein, N. and Handelsman, R. A. *Asymptotic Expansions of Integrals*. Ardent Media, 1975.

Borel, É. Mémoire sur les séries divergentes. In *Annales scientifiques de l'École Normale Supérieure*, volume 16, pages 9–131, 1899. [Memoir on divergent series.].

Boyd, J. P. The Devil's Invention: Asymptotic, Superasymptotic And Hyperasymptotic Series. *Acta Appl. Math.*, 56:1–98, 1999.

Boyer, C. B. *The Rainbow*. Princeton University Press, Princeton, 1959.

Bromwich, T. J. I. *An Introduction to the Theory of Infinite Series*. London Macmillan, 1908.

Bromwich, T. J. I. and MacRobert, T. M. *An Introduction to the Theory of Infinite Series: 2d Ed. Rev., with the Assistance of TM MacRobert*. Macmillan, 1926.

Cannell, D. M. *George Green: Mathematician and Physicist 1793–1841: The Background to his Life and Work*. SIAM, 2001.

Casselman, B. Feature Column from the AMS: The Mathematics of Rainbows, 2009. URL `http://www.ams.org/publicoutreach/feature-column/fcarc-rainbows`.

Casselman, B. Feature Column from the AMS:: The Mathematics of Rainbows, Part II, 2015. URL `http://www.ams.org/publicoutreach/feature-column/fc-2015-02`.

Chaudhry, M. A., Qadir, A., and Tassaddiq, A. Extended Fermi-Dirac and Bose-Einstein Functions with Applications to the Family of Zeta Functions. *preprint, arXiv:1004.0588*, 2010.

Colwell, P. *Solving Keplers Equation Over Three Centuries*. Willmann-Bell Inc., Richmond, VA, 1993.

Copson, E. T. *Asymptotic Expansions*. Cambridge University Press, 2004.

Corradi, M. A Short History of the Rainbow. *Lettera Matematica*, 4:49–57, 2016.

Cosnahan, K. W. Asymptotic Analysis of 1D Path Integrals via Resurgence [electronic resource] . Master's thesis, University of North Carolina Wilmington, Wilmington, North Carolina, 2021.

Costin, O. and Dunne, G. V. Convergence from Divergence. *J. Phys. A: Math.*, 51 (4):04LT01, 2017.

Davis, P. J. Leonhard Euler's Integral: A Historical Profile of the Gamma Function: In Memoriam: Milton Abramowitz. *Am. Math. Mon.*, 66(10):849–869, 1959.

de Moivre, A. *The Doctrine of Chances: or, A Method of Calculating the Probabilities of Events in Play*, volume 200. Chelsea Publishing Company, Incorporated, 1756.

de Moivre, A. *Miscellanea analytica de seriebus et quadraturis*. J. Tonson & J. Watts, London, 1730. [Analytical Miscellaneous Series and Quadratics].

de Morgan, A. On Divergent Series and Various Points of Analysis Connected with Them. *Camb. Phil. Soc., Trans.*, 8, Part II:182–203, 1844.

Deming, W. E. De Moivre's Miscellanea Analytica, and the Origin of the Normal Curve. *Nature*, 132(3340):713–713, 1933.

Descartes, R. *Discours de la méthode pour bien conduire sa raison et chercher la vérité dans les sciences. Plus la Dioptrique. Les Meteores. Et la Geometrie qui sont des essais de cete [sic] methode [par Descartes].* Ian Maire, Leyde, 1637. [Discourse on the method for conducting one's reason well and seeking the truth in science. More Diopter. The Meteors. And Geometry which are tests of this [sic] method [by Descartes]].

Descartes, R. *Descartes: The World and Other Writings*. Cambridge University Press, 1998.

Dingle, R. B. *Asymptotic Expansions: Their Derivation and Interpretation*. Academic Press, London, 1973.

Dingle, R. B. Asymptotic Expansions and Converging Factors I. General Theory and Basic Converging Factors. *Proc. Roy. Soc. Lond. Series A. Math. and Phys. Sci.*, 244(1239):456–475, 1958a.

Dingle, R. B. Asymptotic Expansions and Converging Factors II. Error, Dawson, Fresnel, Exponential, Sine and Cosine, and Similar Integrals. *Proc. Roy. Soc. Lond. Series A. Math. and Phys. Sci.*, 244(1239):476–483, 1958b.

Dingle, R. B. Asymptotic Expansions and Converging Factors III. Gamma, Psi and Polygamma Functions, and Fermi-Dirac and Bose-Einstein integrals. *Proc. Roy. Soc. Lond. Series A. Math. and Phys. Sci.*, 244(1239):484–490, 1958c.

Dingle, R. B. Asymptotic Expansions and Converging Factors V. Lommel, Struve, Modified Struve, Anger and Weber functions, and Integrals of Ordinary and Modified Bessel Functions. *Proc. Roy. Soc. Lond. Series A. Math. and Phys. Sci.*, 249(1257):284–292, 1959a.

Dingle, R. B. Asymptotic Expansions and Converging Factors VI. Application to Physical Prediction. *Proc. Roy. Soc. Lond. Series A. Math. and Phys. Sci.*, 249

(1257):293–295, 1959b.

Dingle, R. B. and Wilkes, M. V. Asymptotic Expansions and Converging Factors IV. Confluent Hypergeometric, Parabolic Cylinder, Modified Bessel, and Ordinary Bessel Functions. *Proc. Roy. Soc. Lond. Series A. Math. and Phys. Sci.*, 249 (1257):270–283, 1959.

DLMF. *NIST Digital Library of Mathematical Functions*. `https://dlmf.nist.gov/`, Release 1.1.10 of 2023-06-15. F. W. J. Olver, A. B. Olde Daalhuis, D. W. Lozier, B. I. Schneider, R. F. Boisvert, C. W. Clark, B. R. Miller, B. V. Saunders, H. S. Cohl, and M. A. McClain, eds.

Dominici, D. Variations on a Theme by James Stirling. *preprint, arXiv math/0603007*, 2006.

Dorigoni, D. An Introduction to Resurgence, Trans-series and Alien Calculus. *Ann. Phys.*, 409:167914, 2019.

Dunne, G. A Beginners Guide to Resurgence and Trans-series in Quantum Theories. `https://dunne.physics.uconn.edu/wp-content/uploads/sites/911/2016/03/dunne_amherst_2016_resurgence.pdf`, 2016. Recent Developments in Semiclassical Probes of Quantum Field Theories UMass Amherst ACFI, March 17-19, 2016.

Dunne, G. Introduction to Resurgence and Non-perturbative Physics . `http://events.kias.re.kr/ckfinder/userfiles/201901/files/LectureNote(Dunne)1.pdf`, 2019. Kavli Pan Asian Winter School Strings, Particles and Cosmology Sogang University, Seoul, January 2019.

Dunne, G. V. and Ünsal, M. Generating Nonperturbative Physics from Perturbation Theory. *Phys. Rev. D*, 89(4):041701, 2014.

Dutka, J. The Early History of the Factorial Function. *Arch. Hist. Exact Sci.*, pages 225–249, 1991.

Dutka, J. On the Early History of Bessel Functions. *Arch. Hist. Exact Sci.*, pages 105–134, 1995.

Dyson, F. J. Divergence of Perturbation Theory in Quantum Electrodynamics. *Phys. Rev.*, 85(4):631–632, 1952.

Écalle, J. *Les fonctions résurgentes:(en trois parties)*, volume 1. Université de Paris-Sud, Département de Mathématique, Bât., 1981. [The resurgent functions: (in three parts)].

Edwards, H. M. *Riemann's Zeta Function*. Academic Press, 1974.

Efthimiou, C. J. Finding Exact Values for Infinite Sums. *Math. Mag.*, 72(1):45–51, 1999.

Efthimiou, C. J. Trigonometric Series via Laplace Transforms. *Math. Mag.*, 79(5): 376–379, 2006.

Erdélyi, A., Magnus, W., Oberhettinger, F., and Tricomi, F. G. *Higher Transcendental Functions*. McGraw-Hill New York, 1953.

Euler, L. Variae observationes circa series infinitas. *Commentarii academiae sci-*

entiarum imperialis Petropolitanae, 9(1737):160–188, 1737. [Several Remarks on Infinite Series.].

Euler, L. De progressionibus transcendentibus seu quarum termini generales algebraice dari nequeunt. *Commentarii academiae scientiarum Petropolitanae*, pages 36–57, 1738. [On Transcendental Progressions That Is, Those Whose General Terms Cannot Be Given Algebraically].

Euler, L. De progressionibus harmonicis observationes. *Commentarii academiae scientiarum Petropolitanae*, 7(1734-35):150–156, 1740a. [Observations on harmonic progressions].

Euler, L. De summis serierum reciprocarum. *Commentarii academiae scientiarum Petropolitanae*, pages 123–134, 1740b. [On the sums of series of reciprocals].

Euler, L. Remarques sur un beau rapport entre les séries des puissances tant directes que réciproques. *Mémoires de l'académie des sciences de Berlin*, pages 83–106, 1768. [Remarks on a beautiful relation between direct as well as reciprocal power series.].

Euler, L. De seriebus divergentibus. [On divergent series] Novi Comm. Acad. Sci. Petrop., 5: 205–237, 1760. *Opera Omnia*, 14(1):585–617, 1760.

Feynman, R. P. Space-time Approach to Quantum Electrodynamics. *The Physical Review*, 76(6):769–789, 1949.

Feynman, R. P. *"Surely You're Joking, Mr. Feynman!": Adventures of a Curious Character*. Bantham, New York, 1985.

Gasper, G. and Rahman, M. *Basic Hypergeometric Series*, volume 96. Cambridge University Press, 2004.

Glasser, M. L. The Summation of Series. *SIAM J. Math. Anal.*, 2(4):595–600, 1971.

Gray, A. and Mathews, G. B. *A Treatise on Bessel Functions and Their Applications to Physics*. Macmillan & Company, 1895.

Green, G. On the Motion of Waves in a Variable Canal of Small Depth and Width. *Trans. Cambridge Philos. Soc.*, 6:457–462, 1838.

Gronau, D. Why is the gamma function so as it is. *Teach. Math. Comput. Sci.*, 1: 43–53, 2003.

Hardy, G. H. *Divergent Series*. Oxford at the Clarendon Press, 1949.

Herman, R. L. The Bailey-Borwein-Plouffe π Formula. Unpublished manuscript, March 26, 2000, 2000.

Herman, R. L. *An Introduction to Fourier Analysis*. Chapman & Hall, CRC, 2016.

Herman, R. L. Laplace Transforms and Greens Functions. In: Martinez-Luaces, V. editor, *Engineering Problems and the Laplace Transform*, Chapter 1, pages 1-112, 2025.

Hirschhorn, M. A New Formula for π. *J. Aust. Math. Soc. Gaz.*, 25:82–83, 1998.

Indorato, L. and Nastasi, P. The 1740 Resolution of the Fermat-Descartes Controversy. *Hist. Math.*, 16(2):137–148, 1989.

Jackson, J. D. From Alexander of Aphrodisias to Young and Airy. *Phys. Rep.*, 320

(1-6):27–36, 1999.

Jeffreys, H. *Operational Methods in Mathematical Physics.* Cambridge at the University Press, 1927.

Khare, V. and Nussenzveig, H. M. Theory of the Rainbow. *Phys. Rev. Lett.*, 33 (16):976, 1974.

Kim, U.-R., Cho, S., and Lee, J. The Art of Schwinger and Feynman Parametrizations. *J. Korean Phys. Soc.*, pages 1–17, 2023.

Kline, M. Euler and infinite series. *Math. Mag.*, 56(5):307–315, 1983.

Kowalenko, V. Euler and Divergent series. *Eur. J. Pure Appl. Math.*, 4(4):370–423, 2011.

Kowalenko, V. Exact Values of the Gamma Function from Stirlings Formula. *Mathematics*, 8(7):1058, 2020.

Lacroix, S. F. *Traité du Calcul Différentiel et du Calcul Intégral*, volume 1. JBM Duprat, 1797. [Treatise on Differential Calculus and Integral Calculus].

Lacroix, S. F. *An Elementary Treatise on the Differential and Integral Calculus.* J. Smith, 1816.

Laven, P. Supernumerary Arcs of Rainbows: Young's Theory of Interference. *Appl. Opt.*, 56(19):G104–G112, 2017.

Lee, R. L. Mie Theory, Airy Theory, and the Natural Rainbow. *Appl. Opt.*, 37(9): 1506–1519, 1998.

Lesko, J. P. and Smith, W. D. A Laplace Transform Technique for Evaluating Infinite Series. *Math. Mag.*, 76(5):394–398, 2003.

Lupas, A. Some BBP-functions. *Publikacije-elektrotehnickog Fakulteta Univerzitet U Beogradu Serija Matematika*, 13:26–29, 2002.

MacRobert, T. M. *Functions of a Complex Variable.* Macmillan & Company, limited, 1917.

Malet, A. Gregorie, Descartes, Kepler, and the Law of Refraction. *Arch. Int. Hist. Sci.*, 40(125):278–304, 1990.

Maor, E. *To Infinity and Beyond: A Cultural History of the Infinite.* Birkhauser.

Mariño, M. Lectures on Non-perturbative Effects in Large N Gauge Theories, Matrix Models and Strings. *Fortschr. Phys.*, 62(5-6):455–540, 2014.

Mariño, M. *Advanced Topics in Quantum Mechanics.* Cambridge University Press, 2021.

Marucho, M. Non Summable Borel Φ^4 Theory in Zero Dimensions and the Generalized Borel Transform. *preprint, arXiv hep-th/0402209*, 2004.

Mas, R. M. Resurgence, A Problem of Missing Exponential Corrections in Asymptotic Expansions. *preprint, arXiv:1904.07217*, 2019.

Mengoli, P. *Novae quadraturae arithmeticae, seu De additione fractionum.* Bononiae: Ex typographia Iacobi Montij, 1650. [New Arithmetic Quadrature, or On the Addition of Fractions].

Merzbacher, E. *Quantum Mechanics.* John Wiley & Sons, 1998.

Miller, W. H. On Spurious Rainbows. *Trans. Camb. Philos. Soc.*, 7:277–286, 1848.

Read Mar. 22, 1841.

Moore, C. N. Summability of Series. *Am. Math. Mon.*, 39(2):62–71, 1932.

Moore, D. H. *Heaviside Operational Calculus: An Elementary Foundation*. American Elsevier Publishing Company, 1971.

Murphy, T. P. *A Study of Bernhard Riemann's 1859 Paper*. Paramount Ridge Press, 2020.

Nahin, P.J. *In Pursuit of Zeta-3*. Princeton University Press, 2021.

Newton, I. *Optice sive de reflexionibus, refractionibus, inflexionibus et coloribus lucis, libri tres*. Sumpt. Marci-Michaelis Bousquet & Sociorum, 1740.

Newton, I. *Opticks, or, a treatise of the reflections, refractions, inflections & colours of light*. Courier Corporation, 1952.

Nicholson, J. W. On the Relation of Airy's Integral to the Bessel Functions. *The London, Edinburgh, and Dublin Phil. Mag. and J. Sci.*, 18(103):6–17, 1909.

Nussenzveig, H. M. The Theory of the Rainbow. *Sci. Am.*, 236(4):116–128, 1977.

Paris, R. B. Smoothing of the Stokes Phenomenon Using Mellin-Barnes Integrals. *J. Comp. and Appl. Math.*, 41(1-2):117–133, 1992.

Paris, R. B. A Note on the Asymptotics of the Modified Bessel Functions on the Stokes Lines. *preprint, arXiv:1708.09656*, 2017.

Paris, R. B. and Kaminski, D. *Asymptotics and Mellin-Barnes Integrals*, volume 85. Cambridge University Press, 2001.

Pathria, R. K. *Statistical Mechanics*. Elsevier, 2016.

Pearson, K. Historical Note on the Origin of the Normal Curve of Errors. *Biometrika*, pages 402–404, 1924.

Potter, R. Mathematical Considerations on the Problem of the Rainbow, Shewing it to Belong to Physical Optics. *Trans. Camb. Philos. Soc.*, 6:141152, 1838.

Riemann, B. Über die Anzahl der Primzahlen unter einer gegebenen Grosse. *Ges. Math. Werke und Wissenschaftlicher Nachlaß*, 2(145-155):2, 1859. [About the Number of Prime Numbers Below a Given Size].

Roy, R. *Sources in the Development of Mathematics: Series and Products from the Fifteenth to the Twenty-first Century*. Cambridge University Press, 2011.

Sandifer, E. Estimating the Basel Problem. *MAA Online, dezembro de*, 2003. `http://eulerarchive.maa.org/hedi/HEDI-2003-12.pdf`.

Sandifer, E. Basel Problem with Integrals. *MAA Online, dezembro de*, 2004. `http://eulerarchive.maa.org/hedi/HEDI-2004-03.pdf`.

Sandifer, E. Gamma the function. *MAA Online, dezembro de*, 2007. `http://eulerarchive.maa.org/hedi/HEDI-2007-09.pdf`.

Schlissel, A. The Initial Development of the WKB Solutions of Linear Second Order Ordinary Differential Equations and Their Use in the Connection Problem. *Hist. Math.*, 4(2):183–204, 1977.

Schwinger, J. On Gauge Invariance and Vacuum Polarization. *Phys. Rev.*, 82(5): 664, 1951.

Schwinger, J. Quantum Mechanics: Symbolism of Atomic Measurements, 2003.

Segur, H., Tanveer, S., and Levine, H. J. *Asymptotics Beyond All Orders*, volume 284 of *NATO ASI Series*. Plenum Press, 1991.

Silverstone, H. J., Nakai, S., and Harris, J. G. Observations on the Summability of Confluent Hypergeometric Functions and on Semiclassical Quantum Mechanics. *Phys. Rev. A*, 32(3):1341, 1985.

Smithies, F. *Cauchy and the Creation of Complex Function Theory*. Cambridge University Press, 1997.

Stillwell, J. *Mathematics and its History*. Springer, 1989.

Stirling, J. *Methodus differentialis: sive tractatus de summatione et interpolatione serierum infinitarum*. 1730. [The differential method: or a treatise on the summation and interpolation of infinite series].

Stokes, G. G. On the Numerical Calculation of a Class of Definite Integrals and Infinite Series. *Trans. Camb. Philos. Soc.*, 9:166–187, 1850. Reprinted in *Mathematical and Physical Papers*, Vol. 2, pp. 329-357. Cambridge Univ. Press, London and New York, 1883.

Stokes, G. G. On the Discontinuity of Arbitrary Constants Which Appear in Divergent Developments. *Trans. Camb. Philos. Soc.*, 10:106–128, 1857. Reprinted in *Mathematical and Physical Papers*, Vol. 4, pp. 77-109. Cambridge Univ. Press, London and New York, 1904.

Stokes, G. G. Supplement to a Paper on the Discontinuity of Arbitrary Constants Which Appear in Divergent Developments. *Trans. Camb. Philos. Soc.*, 11:412–423, 1868.

Tingen, L. L. The Julia and Mandelbrot Sets for the Hurwitz zeta Function [electronic resource] . Master's thesis, University of North Carolina Wilmington, Wilmington, North Carolina, 2009.

Tucciarone, J. The Development of the Theory of Summable Divergent Series from 1880 to 1925. *Arch. Hist. Exact Sci.*, 10(1/2):1–40, 1973.

Van der Poorten, A. A Proof that Euler Missed. *Math. Intell.*, 1(4):195–203, 1979.

Watson, G. N. *A Treatise on the Theory of Bessel Functions*. The University Press, 1922.

Weinberg, S. *The Quantum Theory of Fields: Volume 1, Foundations*. Cambridge University Press, 2005.

Weniger, E. J. Summation of Divergent Power Series by Means of Factorial Series. *Appl. Numer. Math.*, 60(12):1429–1441, 2010.

Wheelon, A. D. On the Summation of Infinite Series in Closed Form. *J. Appl. Phys.*, 25(1):113–118, 1954.

Wheelon, A. D. Note on the Summation of Finite Series. *J. Math. and Phys.*, 34 (1-4):182–186, 1955.

Wheelon, A. D. *Tables of Summable Series and Integrals Involving Bessel Functions*. Holden-Day, 1968.

Whittaker, E. T. and Watson, G. N. *A Course of Modern Analysis: An Introduction to the General Theory of Infinite Series and of Analytic Functions, with an Ac-*

count of the Principal Transcendental Functions. Cambridge at the University Press, 3 edition, 1920.

Young, T. I. The Bakerian Lecture. Experiments and Calculations Relative to Physical Optics. *Phil. Trans. Roy. Soc. Lond.*, 94:1–16, 1804.

Chapter 5

Personal Experience in the Application of the Laplace Transform

Alexander Apelblat*, PhD

Department of Chemical Engineering, Ben-Gurion University of the Negev, Beer Sheva, Israel

Abstract

Integral transform methods, especially the Laplace transform, are some of the most useful and powerful tools of applied mathematics that are applied in various branches of science and engineering. Some examples are presented here based on the author's lengthy experience of using and teaching the basic elements and techniques of operational calculus. Examples include: the solution of nonlinear differential equations (boundary layer flow problems), the solution of partial differential equations (mass transfer coupled with and without a chemical reaction) and different methods for the evaluation of various finite, infinite and convolution integrals of elementary and special functions (Bessel, Kelvin, Struve, Volterra, Whittaker, Mittag-Leffler, Wright, Bateman and Havelock). Special consideration is given to the differentiation of some special functions with respect to the parameters and to the evaluation of the asymptotic relations, limits of series, integrals and integral representations of functions by using the shifted Dirac delta function $\delta(x-1)$. The Efros theorem in the Wlodarski form is applied to evaluate infinite and convolution integrals of the Wright function of the second kind.

* Corresponding Author Email: apelblat@bgu.ac.il

In: The Laplace Transform and Its Applications in Science
Editor: Victor Martinez-Luaces
ISBN: 979-8-89113-747-9

Keywords: Laplace transforms and their inverses, Efros theorem, special functions, evaluation of integrals of elementary and special functions, boundary layer flows, mass transfer with and without chemical reaction

1. Introduction

My first encounter with the Laplace transform occurred when I started my university studies in chemical engineering. I was under the impression that in solving ordinary differential equations in the field of chemical kinetics I was essentially using my high school knowledge of solving algebraic equations, and I was captivated and enchanted by such a method. Later when I encountered partial differential equations associated with heat and mass transfer problems, I realized that this wasn't that simple. I was lacking the mathematical background in complex variable theory, as I found that in order to use the Jordan lemma, the Cauchy residue theorem and the complex Bromwich contours were beyond my knowledge. Unfortunately, I discovered later that even some expert mathematicians had problems with the inversion of Laplace transforms, and they were not always sure whether the final result of complex integration was correct. These mathematicians had a greater interest in the uniqueness rather than the correctness of the solution, and they were even less interested in the physical aspect of the problem. It took me a long time, after studying the McLachlan books [1-3] where a numerous examples of the complex integrations are illustrated, to obtain a degree of confidence that I am also able to perform such integrations, and to have a clear-cut idea that it is better to avoid them if possible. Evidently, in the next step of my mathematical education I found many books dealing with the Laplace transform, and these books were on elementary and more advanced levels [4-33].

It is not always easy to decide on one's career path when someone is equally good in both the humanities field and in mathematics. My parents preferred and suggested that I should pursue some kind of a "practical" profession. Personally, I wanted to be an historian, but I grew up in Poland, during the Soviet-type totalitarian state, and I knew, as people said, "not only is the future dangerous but so is the past", that clearly historical research was not attractive to me. To this day my interest in history is limited to just reading historical books which occupy most shelves of my library. Although I was very good at mathematics, I experienced a setback in high school. My excellent math teacher, Mrs. Sochacka, prepared a small group from my class for the mathematical Olympiads. I had better marks than most other students, but to my surprise I was not included in this group. When I asked for the reason, her answer was "Alex, you know how to solve problems, but you are without the necessary mathematical mind and spirit". Based on her opinion, I figured that I was not good enough to have an academic career in mathematics. At best, I could become a high school teacher which was the last thing I wanted to do. So, I chose chemical engineering, starting my degree in Poland and completing my last year in Israel. In retrospect, I can say it was an excellent choice. Not only were my parents happy, but I ended up flourishing in the discipline as the chemical engineering profession includes a "little" from everything, from mathematics, physics, chemistry and other engineering disciplines. As a chemical engineer and physical chemist, I had the opportunity to publish not just chemistry related papers but also mathematical papers and books [34-60]. My Master thesis was in the irreversible thermodynamics with elements in hydrodynamics, and the PhD thesis in biophysics - dealing with the mathematical model of capillary-fluid exchange. Later, I became a professor, teaching many different courses and doing research, mostly in chemical engineering and physical chemistry. I suppose that what characterized my scientific publications was that numerous experimental results required the usage of many mathematical equations. Not being a mathematics professor had some negative and positive aspects. It is difficult to publish papers in mathematical journals when the author is from the chemical engineering department and not from the math department. Not to mention that the mathematical jargon is often missing, e.g. "small epsilon goes to zero". However, some positive things came about. I did things that expert mathematicians did not even consider attempting. For example, I said to my mathematician friend that by using the Laplace transform, I solved the Blasius problem in hydrodynamics. His reaction was "It is impossible. You cannot apply this transform to nonlinear differential equations". His reply was backed up by many books on integral transforms where it is stated that the operational calculus is only valid to solve linear problems. However, as I will show in this chapter, the Blasius and other boundary layer equations can be solved by using the Laplace transform method in a simple, alternative way. And the applied procedure is even easier than other methods that are used in the literature. One of the benefits when working with the Laplace transform is that one gets familiar with special functions and various groups of finite, infinite or convolution integrals. For me, they included the Bessel and related functions, the lesser known Volterra, Mittag-Leffler, Wright, Bateman and

Havelock functions [51, 52, 54, 55, 59, 60]. My long interest in and involvement with the Laplace transform also led to my published book "Laplace Transforms and Their Applications" [53]. This book is written in different way than many other standard books published on this subject. In it, a strong emphasis is placed on the efficient use of various mathematical manipulations, because as I observed, the ability to apply operational rules in a competent way is inadequate for many users of integral transforms.

The material in this Chapter is some sort of short summary of my quite large involvement with the Laplace transform and it can be divided into two different sections. The first section is dealing with solving physical problems (boundary layer flows and mass transfer in various hydrodynamic situations) and the second section is devoted to mathematical problems, the evaluation of definite, infinite and convolution integrals and some aspects associated with properties of special functions that are linked with direct and inverse transforms derived by rules and theorems of operational calculus.

2. Application of the Laplace Transform to Physical Problems

Applied Notation

The Laplace transform of function $f(t)$ of a real variable t, $t \geq 0$, is defined by the improper integral

$$L\{f(t)\} = \int_0^\infty e^{-st} f(t)\,dt = F(s) \tag{1}$$

$$\operatorname{Re} s > 0$$

where $F(s)$ is an analytical function of variable s. Conditions imposed on the function $f(t)$ and the existence of $F(s)$ are given in Doetsch (1950) [11], Spiegel (1965) [22], Sneddon (1971) [28], and Apelblat (2012) [53]. The inverse transform is usually determined with the help of transform tables and by using the basic properties and operational rules of the Laplace transform. The integral formula which restores the original function $f(t)$ is called the complex Bromwich-Wagner formula

$$L^{-1}\{F(s)\} = f(t) = \begin{cases} \dfrac{1}{2\pi i}\displaystyle\int_{c-i\infty}^{c+i\infty} e^{st}F(s)\,ds & ; \quad t>0 \\ 0 & ; \quad t<0 \end{cases} \tag{2}$$

where c is an arbitrary real number chosen in such a way that all the singularities of $F(s)$ lie to the left of the line $s = c$ in the complex plane. The integration is usually performed by transforming the path of integration of (2) into the Bromwich contour integral [1-3,16,18].

Laminar Boundary Layer Flows

In a boundary layer approximation, the Navier-Stokes equations and the continuity equations for two-dimensional, and unidirectional flow on a semi-infinite plate, together with the boundary conditions are Meksyn (1961) [61]

$$\begin{aligned} & u\frac{\partial u}{\partial x} + \mathrm{v}\frac{\partial u}{\partial y} = -\frac{1}{\rho}\frac{\partial P}{\partial x} + \nu\frac{\partial^2 u}{\partial y^2} \\ & -\frac{1}{\rho}\frac{\partial P}{\partial x} = U_\infty \frac{dU_\infty}{dx} \\ & \frac{\partial u}{\partial x} + \frac{\partial \mathrm{v}}{\partial y} = 0 \\ & u(x,0) = 0 \quad \mathrm{v}(x,0) = 0 \quad ; \quad u(x,\infty) = U_\infty \end{aligned} \tag{3}$$

where u and v are the velocity vector components in the x and y directions, P is the applied pressure, U_∞ denotes the velocity in the main stream just outside the boundary layer, ν is the kinematic viscosity of the incompressible fluid and ρ is its density.

Blasius solved this problem by introducing a new variable η and the stream function ψ in the following way

$$\eta = \frac{1}{2}\sqrt{\frac{U_\infty}{\nu x}}\,y$$
$$\psi(x,y) = \sqrt{\nu U_\infty x}\, f(\eta) \qquad (4)$$
$$u = \frac{\partial \psi}{\partial y} \quad ; \quad \mathrm{v} = -\frac{\partial \psi}{\partial x}$$

and the similarity analysis leads to a solution of the nonlinear third order ordinary differential equation

$$\begin{aligned} & f'''(\eta) + f(\eta) f''(\eta) = 0 \\ & f(0) = f'(0) = 0 \\ & f''(0) = \alpha \\ & f'(\infty) = 1 \end{aligned} \qquad (5)$$

where the primes denote the derivatives with respect to the variable η and the parameter α is determined from the boundary condition at infinity. By expanding $f(\eta)$ into the series at origin, $\eta \to 0$, Blasius derived the following series

$$f(\eta) = \frac{\alpha}{2!}\eta^2 - \frac{\alpha^2}{5!}\eta^5 + \frac{11\alpha^3}{8!}\eta^8 - \frac{375\alpha^4}{11!}\eta^{11} + \frac{27897\alpha^5}{14!}\eta^{14} - \ldots \qquad (6)$$

The series has a finite radius of convergence Meksyn (1961) [61]. From the asymptotic expression, Blasius numerically found that the value of the parameter α is 0.4696.

Even though operational calculus is usually applied to the solution of linear differential equations, the Laplace transform method can also be used for the solution of nonlinear boundary layer equations Apelblat (1967, 1968, 1969) [34-36]. The proposed procedure includes the sequence of successive approximations to obtain the power series expansion (6) and a modification of the Meksyn method (1961) [61] for the evaluation of the parameter α. Denoting the Laplace transform of f(η) by $F(s)$ and introducing it into (5) Apelblat (1967) [34] gives

$$\begin{aligned} & L\left\{f_{n+1}'''(\eta)\right\} = -L\left\{f_n''(\eta) f_n(\eta)\right\} \quad ; \quad n = 0,1,2,3,\ldots \\ & f_0(\eta) = 0 \end{aligned} \qquad (7)$$

And for $n = 0$, it follows from (5) and (7) that

$$s^3 F(s) - \alpha = 0 \tag{8}$$

The inverse of (8) is given by

$$f_1(\eta) = \frac{\alpha}{2!}\eta^2 \tag{9}$$

Calculating the nonlinear term in (7) gives

$$\begin{aligned} s^3 F(s) - \alpha &= -L\left\{\frac{\alpha^2}{2!}\eta^2\right\} = -\frac{\alpha^2}{s^3} \\ F(s) &= \frac{\alpha}{s^3} - \frac{\alpha^2}{s^6} \end{aligned} \tag{10}$$

and the second approximation is

$$f_2(\eta) = \frac{\alpha}{2!}\eta^2 - \frac{\alpha^2}{5!}\eta^5 \tag{11}$$

Following the same process, the third transform and approximation are

$$\begin{aligned} F(s) &= \frac{\alpha}{s^3} - \frac{\alpha^2}{s^6} + \frac{11\alpha^3}{s^9} - \frac{56\alpha^3}{s^{12}} \\ f_3(\eta) &= \frac{\alpha}{2!}\eta^2 - \frac{\alpha^2}{5!}\eta^5 + \frac{11\alpha^3}{8!}\eta^8 - \frac{56\alpha^4}{12!}\eta^{12} \end{aligned} \tag{12}$$

As can be observed, the applied procedure is very simple because the Laplace transforms and their inverses are elementary, they only include operations with the series in ascending powers of η and in descending powers of s. The main part of the necessary calculations is already prepared in previous approximations and each new approximation gives one exact term in the Blasius series. Additional terms in the $f_n(\eta)$ expressions have lower coefficients than those in the power series (6), and the number of terms in each approximation grows considerably.

The power expansions in (6) or (12) satisfy the boundary conditions at the origin $\eta = 0$, and the unknown parameter α should be evaluated from the boundary condition (5) at infinity. In order to perform such an evaluation, Meksyn suggested the following integral (1961) [61]

$$\int_0^\infty e^{-h(\eta;\alpha)} g(\eta;\alpha)\, d\eta = 1$$

$$h(\eta;\alpha) = \sum_{n=0}^{\infty} a_n \eta^{n+1} \tag{13}$$

$$g(\eta;\alpha) = \sum_{n=0}^{\infty} b_n \eta^n$$

where the coefficients a_n and b_n are given functions of the parameter α, and $h(\eta,\alpha)$ is the rapidly decreasing function of η, while $g(\eta,\alpha)$ varies slowly with η. Meksyn asymptotically evaluated integral (13) by the steepest descents method where the low radius of convergence of $f(\eta)$ is of no importance. However, an alternative solution is also available by applying the Laplace transform. If the integral in (13) is replaced with function $I(t)$, which is defined by

$$I(t) = \int_0^\infty e^{-th(\eta;\alpha)} g(\eta;\alpha)\; d\eta \tag{14}$$

then the Laplace transform of $I(t)$ is

$$F(s) = L\{I(t)\} = \int_0^\infty e^{-st} \left[\int_0^\infty e^{-th(\eta;\alpha)} g(\eta;\alpha)\, d\eta \right] dt \tag{15}$$

Changing the order of integration, we have

$$F(s) = \int_0^\infty \frac{g(\eta;\alpha)}{s + h(\eta;\alpha)}\, d\eta \tag{16}$$

If it is possible to evaluate and invert this integral in (16), then the function $I(t)$ becomes a function of t and for $t = 1$, it satisfies the condition required in (13) to determine parameter α.

The Meksyn method includes the substitution of $f(\eta)$ in (5) by the Blasius series in (6), leaving the third and second derivatives of function $f(\eta)$ without change, and formally transforming the nonlinear third order differential equation into a linear first order differential equation

$$[f''(\eta)]' + f(\eta) f''(\eta) = 0 \tag{17}$$

The solution of this differential equation is

$$f''(\eta) = A\,e^{-h(\eta;\alpha)}$$

$$h(\eta;\alpha) = \int_0^{\eta} f(\zeta)d\zeta = \frac{\alpha}{3!}\eta^3 - \frac{\alpha^2}{6!}\eta^6 + \frac{11\alpha^3}{9!}\eta^9 - ... \tag{18}$$

From the boundary condition at η = 0, it follows that the constant of integration A is equal to α and from the condition at infinity in (5) we have

$$f'(\infty) = \alpha \int_0^{\infty} e^{-h(\eta;\alpha)}d\eta = 1 \tag{19}$$

which leads in the present case to $g(\eta,\alpha) = \alpha$.

Returning to the function $I(t)$ in (14), and its Laplace transform in (16), it follows that for the Blasius problem we have

$$I(t) = \alpha \int_0^{\infty} e^{-th(\eta;\alpha)}d\eta$$

$$F(s) = L\{I(t)\} = \alpha \int_0^{\infty} \frac{d\eta}{s + h(\eta;\alpha)} \tag{20}$$

Denoting $g(\eta,\alpha) = \tau$ and inverting the variables η and τ, the Laplace transform in (20) can be expressed by the Lagrange series

$$F(s) = \alpha \sum_{k=0}^{\infty} \frac{A_k}{3} \int_0^{\infty} \frac{\tau^{(k-2)/3}}{s+\tau} d\tau \tag{21}$$

where the coefficients A_k are explicitly given by Meksyn (1961) [61], Whittaker and Watson (1963) [62] and Abramowitz and Stegun (1965) [63]. The first three non-zero coefficients are

$$A_0 = \left(\frac{6}{\alpha}\right)^{1/3}$$

$$A_3 = \frac{1}{15}\left(\frac{6}{\alpha}\right)^{1/3} \tag{22}$$

$$A_6 = -\frac{1}{180}\left(\frac{6}{\alpha}\right)^{1/3}$$

However, all integrals in (21) are divergent, and in order to overcome this difficulty, the rule of repeated differentiation of Laplace transforms should be applied

$$L^{-1}\left\{\frac{d^n F(s)}{ds^n}\right\} = (-1)^n t^n I(t) \quad ; \quad n = 0,1,2,3... \tag{23}$$

From (20) it follows that

$$\frac{d^n F(s)}{ds^n} = (-1)^n n!\alpha \int_0^\infty \frac{d\eta}{[s + h(\eta;\alpha)]^{n+1}} = (-1)^n n!\alpha \sum_{k=0}^{\infty} \frac{A_k}{3} \int_0^\infty \frac{\tau^{(k-2)/3}}{(s+\tau)^{n+1}} d\tau \tag{24}$$

and for $0 < (k+1)/2 < n$, the integrals in (24) are convergent and equal to

$$\int_0^\infty \frac{\tau^{(k-2)/3}}{(s+\tau)^{n+1}} d\tau = \frac{\Gamma\left(\frac{k+1}{3}\right)\Gamma\left(n+1-\frac{k+1}{3}\right)}{n! s^{n+1-(k+1)/3}} \tag{24}$$

Using (25), the inverse of (23) is given by

$$L^{-1}\left\{\frac{d^n F(s)}{ds^n}\right\} = (-1)^n \alpha \sum_{k=0}^{\infty} \frac{A_k}{3} \Gamma\left(\frac{k+1}{3}\right) t^{n-(k+1)/3} \tag{25}$$

and comparing it with (23) we have

$$I(t) = \alpha \sum_{k=0}^{\infty} \frac{A_k}{3} \Gamma\left(\frac{k+1}{3}\right) t^{-(k+1)/3}$$

and the rapidly convergent series for evaluation of the parameter α is

$$I(1) = \alpha \sum_{k=0}^{\infty} \frac{A_k}{3} \Gamma\left(\frac{k+1}{3}\right) = 1 \tag{26}$$

From Abramowitz and Stegun (1965) [63]

$$\Gamma\left(\frac{k+1}{3}\right) = \frac{1.4.7...(3k-2)}{3^k} \Gamma\left(\frac{1}{3}\right)$$

$$\Gamma\left(\frac{1}{3}\right) = 2.6789385... \tag{27}$$

then combining (22), (26) and (27) we have

$$\frac{1}{3}\Gamma\left(\frac{1}{3}\right)(6\alpha^2)^{1/3}\left\{1+\frac{1}{45}-\frac{1}{405}+...\right\}=1 \tag{28}$$

and giving $\alpha = 0.4698$, which is very close to the Blasius evaluated value of $\alpha = 0.4696$.

The Blasius problem is the simplest case of the Laplace transform (equations (7) and (16)) being applied to the solution of boundary layer equations. A more complex hydrodynamic situation will be presented below. For a steady flow in the boundary layer along the wedge with $U_\infty \approx x^m$ (the Falkner and Skan problem), the following nonlinear differential equation should be solved

$$\begin{aligned} & f'''(\eta)+f(\eta)f''(\eta)+\beta(1-[f'(\eta)]^2)=0 \quad ; \quad \beta=\frac{2m}{m+1} \\ & f(0)=f'(0)=0 \\ & f''(0)=\alpha \\ & f'(\infty)=1 \end{aligned} \tag{29}$$

If $m = 0$, equation (29) is reduced to the Blasius problem, and for $m = 1$ to the stagnation-point flow (the Hiemenz problem). Similar to (7), the linear and nonlinear parts of (29) are separated and we have the following sequence of approximations

$$\begin{aligned} & L\left\{f'''_{n+1}(\eta)\right\}=-L\left\{f''_n(\eta)f_n(\eta)+\beta(1-[f'_n(\eta)]^2)\right\} \quad ; \quad n=0,1,2,3,... \\ & f_0(\eta)=0 \end{aligned} \tag{30}$$

The first approximation is the same as in (9)

$$\begin{aligned} & F(s)=\frac{\alpha}{s^3} \\ & f_1(\eta)=\frac{\alpha}{2!}\eta^2 \end{aligned} \tag{31}$$

Introducing the nonlinear part from (30) and using (31), the second approximation is

$$F(s)=\frac{\alpha}{s^3}-\frac{\beta}{s^4}+\frac{\alpha^2(2\beta-1)}{s^4}$$
$$f_2(\eta)=\frac{\alpha}{2!}\eta^2-\frac{\beta}{6}\eta^3+\frac{\alpha^2}{60}\left(\beta-\frac{1}{2}\right)\eta^3 \tag{32}$$

and (32) gives exactly three terms in the series expansion of the solution of the Falkner and Skan problem for $\eta \to 0$ Meksyn (1961) [61]. The third approximation which is presented by Apelblat (1967) [34] already includes eight terms in the series. It is possible to treat the linear part in (30)

$$L\left\{f_{n+1}'''(\eta)+\beta\right\}=-L\left\{f_n''(\eta)f_n(\eta)-\beta[f_n'(\eta)]^2\right\} \quad ; \quad n=0,1,2,3,... \tag{33}$$
$$f_0(\eta)=0$$

and then the first and second approximations are much more advanced

$$f_1(\eta)=\frac{\alpha}{2!}\eta^2-\frac{\beta}{6}\eta^3$$
$$f_1(\eta)=\frac{\alpha}{2!}\eta^2-\frac{\beta}{6}\eta^3+\frac{\alpha^2}{60}\left(\beta-\frac{1}{2}\right)\eta^3-\frac{\alpha\beta}{120}\left(\beta-\frac{2}{3}\right)\eta^6+\frac{\beta^2}{840}\left(\beta-\frac{2}{3}\right)\eta^7 \tag{34}$$

The flows considered above are particular cases of a general differential equation, the so-called equation of similar profiles

$$f'''(\eta)+af(\eta)f''(\eta)+b(1-[f'(\eta)]^2)+c\,f''(\eta)=0$$
$$f(0)=f'(0)=0$$
$$f''(0)=\alpha \tag{35}$$
$$f'(\infty)=1$$

where a, b and c are constants and a suitable choice of them represents the boundary layer flows in convergent and divergent channels, flow with an asymptotic suction profile, the two-dimensional jet and others flows. In this case, the linear part of (35) consists of three terms and therefore it is possible to obtain three sequences of approximate functions

$$L\left\{f_{n+1}'''(\eta)\right\}=-L\left\{af_n''(\eta)f_n(\eta)+b(1-[f_n'(\eta)]^2)+cf_n''(\eta)\right\} \quad ; \quad n=0,1,2,3,...$$
$$L\left\{f_{n+1}'''(\eta)+cf_n''(\eta)\right\}=-L\left\{af_n''(\eta)f_n(\eta)+b(1-[f_n'(\eta)]^2)\right\} \tag{36}$$
$$L\left\{f_{n+1}'''(\eta)+cf_n''(\eta)+b\right\}=-L\left\{af_n''(\eta)f_n(\eta)-b[f_n'(\eta)]^2\right\}$$
$$f_0(\eta)=0$$

The first approximations based on the linear parts only are

$$F(s)=\frac{\alpha}{s^3} \quad ; \quad f_1(\eta)=\frac{\alpha}{2!}\eta^2$$
$$F(s)=\frac{\alpha}{s^2(s+c)} \quad ; \quad f_1(\eta)=\frac{\alpha}{c^2}\left[c\eta-1+e^{-c\eta}\right] \tag{37}$$
$$F(s)=\frac{\alpha}{s^2(s+c)}-\frac{b}{s^3(s+c)} \quad ; \quad f_1(\eta)=\frac{b+c\alpha}{c^2}\left[c\eta-1+e^{-c\eta}\right]-\frac{b}{2!c}\eta^2$$

Evidently, exponentials of η in (37), when expanded into the Taylor series, will give the approximate solutions in the polynomial form. Nonlinear parts of the solution are determined by lengthy algebraic operations, but all inverses of the Laplace transforms are elementary, and they are presented in detail by Apelblat (1967) [34].

The illustrated Laplace transform method can be also applied to sets of nonlinear differential equations. In a first example, the functions representing flow $f(\eta)$, and temperature profiles $\theta(\eta)$, can be solved independently. In the thermal boundary layer in the flow parallel to a semi-infinite plate (the Blasius thermal problem) the following set of simultaneous equations should be solved

$$\begin{aligned}
&f'''(\eta)+f(\eta)f''(\eta)=0\\
&f(0)=f'(0)=0 \quad ; \quad f''(0)=\alpha \quad ; \quad f'(\infty)=1\\
&\theta''(\eta)+\Pr f(\eta)\theta'(\eta)=0\\
&\theta(0)=1 \quad ; \quad \theta'(0)=\beta \quad ; \quad \theta(\infty)=0
\end{aligned} \tag{38}$$

where Pr is the Prandtl number defined as the ratio of kinetic viscosity to thermal diffusivity. Applying the Laplace transforms to (38), the following sequence of approximations is derived

$$\begin{aligned}
&F(s)=L\{f_n(\eta)\} \quad ; \quad G(s)=L\{\theta_{n+1,n}(\eta)\} \quad ; \quad n=0,1,2,3,\ldots\\
&L\{f'''_{n+1}(\eta)\}=-L\{f''_n(\eta)f_n(\eta)\}\\
&L\{\theta''_{n+1,n}(\eta)\}=-\Pr L\{f_n(\eta)\theta'_{n,n-1}(\eta)\}\\
&f_0(\eta)=0 \quad ; \quad \theta_{0,-1}(\eta)=0
\end{aligned}$$
(39)

The first approximation based on only the linear parts in (39) is

$$F(s) = \frac{\alpha}{s^3} \quad ; \quad f_1(\eta) = \frac{\alpha}{2!}\eta^2$$
$$G(s) = \frac{1}{s} + \frac{\beta}{s^2} \quad ; \quad \theta_{1,0}(\eta) = 1 + \beta\eta \tag{40}$$

Introducing (40) into (39) gives the second approximation

$$f_1(\eta) = \frac{\alpha}{2!}\eta^2 - \frac{\alpha^2}{5!}\eta^5$$
$$\theta_{2,1}(\eta) = 1 + \beta\eta - \frac{\alpha\beta \mathrm{Pr}}{4!}\eta^4 \tag{41}$$

and similarly, the third approximation is

$$f_1(\eta) = \frac{\alpha}{2!}\eta^2 - \frac{\alpha^2}{5!}\eta^5 + \frac{11\alpha^3}{8!}\eta^8 - \frac{56\alpha^4}{11!}\eta^{11}$$
$$\theta_{2,1}(\eta) = 1 + \beta\eta - \frac{\alpha\beta \mathrm{Pr}}{4!}\eta^4 + \frac{\alpha^2\beta \mathrm{Pr}(1+10\mathrm{Pr})}{7!}\eta^7 - \frac{56\alpha^3\beta \mathrm{Pr}}{10!}\eta^{10} \tag{42}$$

The constants α and β can be similarly evaluated by the Laplace transform procedure starting from (18). In this case, the equation of motion in (42) does not contain the temperature, and therefore, it can be solved independently. However, the procedure is also valid when both functions $f(\eta)$ and $\theta(\eta)$ do appear simultaneously in the boundary layer differential equations.

The magneto-hydrodynamic equations for the steady, laminar flow of an incompressible, electrically conducting fluid can be written for the Falkner-Skan boundary layer problem as in Apelblat (1969) [36]

$$u\frac{\partial u}{\partial x} + \mathrm{v}\frac{\partial u}{\partial y} = \nu\frac{\partial^2 u}{\partial y^2} - Cx^n + \frac{\mu}{4\pi\rho}\left(H_1\frac{\partial H_1}{\partial x} + H_2\frac{\partial H_1}{\partial y}\right)$$
$$u\frac{\partial H_1}{\partial x} + \mathrm{v}\frac{\partial H_1}{\partial y} - H_1\frac{\partial u}{\partial x} - H_2\frac{\partial u}{\partial y} = \frac{1}{4\pi\sigma\mu}\frac{\partial^2 H_1}{\partial y^2}$$
$$\frac{\partial u}{\partial x} + \frac{\partial \mathrm{v}}{\partial y} = 0 \quad ; \quad \frac{\partial H_1}{\partial x} + \frac{\partial H_2}{\partial y} = 0 \tag{43}$$
$$C = \frac{(n+1)U_\infty^2(S-1)}{2} \quad ; \quad S = \frac{\nu}{4\pi}\left(\frac{H_\infty}{U_\infty}\right)^2$$

u, v and H_1, H_2 are the parallel and normal components of the fluid velocity and magnetic field, respectively. U_∞ and H_∞ denote their values at infinite upstream. σ, ν, μ and ρ are respectively the electrical conductivity, kinematic viscosity, magnetic permeability and density of the fluid. S expresses the ratio of magnetic energy to kinetic energy per unit volume. If C = 0, the equations in (43) are reduced to the magneto-hydrodynamic Blasius problem as given by Apelblat (1968) [35]. By introducing a new set of variables

$$u = U_\infty x^{(n+1)/2} f'(\eta) \quad ; \quad \eta = y\sqrt{\frac{U_\infty(n+3)}{4\nu}}\, x^{1/4(n-1)}$$

$$\text{v} = -\sqrt{\frac{\nu U_\infty}{n+3}}\, x^{1/4(n-1)}[(n+3)f(\eta)+(n-1)\eta f'(\eta)] \tag{44}$$

$$H_1 = H_\infty x^{1/2(n+1)} g'(\eta)$$

$$H_2 = -H_\infty\sqrt{\frac{\nu}{4(n+3)U_\infty}}\, x^{1/4(n-1)}[(n+3)g(\eta)+(n-1)\eta g'(\eta)]$$

equations in (43), together with the boundary conditions of the Falkner-Skan magneto-hydrodynamic problem, are transformed to the following form Apelblat (1969) [36]

$$f'''(\eta)+f(\eta)f''(\eta)+\beta(1-[f'(\eta)]^2) = S[g(\eta)g''(\eta)-\beta(1-[g'(\eta)]^2)]$$

$$g'''(\eta)+\frac{1}{\lambda}[f(\eta)g''(\eta)-g(\eta)f''(\eta)] = 0$$

$$f(0)=f'(0)=0 \quad ; \quad f''(0)=\alpha \quad ; \quad f'(\infty)=1 \tag{45}$$

$$g(0)=g'(0)=0 \quad ; \quad g''(0)=\gamma \quad ; \quad g'(\infty)=1$$

$$\beta = \frac{2(n+1)}{n+3} \quad ; \quad \lambda = \frac{1}{4\pi\nu\sigma\mu}$$

If $\beta = 0$, equations in (45) are reduced to the magneto-hydrodynamic Blasius problem. When $\beta = 1$ and $0 < \beta < 2$ they represent the flow near the front stagnation point and the flow past a wedge placed symmetrically in a stream. In magneto-hydrodynamic boundary layers we have $S < 1$ and $\lambda >> 1$.

If linear and nonlinear terms in (45) are separated, the sequence of approximations using the Laplace transform method is

$$F(s)=L\{f_{n+1,n}(\eta)\} \;;\; G(s)=L\{\theta_{n+1,n}(\eta)\} \;;\; n=1,2,3,...$$
$$L\{f'''_{n,n+1}(\eta)+\beta(1-S)\}=L\{\beta[f'_{n,n-1}(\eta)]^2-f_{n,n-1}(\eta)f''_{n,n-1}(\eta)+$$
$$S[g_{n,n-1}g''_{n,n-1}(\eta)-\beta[g'_{n,n-1}(\eta)]^2)]\} \tag{46}$$
$$L\{g'''_{n,n+1}(\eta)\}+\frac{1}{\lambda}L\{f''_{n,n-1}(\eta)g_{n,n-1}-f_{n,n-1}g''_{n,n-1}(\eta)\}$$
$$f_{1,0}(\eta)=0 \;;\; g_{1,0}(\eta)=0$$

The Laplace transforms and their inverses of the linear terms in (46) are

$$F(s)=\frac{\alpha}{s^3}-\frac{\beta((1-S)}{s} \;;\; f_{2,1}(\eta)=\frac{\alpha}{2!}\eta^2-\frac{\beta((1-S)}{6}\eta^3$$
$$G(s)=\frac{\gamma}{s^3} \;;\; g_{2,1}(\eta)=\frac{\gamma}{2!}\eta^2 \tag{47}$$

and introducing (47) into the nonlinear part of (46), the second approximation is given by

$$f_{3,2}(\eta)=\frac{\alpha}{2!}\eta^2-\frac{\beta((1-S)}{6}\eta^3+\frac{(\alpha^2-S\gamma^2)(\beta-\frac{1}{2})}{60}\eta^5-$$
$$\frac{\alpha\beta(1-S)(\beta-\frac{2}{3})}{120}\eta^6+\frac{\beta^2(1-S)^2(\beta-\frac{2}{3})}{840}\eta^7 \tag{48}$$
$$g_{3,2}(\eta)=\frac{\gamma}{2!}\eta^2-\frac{\beta\gamma((1-S)}{360\lambda}\eta^6+\frac{\gamma(\alpha^2-S\gamma^2)(\beta-\frac{1}{2})}{2240\lambda}\eta^8-$$
$$\frac{\alpha\beta\gamma(1-S)(\beta-\frac{2}{3})}{4320\lambda}\eta^9+\frac{\beta^2\gamma(1-S)^2(\beta-\frac{2}{3})}{30240\lambda}\eta^{10}$$

The values of α and γ can be evaluated by applying the procedure similar to that given in the set of equations (18) - (28), and is described in detail by Apelblat (1969) [36]. Values derived by this method agree well with literature results.

Mass Transfer Coupled with Chemical Reaction of the First Order

In chemical engineering operations in two phase systems (absorption, solvent extraction, and distillation), mass transfer is frequently accompanied by a chemical reaction. The case of a fluid being at the boundary with constant concentration c_0 and soluble material of concentration $c(x,y)$ is dispersed by the combined action of diffusion and the first order irreversible chemical reaction is discussed below. In the mathematical description, the problem includes the first order homogenous (in a single phase) chemical reaction or the heterogeneous (at an interface) chemical reaction. The mass transfer problems for a two-dimensional formulation, and for three hydrodynamic situations: the uniform flow when fluid is moving with constant velocity U_0 (plug flow), the semi-infinite Couette flow (velocity with a constant gradient a in the y direction) and fully developed boundary layer flow, are solved by applying the Laplace transform methods Apelblat (1980,1982) [37, 39].

The inverse operations of the Laplace transforms are performed by using the Bromwich contour where contributions to the complex integral (2) come by considering, the poles of $F(s)$ at s_n, the branch point at the origin, and two lines (cuts) McLachlan (1962, 1963) [2,3]

$$f(t)=L^{-1}\{F(s)\}=\sum_{n}\mathrm{Res}\left[e^{s_n t}F(s_n)\right]+$$

$$\frac{1}{2\pi}\lim_{r\to 0}\left[\int_{-\pi}^{\pi}e^{tre^{i\theta}}F(re^{i\theta})re^{i\theta}\,d\theta\right]+\frac{1}{2\pi i}\int_{0}^{\infty}e^{-t\xi}\left[F(\xi e^{-\pi i})-F(\xi e^{\pi i})\right]d\xi \tag{49}$$

Mass transfer from boundaries into an adjoining steady, laminar flow, coupled with chemical reaction of the first order (diffusion is neglected in the direction of the convective transport) is governed by the following partial differential equation

$$(U_0+ay)\frac{\partial c}{\partial x}=D\frac{\partial^2 c}{\partial y^2}-kc$$

$$c=0\quad;\quad x=0\quad;\quad y>0 \tag{50}$$

$$c=c_0\quad;\quad x\geq 0\quad;\quad y=0$$

It is assumed that concentration $c(x,y)$ is finite for $x > 0$ and $y \to \infty$. D denotes the diffusion coefficient and k is the rate constant of the chemical reaction for the considered component. In the case of a surface reaction, equation in (50) takes the following form

$$(U_0 + ay)\frac{\partial c}{\partial x} = D\frac{\partial^2 c}{\partial y^2}$$

$$c = c_0 \quad ; \quad x = 0 \quad ; \quad y \geq 0 \qquad (51)$$

$$D\frac{\partial c}{\partial y} = k_s c \quad ; \quad x > 0 \quad ; \quad y = 0$$

Applying the Laplace transform method, and inverting the Laplace transforms with the help of (49) (for details see Apelblat (1980) [37] and Apelblat (2012) [53]), the derived solutions of the partial differential equations in (50) and (51) are presented in Table 1. They include the following cases: $a = 0$ and $U_0 \neq 0$ (plug flow); $a \neq 0$ and $U_0 = 0$ (Couette flow), and $a \neq 0$ and $U_0 \neq 0$ (Couette flow with moving interface); with and without chemical reactions. The last case is called the Lévéque problem.

Table 1. Normalized distribution of concentrations in various hydrodynamic situations with and without a chemical reaction.

Solutions of partial differential equations in (50) and (51).
$\frac{c(X,Y)}{c_0} = 1 - erf\left(\frac{Y}{2\sqrt{X}}\right)$ $X = \frac{x}{U_0} \;;\; Y = \frac{y}{\sqrt{D}} \;;\; U_0 \neq 0 \;;\; a = 0 \;;\; k = 0$
$\frac{c(X,Y)}{c_0} = \frac{1}{2}\left[e^{-\sqrt{k}Y} erfc\left(\frac{Y}{2\sqrt{X}} - \sqrt{kX}\right) + e^{\sqrt{k}Y} erfc\left(\frac{Y}{2\sqrt{X}} + \sqrt{kX}\right)\right]$ $X = \frac{x}{U_0} \;;\; Y = \frac{y}{\sqrt{D}} \;;\; U_0 \neq 0 \;;\; a = 0 \;;\; k \neq 0$
$\frac{c(X,Y)}{c_0} = e^{-\beta Y + \beta^2 X} erfc\left(\frac{Y}{2\sqrt{X}} + \beta\sqrt{X}\right) + erf\left(\frac{Y}{2\sqrt{X}}\right) \;;\; \beta = \frac{k_s}{\sqrt{D}}$ $X = \frac{x}{U_0} \;;\; Y = \frac{y}{\sqrt{D}} \;;\; U_0 \neq 0 \;;\; a = 0 \;;\; k_s \neq 0$

Solutions of partial differential equations in (50) and (51).
$\frac{c(X,Y)}{c_0}=1-\frac{3^{1/6}\Gamma\left(\frac{2}{3}\right)\sqrt{Y}}{2\pi}\int_0^\infty \frac{e^{-\xi X}}{\xi^{5/6}}J_{1/3}(\phi)\,d\xi \;;\; \phi=\frac{2Y^{3/2}\sqrt{\xi}}{3}$ $X=\frac{x}{a\sqrt{D}} \;;\; Y=\frac{y}{\sqrt{D}} \;;\; U_0=0 \;;\; a\neq 0 \;;\; k=0$
$\frac{c(X,Y)}{c_0}=e^{-Y}-\frac{1}{\pi}\int_0^\infty \frac{e^{-\xi X}}{\xi}G(\phi,\phi_0)\,d\xi \;;\; \phi=\frac{1-\xi Y}{\xi^{2/3}} \;;\; \phi_0=\frac{1}{\xi^{2/3}}$ $G(\phi,\phi_0)=\left[\frac{Ai(\phi)Bi(\phi_0)-Bi(\phi)Ai(\phi_0)}{[Ai(\phi_0)]^2+[Bi(\phi_0)]^2}\right]$ $X=\frac{k^{3/2}x}{a\sqrt{D}} \;;\; Y=y\sqrt{\frac{k}{D}} \;;\; U_0=0 \;;\; a\neq 0 \;;\; k\neq 0$
$\frac{c(X,Y)}{c_0}=\frac{3^{5/3}\Gamma\left(\frac{2}{3}\right)}{4\pi}\int_0^\infty \frac{e^{-\xi X}}{\xi}G(\phi,\xi)\,d\xi \;;\; \phi=\xi Y$ $G(\phi,u)=\left[\frac{\sqrt{3}Ai(-\phi)(1+r\xi)+Bi(-\phi)(-1+r\xi)}{1+r\xi+(r\xi)^2}\right] \;;\; r=\frac{3^{1/3}\Gamma\left(\frac{2}{3}\right)\sqrt{D}}{\Gamma\left(\frac{1}{3}\right)k_s}$ $X=\frac{x}{a\sqrt{D}} \;;\; Y=\frac{y}{\sqrt{D}} \;;\; U_0=0 \;;\; a\neq 0 \;;\; k_s\neq 0$
$\frac{c(X,Y)}{c_0}=1-\frac{1}{\pi}\int_0^\infty \frac{e^{-\xi X}}{\xi}G(\phi,\phi_0)\,d\xi \;;\; \phi=\xi^{1/3}(1+Y) \;;\; \phi_0=\xi^{1/3}$ $G(\phi,\phi_0)=\left[\frac{Ai(-\phi)Bi(-\phi_0)-Bi(-\phi)Ai(-\phi_0)}{[Ai(-\phi_0)]^2+[Bi(-\phi_0)]^2}\right]$ $X=\frac{Da^2}{U_0^3}x \;;\; Y=\frac{a}{U_0}y \;;\; U_0\neq 0 \;;\; a\neq 0 \;;\; k=0$

Table 1. (Continued)

Solutions of partial differential equations in (50) and (51).
$\frac{c(X,Y)}{c_0} = e^{-\sqrt{\beta}Y} - \frac{1}{\pi}\int_0^{\infty} \frac{e^{-\xi X}}{\xi} G(\phi,\phi_0)\, d\xi$; $\phi = \frac{\beta - \xi(1+Y)}{\xi^{2/3}}$; $\phi_0 = \frac{\beta - \xi}{\xi^{2/3}}$ $G(\phi,\phi_0) = \left[\frac{Ai(\phi)Bi(\phi_0) - Bi(\phi)Ai(\phi_0)}{[Ai(\phi_0)]^2 + [Bi(\phi_0)]^2}\right]$; $\beta = \frac{k}{D}\left(\frac{U_0}{a}\right)^2$ $X = \frac{Da^2}{U_0^3}x$; $Y = \frac{a}{U_0}y$; $U_0 \neq 0$; $a \neq 0$; $k \neq 0$
$\frac{c(X,Y)}{c_0} = \frac{\beta}{\pi}\int_0^{\infty} \frac{e^{-\xi X}}{\xi}\left[\frac{\xi^{1/3}G(\phi,\phi_0) + \beta H(\phi,\phi_0)}{M(\phi_0)}\right] d\xi$; $\beta = \frac{k_s U_0}{Da}$ $G(\phi,\phi_0) = [Ai(\phi)Bi'(\phi_0) - Bi(\phi)Ai'(\phi_0)]$; $\phi = -\xi^{1/3}(1+Y)$; $\phi_0 = -\xi^{1/3}$ $H(\phi,\phi_0) = [Ai(\phi)Bi(\phi_0) - Bi(\phi)Ai(\phi_0)]$ $M(\phi_0) = [\xi^{1/3}Ai'(\phi_0) + \beta Ai(\phi)]^2 + [\xi^{1/3}Bi'(\phi_0) + \beta Bi(\phi)]^2$ $X = \frac{Da^2}{U_0^3}x$; $Y = \frac{a}{U_0}y$; $U_0 \neq 0$; $a \neq 0$; $k_s \neq 0$

The solutions in Table 1 are valid for the corresponding heat transfer problems where concentrations $c(x,y)$ can be replaced by temperatures $T(x,y)$. In order to simplify notation, all variables in the solutions of the partial differential equations in (50) and (51) are dimensionless and are expressed in terms of the error functions *erf*(z), *erfc*(z), the Bessel functions of the first kind $J_\nu(z)$ and the Airy functions $Ai(z)$ and $Bi(z)$ which are defined in Abramowitz and Stegun (1965) [63].

From a practical point of view, a very important case is the mass transfer in a boundary layer with a surface reaction, the so-called the Lighthill-Levich problem. Mathematically this is a most complex problem. The convective diffusion equation and the boundary conditions of this case are governed by

$$u\frac{\partial c}{\partial x} + \text{v}\frac{\partial c}{\partial y} = D\frac{\partial^2 c}{\partial y^2}$$
$$c = c_0 \quad ; \quad x = 0 \quad ; \quad y \geq 0$$
$$c = c_0 \quad ; \quad x > 0 \quad ; \quad y \to \infty \tag{52}$$
$$D\frac{\partial c}{\partial y} = k_s c \quad ; \quad x > 0 \quad ; \quad y = 0$$

Using the first term from the Blasius series (6), and after several steps and transforms (for details see Apelblat (1980) [37]), it is possible to present (52) in the following form

$$Y\frac{\partial C}{\partial X} = \frac{\partial^2 C}{\partial Y^2} \quad ; \quad C(X,Y) = c(x,y) - c_0$$
$$C = 0 \quad ; \quad X = 0 \quad ; \quad Y \geq 0$$
$$C = 0 \quad ; \quad X > 0 \quad ; \quad Y \to \infty \tag{53}$$
$$\frac{\partial C}{\partial Y} = \beta X^{1/3} C(X,Y) + \beta X^{1/3} c_0 \quad ; \quad X > 0 \quad ; \quad Y = 0$$

$$X = \frac{4D\sqrt{2\lambda}}{3} x^{3/4} \quad ; \quad Y = 2^{1/6}\sqrt{\lambda}\left(\frac{y}{x^{1/4}}\right) \quad ; \quad \lambda = \frac{\alpha}{4}\sqrt{\frac{U_\infty^3}{\nu}}$$

Applying the Laplace transform to the partial differential equation in (53)

$$F(s,Y) = L\{C(X,Y)\} = \int_0^\infty e^{-sX} C(X,Y)\,dX \tag{54}$$

and solving the ordinary differential equation we have that

$$F(s,Y) = A(s)Ai(s^{1/3}Y) + B(s)Bi(s^{1/3}Y) \tag{55}$$

where $A(s)$ and $B(s)$ are the integration constants which are unknown functions of the variable s. However, $B(s) = 0$, because the Airy function $Bi(s^{1/3}Y)$ is not finite for $Y \to \infty$, and $F(s,Y)$ can be written as the product

$$F(s,Y) = A(s)Ai(s^{1/3}Y) = G(s)\cdot H(s,Y)$$
$$G(s) = A(s) \quad ; \quad H(s,Y) = Ai(s^{1/3}Y) \tag{56}$$

Therefore, by using the convolution theorem we have

$$g(X) = L^{-1}\{G(s)\} \quad ; \quad h(X,Y) = L^{-1}\left\{Ai(s^{1/3}Y)\right\}$$

$$C(X,Y) = \int_0^X g(\xi)h(X-\xi)d\xi \tag{57}$$

The function $h(X,Y)$ can be derived from the corresponding solution of the Lévéque problem (see Table 1)

$$h(X,Y) = \frac{\sqrt{Y}}{2\sqrt{3}\pi}\int_0^\infty \xi^{1/6} e^{-\xi X} J_{1/3}(\phi)d\xi \quad ; \quad \phi = \frac{2Y^{3/2}\sqrt{\xi}}{3}$$

$$C(X,Y) = \int_0^X g(\zeta)\left[\frac{\sqrt{Y}}{2\sqrt{3}\pi}\int_0^\infty \xi^{1/6} e^{-\xi(X-\zeta)} J_{1/3}(\phi)d\xi\right]d\zeta \tag{58}$$

and the function $g(X)$ comes from the boundary condition at Y = 0 in (53).

$$\frac{d}{dY}\left[\sqrt{Y}J_{1/3}\left(\frac{2Y^{3/2}\sqrt{\xi}}{3}\right)\right]_{Y=0} = \frac{3^{2/3}\xi^{1/6}}{\Gamma\left(\frac{1}{3}\right)} \tag{59}$$

$$h(X,0) = 0$$

Using (58) and (59), the following integral equation is derived

$$\int_0^X \frac{g(\zeta)}{(X-\zeta)^{4/3}}d\zeta = 3^{5/6}2\pi\beta c_0 \tag{60}$$

and its solution is

$$g(X) = -\frac{\pi\Gamma\left(\frac{1}{3}\right)\beta c_0 X^{2/3}}{3^{1/6}\left[\Gamma\left(\frac{2}{3}\right)\right]^2} \tag{61}$$

Finally, from (58) and (61), the normalized distribution of concentrations in a laminar boundary layer flow with a surface chemical reaction of the first order (the Lighthill-Levich problem) is given by

$$\frac{C(X,Y)}{c_0} = 1 - \lambda \int_0^X \frac{(X-\zeta)^{2/3}}{\zeta^{4/3}} e^{-Y^3/9\zeta} d\zeta$$

$$\lambda = \left(\frac{2\nu}{9\alpha^2 D}\right)^{1/3} \frac{\Gamma\left(\frac{1}{3}\right)}{\left[\Gamma\left(\frac{2}{3}\right)\right]^2} \left(\frac{k_s}{DU_\infty}\right) \tag{62}$$

where the integral in (62) can be reduced to the Whittaker function or confluent hypergeometric function (for details see Apelblat (1982) [39]).

The normalized concentration distributions, presented in Table 1 for various hydrodynamic situations, coupled with and without a chemical reaction of the first order, were derived by neglecting diffusion in the direction of the convective transport, and the axial diffusion effect is omitted. Such assumptions are often valid for relatively long absorption channels. However, there are situations where the estimated mass transfer or corresponding heat transfer are evaluated with a significant error Apelblat (1982)[39]. If an additional term of the axial diffusion is included, then the mathematical formulation of the mass transfer problems is similar to the ones presented in equations (50) and (51). In this case, the partial differential equations can also be solved in closed form by applying the Laplace transform method with the inversion operations performed with the help of contributions given in (49).

The mass conservation equations which express the dissolution and movement of solid material in the flow (diffusion in both the x and y directions) and in the presence of a chemical reaction of the first order, take the following form

$$(U_0 + ay)\frac{\partial c}{\partial x} = D\left(\frac{\partial^2 c}{\partial x^2} + \frac{\partial^2 c}{\partial y^2}\right) - kc$$

$$c = 0 \quad ; \quad x = 0 \quad ; \quad y > 0$$

$$\frac{\partial c}{\partial x} = 0 \quad ; \quad x = 0 \quad ; \quad y > 0 \tag{63}$$

$$c = c_0 \quad ; \quad x > 0 \quad ; \quad y = 0$$

$$c(x,y) \quad \text{is finite for } x \geq 0 \quad ; \quad y \to \infty$$

and in the case of a chemical surface reaction of the first order

$$
\begin{aligned}
&(U_0 + ay)\frac{\partial c}{\partial x} = D\left(\frac{\partial^2 c}{\partial x^2} + \frac{\partial^2 c}{\partial y^2}\right) \\
&c = c_0 \quad ; \quad x = 0 \quad ; \quad y \geq 0 \\
&\frac{\partial c}{\partial x} = 0 \quad ; \quad x = 0 \quad ; \quad y \geq 0 \\
&D\frac{\partial c}{\partial y} = k_s c \quad ; \quad x > 0 \quad ; \quad y = 0 \\
&c = c_0 \quad ; \quad x > 0 \quad ; \quad y \to \infty
\end{aligned} \tag{64}
$$

Solutions of the partial differential equations subject to the boundary conditions (63) and (64) for various hydrodynamic situations, with and without chemical reactions are presented in Table 2. The case of simultaneous homogenous and heterogeneous chemical reaction of the first order in uniform (plug) flow is also included.

Table 2. Normalized distribution of concentrations in various hydrodynamic situations, with and without a chemical reaction, and with included the axial diffusion effect.

Solutions of partial differential equations in (63) and (64).
$\frac{c(X,Y)}{c_0} = 1 - \frac{1}{\pi}\int_0^\infty \frac{e^{-\xi X}}{\xi} \sin\left[Y\sqrt{\xi(1+\xi)}\right] d\xi$ $X = \frac{U_0}{D}x \quad ; \quad Y = \frac{U_0}{D}y \quad ; \quad U_0 \neq 0 \quad ; \quad a = 0 \quad ; \quad k = 0$
$\frac{c(X,Y)}{c_0} = e^{-\sqrt{\beta}Y} - \frac{1}{\pi}\int_0^\infty \frac{e^{-\xi X}}{\xi} \sin\left[Y\sqrt{\xi^2 + \xi - \beta}\right] d\xi \quad ; \quad \beta = \frac{kD}{U_0^2}$ $X = \frac{U_0}{D}x \quad ; \quad Y = \frac{U_0}{D}y \quad ; \quad U_0 \neq 0 \quad ; \quad a = 0 \quad ; \quad k \neq 0$

Solutions of partial differential equations in (63) and (64).
$\frac{c(X,Y)}{c_0}=\frac{\beta_s}{\pi}\int_0^{\infty}\frac{e^{-\xi X}}{\xi}\left[\frac{\beta_s \sin(Y\phi)+\phi\cos(Y\phi)}{\xi^2+\xi+\beta_s^2}\right]d\xi$ $\phi=\sqrt{\xi(1+\xi)}\ ;\ \beta_s=\frac{k_s}{U_0}$ $X=\frac{U_0}{D}x\ ;\ Y=\frac{U_0}{D}y\ ;\ U_0\neq 0\ ;\ a=0\ ;\ k_s\neq 0$
$\frac{c(X,Y)}{c_0}=1-\frac{1}{\pi}\int_0^{\infty}\frac{e^{-\xi X}}{\xi}G(\phi,\phi_0)d\xi\ ;\ \phi=\xi^{1/3}(Y+\xi)\ ;\ \phi_0=\xi^{4/3}$ $G(\phi,\phi_0)=\left[\frac{Ai(-\phi)Bi(-\phi_0)-Bi(-\phi)Ai(-\phi_0)}{[Ai(-\phi_0)]^2+[Bi(-\phi_0)]^2}\right]$ $X=\sqrt{\frac{a}{D}}x\ ;\ Y=\sqrt{\frac{a}{D}}y\ ;\ U_0=0\ ;\ a\neq 0\ ;\ k=0$
$\frac{c(X,Y)}{c_0}=e^{-\sqrt{\beta}Y}-\frac{1}{\pi}\int_0^{\infty}\frac{e^{-\xi X}}{\xi}G(\phi,\phi_0)d\xi\ ;\ \beta=\frac{k}{a}$ $\phi=\xi^{1/3}\left(Y+\xi+\frac{\beta}{\xi}\right)\ ;\ \phi_0=\xi^{1/3}\left(\xi+\frac{\beta}{\xi}\right)$ $G(\phi,\phi_0)=\left[\frac{Ai(-\phi)Bi(-\phi_0)-Bi(-\phi)Ai(-\phi_0)}{[Ai(-\phi_0)]^2+[Bi(-\phi_0)]^2}\right]$ $X=\sqrt{\frac{a}{D}}x\ ;\ Y=\sqrt{\frac{a}{D}}y\ ;\ U_0=0\ ;\ a\neq 0\ ;\ k\neq 0$

Table 2. (Continued)

Solutions of partial differential equations in (63) and (64).
$\frac{c(X,Y)}{c_0}=\frac{\beta_s}{\pi}\int_0^\infty \frac{e^{-\xi X}}{\xi}G(\phi,\phi_0)\,d\xi \quad ; \quad \phi=\xi^{1/3}(1+Y) \quad ; \quad \phi_0=\xi^{4/3}$ $G(\phi,\phi_0)=\left[\frac{Ai(-\phi)M_2(-\phi_0)-Bi(-\phi)M_1(-\phi_0)}{[M_1(-\phi_0)]^2+[M_2(-\phi_0)]^2}\right] \quad ; \quad \beta_s=\frac{k_s}{\sqrt{aD}}$ $M_1(-\phi_0)=[\xi^{1/3}Ai'(-\phi_0)+\beta_s Ai(-\phi_0)]$ $M_2(-\phi_0)=[\xi^{1/3}Bi'(-\phi_0)+\beta_s Bi(-\phi_0)]$ $X=\sqrt{\frac{a}{D}}x \quad ; \quad Y=\sqrt{\frac{a}{D}}y \quad ; \quad U_0=0 \quad ; \quad a\neq 0 \quad ; \quad k_s\neq 0$
$\frac{c(X,Y)}{c_0}=1-\frac{1}{\pi}\int_0^\infty \frac{e^{-\xi X}}{\xi}G(\phi,\phi_0)\,d\xi \quad ; \quad \beta=\frac{aD}{U_0^2}$ $\phi=\left(\frac{\xi}{\beta}\right)^{1/3}(1+Y+\beta\xi) \quad ; \quad \phi_0=\left(\frac{\xi}{\beta}\right)^{1/3}(1+\beta\xi)$ $G(\phi,\phi_0)=\left[\frac{Ai(-\phi)Bi(-\phi_0)-Bi(-\phi)Ai(-\phi_0)}{[Ai(-\phi_0)]^2+[Bi(-\phi_0)]^2}\right]$ $X=\frac{a}{U_0}x \quad ; \quad Y=\frac{a}{U_0}y \quad ; \quad U_0\neq 0 \quad ; \quad a\neq 0 \quad ; \quad k=0$
$\frac{c(X,Y)}{c_0}=e^{-\sqrt{\lambda/\beta}\,Y}-\frac{1}{\pi}\int_0^\infty \frac{e^{-\xi X}}{\xi}G(\phi,\phi_0)\,d\xi \quad ; \quad \beta=\frac{aD}{U_0^2} \quad ; \quad \lambda=\frac{k}{a}$ $\phi=\left(\frac{\xi}{\beta}\right)^{1/3}[1+Y+\beta-\frac{\lambda}{\xi}] \quad ; \quad \phi_0=\left(\frac{\xi}{\beta}\right)^{1/3}[1+\beta-\frac{\lambda}{\xi}]$ $G(\phi,\phi_0)=\left[\frac{Ai(-\phi)Bi(-\phi_0)-Bi(-\phi)Ai(-\phi_0)}{[Ai(-\phi_0)]^2+[Bi(-\phi_0)]^2}\right]$ $X=\frac{a}{U_0}x \quad ; \quad Y=\frac{a}{U_0}y \quad ; \quad U_0\neq 0 \quad ; \quad a\neq 0 \quad ; \quad k\neq 0$

Solutions of partial differential equations in (63) and (64).
$\frac{c(X,Y)}{c_0} = \frac{\lambda}{\pi}\int_0^\infty \frac{e^{-\xi X}}{\xi} G(\phi,\phi_0)d\xi$; $\phi = \xi^{1/3}(1+Y-\beta\xi)$; $\phi_0 = \xi^{1/3}(1-\beta\xi)$ $G(\phi,\phi_0) = \left[\frac{Ai(-\phi)M_2(-\phi_0) - Bi(-\phi)M_1(-\phi_0)}{[M_1(-\phi_0)]^2 + [M_2(-\phi_0)]^2}\right]$; $\beta = \frac{aD}{U_0^2}$; $\lambda = \beta^{1/3}\frac{k_s U_0}{aD}$ $M_1(-\phi_0) = [\xi^{1/3}Ai'(-\phi_0) + \lambda Ai(-\phi_0)]$ $M_2(-\phi_0) = [\xi^{1/3}Bi'(-\phi_0) + \lambda Bi(-\phi_0)]$ $X = \frac{a}{U_0}x$; $Y = \frac{a}{U_0}y$; $U_0 \neq 0$; $a \neq 0$; $k_s \neq 0$
$\frac{c(X,Y)}{c_0} = \frac{\lambda}{\pi}\int_\zeta^\infty e^{-\xi X}(1+\xi)\left[\frac{\lambda\sin(Y\sqrt{\phi}) + \sqrt{\phi}\cos(Y\sqrt{\phi})}{\phi(\phi+\lambda^2)}\right]d\xi$ $\phi = \xi^2 + \xi - \beta$; $\zeta = \frac{-1+\sqrt{1+4\beta}}{2}$; $\beta = \frac{kD}{U_0^2}$; $\lambda = \frac{k_s}{U_0}$ $X = \frac{U_0}{D}x$; $Y = \frac{U_0}{D}y$; $U_0 \neq 0$; $a = 0$; $k \neq 0$; $k_s \neq 0$

For the case of simultaneous homogenous and heterogeneous chemical reactions in uniform flow (the last case in Table 2), if the axial diffusion term is neglected, the solution has nearly the same form Apelblat (1982) [39]

$$\frac{c(X,Y)}{c_0} = \frac{\lambda}{\pi}\int_\beta^\infty e^{-\xi X}\left[\frac{\lambda\sin(Y\sqrt{\phi}) + \sqrt{\phi}\cos(Y\sqrt{\phi})}{\phi\,(\phi+\lambda^2)}\right]d\xi; \phi = \xi - \beta$$
$$\beta = \frac{kD}{U_0^2} \;;\; \lambda = \frac{k_s}{U_0} \;;\; X = \frac{U_0}{D}x \;;\; Y = \frac{U_0}{D}y \qquad (65)$$

3. Application of the Laplace Transform to the Properties of Elementary and Special Functions.

The evaluation of finite, infinite and convolution integrals of elementary and special functions by applying integration and differentiation operations in a combination with the rules and theorems of the operational calculus is a standard technique in the Laplace transform and in other integral transforms [2, 5, 7, 21, 22, 31, 32, 53]. Several methods, developed by the present author to particular groups of integrals, the repeated use of different integral transforms and the Laplace transforms expressed in terms of the logarithmic functions, are presented in Apelblat (1981) [38], Apelblat (1985) [40],

Apelblat and Kravitsky (1988) [42], Apelblat (1994) [45], Apelblat (1998) [46] and Apelblat (2012) [53]. The integrals evaluated by the Laplace transform methods are tabulated in Apelblat (1996) [64]. The Laplace transform is also very useful in the derivations and the operations associated with the properties of special functions. This is illustrated in the case of Bessel, Kelvin, Struve, Volterra, Whittaker, Mittag-Leffler, Wright, Bateman and Havelock functions [41, 43, 44, 54-60].

Evaluation of Logarithmic Integrals Using Direct Laplace Transforms.

If function $f(t)$ possess the Laplace transform $F(s)$, then by the changing the integration variable we have

$$L\{f(t)\} = \int_0^\infty e^{-st} f(t)dt = F(s) \quad ; \quad x = e^{-st}$$
$$\int_0^1 f\left(-\frac{1}{s}\ln x\right)dx = \int_0^1 f\left[\frac{1}{s}\ln\left(\frac{1}{x}\right)\right]dx = sF(s) \tag{66}$$

and if it is in the forms $f(t) = u(t-a)h(t)$ or $f(t) = [u(t)-u(t-a)]\ h(t)$, where $u(t)$ and $u(t-a)$ are the step and shifted step functions, then

$$\int_0^a h\left(-\frac{1}{s}\ln x\right)dx = sF(s) \quad ; \quad a > 0$$
$$\int_a^1 h\left(-\frac{1}{s}\ln x\right)dx = sF(s) \tag{67}$$

Since extensive tables of the function-transform pairs are available in the literature [1, 65-67], many logarithmic and other integrals can be derived from (66) and combining with operational rules of the Laplace transform. The complete list of equivalent logarithmic forms of these rules is given by Apelblat (2012) [53]. Two simple examples will illustrate the procedure, but many other problems of much more complex integrals are presented by Apelblat (1994, 1998) [45, 46]. From the following transform-inverse pairs

$$L\left\{t^{\nu}\right\}=\frac{\Gamma(\nu+1)}{s^{\nu+1}} \quad ; \quad \mathrm{Re}\,\nu>-1 \quad ; \quad \mathrm{Re}\,s>0$$

$$L\begin{Bmatrix} 0 & ; & 0<t<a \\ t^{\nu} & ; & t>a \end{Bmatrix}=\frac{\Gamma(\nu+1, as)}{s^{\nu+1}}$$

$$L\begin{Bmatrix} t^{\nu} & ; & 0<t<a \\ 0 & ; & t>a \end{Bmatrix}=\frac{\gamma(\nu+1, as)}{s^{\nu+1}}$$

(68)

we have by using (66) and (67)

$$\int_0^1\left[\ln\left(\frac{1}{x}\right)\right]^{\nu} dx=\Gamma(\nu+1) \quad ; \quad \mathrm{Re}\,\nu>-1$$

$$\int_0^a\left[\ln\left(\frac{1}{x}\right)\right]^{\nu} dx=\Gamma(\nu+1, \lambda) \quad ; \quad \lambda=\ln\left(\frac{1}{a}\right) \tag{69}$$

$$\int_a^1\left[\ln\left(\frac{1}{x}\right)\right]^{\nu} dx=\gamma(\nu+1, \lambda) \quad ; \quad 0<a<1$$

where $\Gamma(a,z)$ and $\gamma(a,z)$ are the incomplete gamma functions Abramowitz and Stegun (1965) [63].

In the case of trigonometric functions, the Laplace transforms are

$$L\{\sin(at)\}=\frac{a}{s^2+a^2}$$

$$L\{\cos(at)\}=\frac{s}{s^2+a^2} \tag{70}$$

and by denoting $\lambda = a/s$, it follows from (66) that

$$\int_0^1 \sin(\lambda \ln x)\, dx=-\frac{\lambda}{1+\lambda^2}$$

$$\int_0^1 \cos(\lambda \ln x)\, dx=\frac{1}{1+\lambda^2} \tag{71}$$

A number of integrals can be enlarged by applying various mathematical operations. For example, by a direct differentiation with respect to the parameter λ of integrals in (71) we have

$$\int_0^1 \ln x \cos(\lambda \ln x)\,dx = \frac{2\lambda}{(1+\lambda^2)^2}$$
$$\int_0^1 \ln x \sin(\lambda \ln x)\,dx = \frac{1-\lambda^2}{(1+\lambda^2)^2} \quad (72)$$

Closely related to the Laplace transform are the Mellin, Stieltjes and generalized Stieltjes transforms and they also provide by a change of integration variable many logarithmic integrals (for details see Apelblat (2012) [53]).

Integrals Associated with the Inverse Laplace Transforms of Logarithmic Functions.

If as a result of performing a complex integration by using (2), the original function $f(t)$ can be presented in the form Apelblat and Kravitsky (1988) [42]

$$f(t) = L^{-1}\{F(s)\} = \int_0^\infty g(t,\xi)\,d\xi + h(t) \quad (73)$$

then it is possible to derive the following infinite integral from the knowledge of the Laplace transforms of $g(t,\xi)$ and $h(t)$ functions

$$\int_0^\infty G(s,\xi)\,d\xi = F(s) - H(s)$$
$$G(s,\xi) = L\{g(t,\xi)\} \quad ; \quad H(s) = L\{h(t)\} \quad (74)$$

The functional form of $g(t,\xi)$ and $h(t)$ functions depends on the chosen contour in the complex integration and therefore it permits obtaining a number of different integrals using the same Laplace transform $F(s)$. The procedure presented in (73) and (74) is illustrated by the inverse transforms of the following logarithmic functions

$$f(t) = L^{-1}\{F(s)\} = \frac{1}{2\pi i}\int_{c-i\infty}^{c+i\infty} \frac{e^{st}}{\ln(s^2+a^2)} ds \quad ; \quad t > 0 \quad ; \quad a \geq 0$$

$$f(t) = L^{-1}\{F(s)\} = \frac{1}{2\pi i}\int_{c-i\infty}^{c+i\infty} \frac{e^{st}}{[\ln(s^2+a^2)]^2} ds$$

(75)

In the inversion of the Laplace transforms in (75), the locations of branch points and removable singularities must be considered, and that four different Bromwich contours exist for various values of the parameter a ($a > 0$, $a = 1$, $0 < a < 1$ and $a = 0$, for details see Apelblat and Kravitsky (1988) [42]). In the case of $0 < a < 1$, the final results of complex integration of (75) are

$$f(t) = 2\int_0^\infty \frac{\sin(t\xi)}{\pi^2 + [\ln(\xi^2 - a^2)]^2} d\xi + \frac{\sinh(t\sqrt{1-a^2})}{\sqrt{1-a^2}} \quad ; \quad 0 < a < 1$$

$$f(t) = 4\int_0^\infty \frac{\sin(t\xi)}{\left[\pi^2 + [\ln(\xi^2 - a^2)]^2\right]^2} d\xi + \frac{(1-2a^2)\sinh(t\sqrt{1-a^2})}{2(1-a^2)^{3/2}} \tag{76}$$

$$\frac{t\cosh(t\sqrt{1-a^2})}{2(1-a^2)}$$

The Laplace transforms of $f(t)$ in (76) have the form of (73). By changing the variable of integration and by denoting $s^2 + a^2 = \alpha$, it follows from (74) that

$$\int_0^\infty \frac{d\xi}{(\alpha+\xi)[\pi^2 + (\ln\xi)^2]} = \frac{1}{\ln\alpha} - \frac{1}{\alpha - 1} \quad ; \quad \alpha \neq 1$$

$$\int_0^\infty \frac{\ln\xi}{(\alpha+\xi)[\pi^2 + (\ln\xi)^2]^2} d\xi = \frac{1}{2}\left[\frac{1}{(\ln\alpha)^2} - \frac{\alpha}{(\alpha-1)^2}\right] \tag{77}$$

Using simple mathematical operations, and repeated differentiation of the integrands with respect to the parameter α in (77), it is possible to derive many infinite logarithmic integrals Apelblat and Kravitsky (1988) [42], Apelblat (1996) [64]. For example, the first differentiation of integral in (77) gives

$$\int_0^\infty \frac{d\xi}{(\alpha+\xi)^2[\pi^2+(\ln\xi)^2]} = \frac{1}{\alpha(\ln\alpha)^2} - \frac{1}{(\alpha-1)^2} \quad ; \quad \alpha \neq 1$$

$$\int_0^\infty \frac{\ln\xi}{(\alpha+\xi)^2[\pi^2+(\ln\xi)^2]^2} d\xi = \frac{1}{\alpha(\ln\alpha)^3} - \frac{\alpha+1}{2(\alpha-1)^3} \tag{78}$$

The integral in (77) with $\alpha = 0$, is especially important, because it is associated with the representation of the Volterra function $\nu(t)$

$$\nu(t) = e^t - \int_0^\infty \frac{e^{-t\xi}}{\xi[\pi^2+(\ln\xi)^2]} d\xi \quad ; \quad t > 0$$

$$\nu(t) = \int_0^\infty \frac{t^\xi}{\Gamma(\xi+1)} d\xi \quad ; \quad L\{\nu(t)\} = \frac{1}{s\ln s} \quad ; \quad \operatorname{Re} s > 1 \tag{79}$$

This result was derived by Ramanujan [51, 52, 68, 69].

If the Laplace transforms $G(s) = L\{g(t)\}$ and $R(s) = L\{\nu(t)g(t)\}$ are known, similarly as in (74), we have

$$\int_0^\infty \frac{G(s+\xi)}{\xi[\pi^2+(\ln\xi)^2]} d\xi = G(s-1) - R(s) \tag{80}$$

For example by introducing $g(t) = \sinh(at)$, it is possible after a few steps obtain the following infinite integral Apelblat and Kravitsky (1988) [42]

$$\int_0^\infty \frac{d\xi}{(\alpha+\xi)(\beta+\xi)[\pi^2+(\ln\xi)^2]} = \frac{1}{\alpha\beta-\alpha-\beta+1} - \frac{1}{\alpha-\beta}\left[\frac{1}{\beta\ln\beta} - \frac{1}{\alpha\ln\alpha}\right] \tag{81}$$

$\alpha \neq \beta$

A simple form of the Laplace transform, which is presented in (79), permits performing various integrations of the Volterra and other special functions Apelblat (2008,2009) [51, 52].

Integrals of Gamma, Polygamma and Volterra Functions.

Indefinite, finite, infinite and convolution integrals of the gamma, polygamma (psi) and the Volterra functions can be evaluated by applying the substitution formulas of the Laplace transform Humbert and Colombo (1965) [23]. The Volterra functions are defined by the improper integrals and their Laplace transforms are expressed in terms of logarithmic functions of variable s [23, 51, 52, 69]

$$\nu(t)=\int_0^\infty \frac{t^\xi}{\Gamma(\xi+1)}d\xi \quad ; \quad L\{\nu(t)\}=\frac{1}{s\ln s} \quad ; \quad \mathrm{Re}\, s>1$$

$$\nu(t,\alpha)=\int_0^\infty \frac{t^{\xi+\alpha}}{\Gamma(\xi+\alpha+1)}d\xi \quad ; \quad L\{\nu(t,\alpha)\}=\frac{1}{s^{\alpha+1}\ln s} \quad ; \quad \mathrm{Re}\,\alpha>-1 \tag{82}$$

$$\mu(t,\beta)=\int_0^\infty \frac{t^\xi \xi^\beta}{\Gamma(\beta+1)\Gamma(\xi+1)}d\xi \quad ; \quad L\{\mu(t,\beta)\}=\frac{1}{s(\ln s)^{\beta+1}} \quad ; \quad \mathrm{Re}\,\beta>-1$$

$$\mu(t,\beta,\alpha)=\int_0^\infty \frac{t^{\xi+1}\xi^\beta}{\Gamma(\beta+1)\Gamma(\xi+\alpha+1)}d\xi \quad ; \quad L\{\mu(t,\beta,\alpha)\}=\frac{1}{s^{\alpha+1}(\ln s)^{\beta+1}}$$

Replacing the Laplace variable *s* by ln*s*, the corresponding substitution formulas can be derived

$$L\left\{\int_0^\infty f(\xi)\nu(t,\xi)\,d\xi\right\}=\frac{F(\ln s)}{s\ln s} \quad ; \quad L\{f(t)\}=F(s) \quad ; \quad \mathrm{Re}\, s>1$$

$$L\left\{\int_0^\infty f(\xi)\nu(t,\alpha+\xi)\,d\xi\right\}=\frac{F(\ln s)}{s^{\alpha+1}\ln s} \quad ; \quad \mathrm{Re}\,\alpha>-1 \tag{83}$$

$$L\left\{\int_0^\infty f(\xi)\,\mu(t,\beta,\xi)\,d\xi\right\}=\frac{F(\ln s)}{s(\ln s)^{\beta+1}} \quad ; \quad \mathrm{Re}\,\beta>-1$$

$$L\left\{\int_0^\infty f(\xi)\,\mu(t,\beta,\alpha+\xi)\,d\xi\right\}=\frac{F(\ln s)}{s^{\alpha+1}(\ln s)^{\beta+1}}$$

Using the Laplace transform of the first derivative of the Dirac delta function $\delta'(t)$ and (83), it is possible to obtain infinite integrals of the psi (digamma) function Apelblat (1985) [40]

$$\int_0^\infty \frac{\psi(\alpha+\xi+1)}{\Gamma(\alpha+\xi+1)}d\xi=\frac{1}{\Gamma(\alpha+1)} \quad ; \quad \psi(z)=\frac{\Gamma'(z)}{\Gamma(z)}$$

$$\int_0^\infty \frac{t^{\alpha+\xi}\,\psi(\alpha+\xi+1)}{\Gamma(\alpha+\xi+1)}d\xi=\frac{t^\alpha}{\Gamma(\alpha+1)}+\nu(t,\alpha)\ln t \quad ; \quad t\neq 1 \tag{84}$$

and integrals of the Volterra functions

$$\int_0^\infty \xi^{\beta-1}\nu(t,\alpha+\xi)\,d\xi = \Gamma(\beta)\,\mu(t,\beta,\alpha) \quad ; \quad \mathrm{Re}\,\alpha > -1 \quad ; \quad \mathrm{Re}\,\beta > 0$$

$$\int_0^\infty \xi^{\lambda-1}\mu(t,\beta,\alpha+\xi)\,d\xi = \Gamma(\beta)\,\mu(t,\beta+\lambda,\alpha) \quad ; \quad \mathrm{Re}\,\lambda > 0$$

(85)

Applying various substitution formulas which include the Volterra functions, the infinite integrals with the modified Bessel functions of the second kind $K_\nu(z)$, the complementary error functions $erfc(z)$ and the parabolic cylinder functions $D_\nu(z)$ can be evaluated (for details see Apelblat (1985) [40]).

In the case of convolution integrals, in the simplest case of $f(t) = \nu(t)$, using the product theorem of the Laplace transform we have

$$\int_0^t \nu(\xi)\nu(t-\xi)\,d\xi = L^{-1}\left\{\frac{1}{(s\ln s)^2}\right\} = \mu(t,1,1) \qquad (86)$$

and from the following Laplace transform

$$L\left\{t^{\lambda-1}\ln t\right\} = \frac{\Gamma(\lambda)}{s^\lambda}[\psi(\lambda)-\ln s] \quad ; \quad \mathrm{Re}\,\lambda > 0 \qquad (87)$$

the convolution integral of the Volterra function is

$$\int_0^t (t-\xi)^{\lambda-1}\ln(t-\xi)\nu(\xi,\alpha)\,d\xi = \Gamma(\lambda)\left[\psi(\lambda)\nu(t,\alpha+\lambda) - \frac{t^{\alpha+\lambda}}{\Gamma(\alpha+\lambda+1)}\right] \qquad (88)$$

For $\lambda = 1$ and $\lambda = 1/2$, the integral in (88) becomes

$$\int_0^t \ln(t-\xi)\nu(\xi,\alpha)\,d\xi = -\left[\gamma\,\nu(t,\alpha+1) + \frac{t^{\alpha+1}}{\Gamma(\alpha+2)}\right] \quad ; \quad \gamma = -\psi(1)$$

$$\int_0^t \frac{\ln(t-\xi)}{\sqrt{t-\xi}}\nu(\xi,\alpha)\,d\xi = -(\gamma+2\ln 2)\nu(t,\alpha+\frac{1}{2}) - \frac{t^{\alpha+1/2}}{\Gamma\left(\alpha+\frac{3}{2}\right)} \qquad (89)$$

where $\gamma = 0.57721566\ldots$ is the Euler constant.

Indefinite integrals of the Volterra function can be evaluated successively by applying the following procedure

$$\int_0^t (t-\xi)^{n-1}\nu(\xi)\,d\xi = (n-1)!\,\nu(t,n) \quad ; \quad n = 1,2,3,\ldots$$

$$L\left\{\int_0^t \nu(\xi)\,d\xi\right\} = \frac{1}{s^2 \ln s} \tag{90}$$

$$\int_0^t \nu(\xi)\,d\xi = \nu(t,1)$$

and for example, for $n = 2$ and $n = 3$, from (90) we have

$$\begin{aligned} &\int_0^t \xi\,\nu(\xi)\,d\xi = t\nu(t,1) - \nu(t,2) \\ &\int_0^t \xi^2\nu(\xi)\,d\xi = 2\nu(t,3) - 2t\nu(t,2) + t^2\nu(t,1) \end{aligned} \tag{91}$$

Similarly, the indefinite integrals in (90) and (91), can be derived for other Volterra functions which are defined in (82).

Evaluation of Integrals by Repeated Usage of Integral Transforms.

A very simple generation of infinite integrals can be performed by repeated usage of the various direct integral transforms which are available in literature tables of transforms. In this context, the operational rules of linearity, scale change and translation, and the change in the order of applied transforms are also useful Apelblat (1981, 1996, 2012) [38, 53, 64]. The procedure is illustrated by applying the Laplace transform and the Fourier sine transform of the function $f(x)$

$$\begin{aligned} &F_s(y) = \int_0^\infty \sin(yx) f(x)\,dx \\ &G(s) = \int_0^\infty e^{-sy}\left[\int_0^\infty \sin(yx) f(x)\,dx\right] dy \end{aligned} \tag{92}$$

Formally, by changing the order of integration in (92) (the conditions for this operation are discussed by McLachlan (1962) [2]) and by performing the integration we have a new infinite integral $G(s)$

$$G(s) = \int_0^\infty f(x)\left[\int_0^\infty e^{-sy}\sin(yx)\,dy\right]dx$$
$$G(s) = \int_0^\infty \frac{x f(x)}{s^2+x^2}\,dx \tag{93}$$

In particular, the Fourier sine transform

$$F_s(y) = \int_0^\infty \frac{\sin(yx)}{x(a^2+x^2)}\,dx = \frac{\pi}{2a^2}\left[1-e^{-ay}\right]$$
$$G(s) = \int_0^\infty e^{-sy}\left[\frac{\pi}{2a^2}\left(1-e^{-ay}\right)\right]dy = \frac{\pi}{2a^2}\left(\frac{1}{s}-\frac{1}{s+a}\right) \tag{94}$$

from (94) and denoting $s = b$, it follows from (93) that

$$\int_0^\infty \frac{dx}{(a^2+x^2)(b^2+x^2)} = \frac{\pi}{2ab(a+b)} \tag{96}$$

In general case, if integral transforms $F_1(\xi)$ and $F_2(\xi)$ of a function $f(t)$ with respect to the kernels $K_1(t,\xi)$ and $K_2(t,\xi)$ are

$$F_1(\xi) = \int_0^\infty K_1(t,\xi) f(t)\,dt$$
$$F_2(\xi) = \int_0^\infty K_2(t,\xi) f(t)\,dt \tag{96}$$

and they are integrable over $(0,\infty)$ interval, then it is possible to introduce the integral transform $F(s)$ of the function $F_1(\xi)$ with respect to the kernel $K_2(t,\xi)$ and integral transform $G(s)$ of the function $F_2(\xi)$ with respect to the kernel $K_1(t,\xi)$

$$F(s) = \int_0^\infty K_2(\xi, s) F_1(\xi)\, d\xi$$
$$G(s) = \int_0^\infty K_1(\xi, s) F_2(\xi)\, d\xi \qquad (97)$$

Denoting

$$g(t,s) = \int_0^\infty K_1(t, \xi) K_2(\xi, s)\, d\xi$$
$$h(t,s) = \int_0^\infty K_2(t, \xi) K_1(\xi, s)\, d\xi \qquad (98)$$

it is possible to express (97) by

$$F(s) = \int_0^\infty f(t) g(t,s)\, dt$$
$$G(s) = \int_0^\infty f(t) h(t,s)\, dt \qquad (99)$$

Many infinite integrals derived by applying (99) are presented by Apelblat (1981, 1996) [38, 64]. For various integral transforms, the functions $g(t,s)$ and $h(t,s)$ can be evaluated in the closed form, and they are reported by Apelblat (1981, 2012) [38, 53]. These functions are usually not the same if the applied integral transforms are interchanged. Table 3 presents the $g(t,s)$ and $h(t,s)$ functions when the Laplace transform is associated with the sine and cosine Fourier transforms, the Stieltjes and generalized Stieltjes transforms, the Bessel J, Y and K transforms, and the Hankel transforms. For the Bessel transforms, these functions are expressed in terms of the Legendre functions Abramowitz and Stegun (1965) [63].

Table 3. The kernels $K_1(t,\xi)$ and $K_2(\xi,s)$ and functions $g(t,s)$ and $h(t,s)$ of different integral transforms.

$K_1(t,\xi)$	$K_2(\xi,s)$	$g(t,s)$ and $h(t,s)$
$\sin(t\xi)$	$e^{-s\xi}$	$\frac{t}{s^2+t^2}$
$e^{-t\xi}$	$\sin(s\xi)$	$\frac{s}{s^2+t^2}$
$\cos(t\xi)$	$e^{-s\xi}$	$\frac{s}{s^2+t^2}$
$e^{-t\xi}$	$\cos(s\xi)$	$\frac{t}{s^2+t^2}$
$\frac{1}{t+\xi}$	$e^{-s\xi}$	$-e^{st}Ei(-st)$
$e^{-t\xi}$	$e^{-s\xi}$	$\frac{1}{s+t}$
$e^{-t\xi}$	$\frac{1}{s+\xi}$	$-e^{st}Ei(-st)$
$\frac{1}{(t+\xi)^{\rho}}$	$e^{-s\xi}$	$s^{\rho-1}e^{st}\Gamma(1-\rho,st)$
$e^{-t\xi}$	$\frac{1}{(s+\xi)^{\rho}}$	$t^{\rho-1}e^{st}\Gamma(1-\rho,st)$
$\frac{1}{\sqrt{t+\xi}}$	$e^{-s\xi}$	$\sqrt{\frac{\pi}{s}}\,e^{st}\,erfc(\sqrt{st})$
$e^{-t\xi}$	$\frac{1}{\sqrt{s+\xi}}$	$\sqrt{\frac{\pi}{t}}\,e^{st}\,erfc(\sqrt{st})$
$\frac{1}{(t+\xi)^{3/2}}$	$e^{-s\xi}$	$\frac{2}{\sqrt{t}}-2\sqrt{\pi s}\,e^{st}\,erfc(\sqrt{st})$
$e^{-t\xi}$	$\frac{1}{(s+\xi)^{3/2}}$	$\frac{2}{\sqrt{s}}-2\sqrt{\pi t}\,e^{st}\,erfc(\sqrt{st})$

$K_1(t,\xi)$	$K_2(\xi,s)$	$g(t,s)$ and $h(t,s)$
$\sqrt{t\xi}\,J_\nu(t\xi)$	$e^{-s\xi}$	$\dfrac{\Gamma\left(\nu+\frac{3}{2}\right)\sqrt{t}}{(s^2+t^2)^{3/4}}P_{1/2}^{-\nu}\left[\dfrac{s}{\sqrt{s^2+t^2}}\right]$; $\mathrm{Re}\,\nu>-\dfrac{3}{2}$
$e^{-t\xi}$	$\sqrt{s\xi}\,J_\nu(s\xi)$	$\dfrac{\Gamma\left(\nu+\frac{3}{2}\right)\sqrt{s}}{(s^2+t^2)^{3/4}}P_{1/2}^{-\nu}\left[\dfrac{t}{\sqrt{s^2+t^2}}\right]$; $\mathrm{Re}\,\nu>-\dfrac{3}{2}$
$\sqrt{t\xi}\,Y_\nu(t\xi)$	$e^{-s\xi}$	$-\dfrac{2\Gamma\left(\nu+\frac{3}{2}\right)\sqrt{t}}{\pi(s^2+t^2)^{3/4}}Q_{1/2}^{-\nu}\left[\dfrac{s}{\sqrt{s^2+t^2}}\right]$; $\lvert\mathrm{Re}\,\nu\rvert<\dfrac{3}{2}$
$e^{-t\xi}$	$\sqrt{s\xi}\,Y_\nu(s\xi)$	$-\dfrac{2\Gamma\left(\nu+\frac{3}{2}\right)\sqrt{s}}{\pi(s^2+t^2)^{3/4}}Q_{1/2}^{-\nu}\left[\dfrac{t}{\sqrt{s^2+t^2}}\right]$; $\lvert\mathrm{Re}\,\nu\rvert<\dfrac{3}{2}$
$\sqrt{t\xi}\,K_\nu(t\xi)$	$e^{-s\xi}$	$\dfrac{\Gamma\left(\frac{3}{2}-\nu\right)\sec(\pi\nu)\sqrt{t}}{(s^2-t^2)^{3/4}}Q_{1/2}^{\nu}\left[\dfrac{s}{\sqrt{s^2-t^2}}\right]$ $\mathrm{Re}\left(\frac{1}{2}\pm\nu\right)>-1$
$e^{-t\xi}$	$\sqrt{s\xi}\,K_\nu(s\xi)$	$\dfrac{\Gamma\left(\frac{3}{2}-\nu\right)\sec(\pi\nu)\sqrt{s}}{(s^2-t^2)^{3/4}}Q_{1/2}^{\nu}\left[\dfrac{t}{\sqrt{s^2-t^2}}\right]$ $\mathrm{Re}\left(\frac{1}{2}\pm\nu\right)>-1$
$\sqrt{t\xi}\,H_{1/2}(t\xi)$	$e^{-s\xi}$	$\sqrt{\dfrac{2}{\pi}}\dfrac{t^2}{s(s^2+t^2)}$
$e^{-t\xi}$	$\sqrt{s\xi}\,H_{1/2}(s\xi)$	$\sqrt{\dfrac{2}{\pi}}\dfrac{s^2}{t(s^2+t^2)}$
$\sqrt{t\xi}\,H_{-1/2}(t\xi)$	$e^{-s\xi}$	$\sqrt{\dfrac{2}{\pi}}\dfrac{t}{(s^2+t^2)}$
$e^{-t\xi}$	$\sqrt{s\xi}\,H_{-1/2}(s\xi)$	$\sqrt{\dfrac{2}{\pi}}\dfrac{s}{(s^2+t^2)}$
$\sqrt{t\xi}\,H_{-1/2}(t\xi)$	$e^{-s\xi}$	$\sqrt{\dfrac{2}{\pi}}\dfrac{t}{(s^2+t^2)}$
$e^{-t\xi}$	$\sqrt{s\xi}H_{3/2}(s\xi)$	$\dfrac{s}{\sqrt{2\pi}}\left[\dfrac{s^2-t^2}{t^2(s^2+t^2)}+\dfrac{1}{s^2}\ln\left(\dfrac{s^2+t^2}{t^2}\right)\right]$

Table 3. (Continued)

$K_1(t,\xi)$	$K_2(\xi,s)$	$g(t,s)$ and $h(t,s)$
$\sqrt{t\xi}\,H_{3/2}(t\xi)$	$e^{-s\xi}$	$\frac{t}{\sqrt{2\pi}}\left[\frac{s^2-t^2}{s^2(s^2+t^2)}+\frac{1}{t^2}\ln\left(\frac{s^2+t^2}{s^2}\right)\right]$
$e^{-t\xi}$	$\sqrt{s\xi}H_{-3/2}(s\xi)$	$\sqrt{\frac{2}{\pi}}\left[\frac{t}{s^2+t^2}-\frac{1}{s}\tan^{-1}\left(\frac{s}{t}\right)\right]$
$\sqrt{t\xi}\,H_{-3/2}(t\xi)$	$e^{-s\xi}$	$\sqrt{\frac{2}{\pi}}\left[\frac{s}{s^2+t^2}-\frac{1}{t}\tan^{-1}\left(\frac{t}{s}\right)\right]$

Additional integrals linked with the Laplace transforms and special functions will be considered in the next sections of this Chapter.

Derivatives and Integrals with Respect to the Order of the Bessel and Related Functions.

Although there is a vast literature devoted to the properties of the Bessel and related functions, the topic of mathematical operations with respect to the order of these functions, is less studied. Differentiation and integration with respect to the order of the Bessel functions can be performed by various methods Apelblat (2020) [56], these which are associated with the Laplace transform are presented here Apelblat and Kravitsky (1981) [41], Apelblat (1989, 1991) [43, 44].

The Laplace transform of the Bessel function of the first kind is Oberhettinger and Badii (1973) [67]

$$L\{J_\nu(t)\}=\frac{1}{\sqrt{s^2+1}\,[s+\sqrt{s^2+1}]^\nu} \quad ; \quad \mathrm{Re}\,s>0 \quad ; \quad \mathrm{Re}\,\nu>-1$$

(100)

If the integration under the sign of the Laplace transform (100) is performed, then we have

$$F(s,\nu)=L\left\{\int_\nu^\infty J_\mu(t)\,d\mu\right\}=\int_\nu^\infty \frac{d\mu}{\sqrt{s^2+1}\,[s+\sqrt{s^2+1}]^\mu} \tag{101}$$

Using the complex integral in (2), deforming the Bromwich contour, and taking into account that the integrand in (101) has two conjugate branch points at $s = \pm i$, *and* a pole at the origin, the final result of complex integration performed with the help of (49) is Apelblat and Kravitsky (1981) [41]

$$I(t,\nu) = L^{-1}\{F(s,\nu)\} = \int_{\nu}^{\infty} J_{\mu}(t)\,d\mu = \frac{1}{2} + \frac{1}{\pi}\int_{0}^{\infty} \sin(t\sin\xi - \nu\xi)\frac{d\xi}{\xi} + \int_{0}^{\infty} \frac{e^{-\nu\xi - t\sinh\xi}}{\pi^2+\xi^2}\left[\cos(\pi\nu) + \frac{\xi}{\pi}\sin(\pi\nu)\right]d\xi \qquad (102)$$

Thus, the integral with respect to the order of the Bessel function is expressed in terms of two integrals of elementary functions. For $\nu = 0$, this integral is reduced to

$$I(t,0) = \int_{0}^{\infty} J_{\mu}(t)\,d\mu = \frac{1}{2} + \frac{1}{\pi}\int_{0}^{\infty} \sin(t\sin\xi)\frac{d\xi}{\xi} + \int_{0}^{\infty} \frac{e^{-t\sinh\xi}}{\pi^2+\xi^2}d\xi \qquad (103)$$

The operational rule of the Laplace transform permits deriving the limiting expressions for the integral in (103) for small and large values of variable t by using the limiting forms of the transform $F(s,0)$

$$F(s \to 0,0) \to \frac{1}{\ln(s+1)} \to \frac{1}{s}$$
$$F(s \to \infty,0) \to \frac{1}{s\ln(2s)} \qquad (104)$$

and the inverses of (104) are Apelblat and Kravitsky (1981) [41]

$$I(t \to \infty,0) \to 1$$
$$I(t \to 0,0) \to \nu\left(\frac{t}{2}\right) \to \frac{1}{\tau}\left[1 - \frac{\psi(1)}{\tau} + \frac{\psi(1)^2 - \psi^{(1)}(1)}{\tau^2} - \ldots\right] \quad ; \quad \tau = \ln\left(\frac{2}{t}\right)$$

(105)

Differentiation of (100) with respect to the order ν gives the Laplace transform of the derivative of the Bessel function

$$L\left\{\frac{\partial J_{\nu}(t)}{\partial \nu}\right\} = -\frac{\ln[s+\sqrt{s^2+1}]}{\sqrt{s^2+1}\,[s+\sqrt{s^2+1}]^{\nu}} = -\frac{\ln[s+\sqrt{s^2+1}]}{\sqrt{s^2+1}}\cdot\frac{1}{[s+\sqrt{s^2+1}]^{\nu}} \qquad (106)$$

But these inverse transforms are known Oberhettinger and Badii (1973) [67]

$$L^{-1}\left\{\frac{\ln[s+\sqrt{s^2+1}]}{\sqrt{s^2+1}}\right\}=-\frac{\pi}{2}Y_0(t)$$

$$L^{-1}\left\{\frac{1}{[s+\sqrt{s^2+1}]^{\nu}}\right\}=\frac{\nu}{t}J_{\nu}(t) \quad ; \quad \mathrm{Re}\,\nu>0 \tag{107}$$

and from the convolution theorem we have van der Pol and Bremmer (1964) [21]

$$\frac{\partial J_{\nu}(t)}{\partial \nu}=\frac{\pi\nu}{2}Y_0(t)*\frac{J_{\nu}(t)}{t}=\frac{\pi\nu}{2}\int_0^t Y_0(t-\xi)\frac{J_{\nu}(\xi)}{\xi}d\xi \tag{108}$$

Introducing a new variable $\xi = t(\cos\theta)^2$ into (108), the derivative of the Bessel function with respect to the order can be expressed by the finite integral Apelblat and Kravitsky (1981) [41]

$$\frac{\partial J_{\nu}(t)}{\partial \nu}=\pi\nu\int_0^{\pi/2}\tan\theta\, Y_0[t(\sin\theta)^2]J_{\nu}[t(\cos\theta)^2]d\theta \quad ; \quad \mathrm{Re}\,\nu>0 \tag{109}$$

The next differentiation of (109) gives the following product of the Laplace transforms

$$L\left\{\frac{\partial^2 J_{\nu}(t)}{\partial \nu^2}\right\}=\frac{\left[\ln[s+\sqrt{s^2+1}]\right]^2}{\sqrt{s^2+1}\,[s+\sqrt{s^2+1}]^{\nu}}=$$

$$\frac{\left[\ln[s+\sqrt{s^2+1}]\right]^2}{s}\cdot\frac{s}{\sqrt{s^2+1}[s+\sqrt{s^2+1}]^{\nu}} \tag{110}$$

In this case, these inverses are known and they can be expressed in terms of the Bessel functions and the integral Bessel function Oberhettinger and Badii (1973) [67]

$$L^{-1}\left\{\frac{1}{s}\left[\ln[s+\sqrt{s^2+1}]\right]^2\right\}=-\pi\, Yi_0(t) \quad ; \quad Yi_0(t)=-\int_t^{\infty}\frac{Y_0(\xi)}{\xi}d\xi$$

$$L^{-1}\left\{\frac{s}{\sqrt{s^2+1}[s+\sqrt{s^2+1}]^{\nu}}\right\}=\frac{d}{dt}\{J_{\nu}(t)\}=\frac{J_{\nu-1}(t)-J_{\nu+1}(t)}{2} \tag{111}$$

Therefore, by using the convolution theorem, the second derivative of the Bessel function with respect to the order is

$$\frac{\partial^2 J_\nu(t)}{\partial \nu^2} = -\frac{\pi}{2}\int_0^t Yi_0(t-\xi)[J_{\nu-1}(\xi)-J_{\nu+1}(\xi)]d\xi \tag{112}$$

Similarly, in the case of the modified Bessel function of the first kind, the corresponding expression to that in (101) is Apelblat and Kravitsky (1981) [41]

$$F(s,\nu) = L\left\{\int_\nu^\infty I_\mu(t)\,d\mu\right\} = \int_\nu^\infty \frac{d\mu}{\sqrt{s^2-1}\,[s+\sqrt{s^2-1}]^\mu} \tag{113}$$

$$\operatorname{Re}\mu > -1 \quad ; \quad \operatorname{Re}s > 1$$

The complex inversion can be performed by considering that the integrand in (113) has two branch points at $s = \pm\, 1$, and the final result of complex integration is

$$I(t,\nu) = \int_\nu^\infty I_\mu(t)\,d\mu = \frac{e^t}{2} - \frac{1}{\pi}\int_0^\infty e^{t\cos\xi}\sin(\nu\xi)\frac{d\xi}{\xi} + \\ \int_0^\infty \frac{e^{-\nu\xi - t\cosh\xi}}{\pi^2+\xi^2}[\xi\sin(\pi\nu)+\pi\cos(\pi\nu)]\,d\xi \tag{114}$$

Similarly as in (103) - (105) we have

$$I(t,0) = \int_0^\infty I_\mu(t)\,d\mu = \frac{e^t}{2} - \int_0^\infty \frac{e^{-t\cosh\xi}}{\pi^2+\xi^2}\,d\xi$$

$$I(t\to\infty,0) \to \frac{e^t}{2} \tag{115}$$

$$I(t\to 0,0) \to \frac{1}{\tau}\left[1-\frac{\psi(1)}{\tau}+\frac{\psi(1)^2-\psi^{(1)}(1)}{\tau^2}-\ldots\right] \quad ; \quad \tau = \ln\left(\frac{2}{t}\right)$$

The mathematical operations comparable to those performed in (107) - (109) and (112) give the following results

$$\frac{\partial I_\nu(t)}{\partial \nu} = -K_0(t) * \frac{\nu J_\nu(t)}{t} = -\nu \int_0^t K_0(t-\xi)\frac{I_\nu(\xi)}{\xi}d\xi \quad ; \quad \mathrm{Re}\,\nu > 0$$

$$Ki_0(t) = -\int_t^\infty \frac{K_0(\xi)}{\xi}d\xi$$

$$\frac{\partial I_\nu(t)}{\partial \nu} = -2\nu \int_0^{\pi/2} \tan\theta\, K_0[t(\sin\theta)^2]I_\nu[t(\cos\theta)^2]d\theta$$

$$\frac{\partial^2 I_\nu(t)}{\partial \nu^2} = \int_0^t [Ki_0(t-\xi) - \frac{\pi^2}{8}][I_{\nu-1}(\xi) + I_{\nu+1}(\xi)]d\xi$$

(116)

The Anger functions $\mathbf{J}_\nu(t)$ are closely related to the Bessel functions Abramowitz and Stegun (1965) [63]

$$\mathrm{J}_\nu(t) = J_\nu(t) + \frac{\sin(\pi\nu)}{\pi}\int_0^\infty e^{-\nu\xi - t\sinh\xi}d\xi = \frac{1}{\pi}\int_0^\pi \cos(\nu\zeta - t\sin\zeta)d\zeta \tag{117}$$

$$L\{\mathrm{J}_\nu(t)\} = L\{J_\nu(t)\} + \frac{\sin(\pi\nu)}{\pi}\int_0^\infty \frac{e^{-\nu\xi}}{s+\sinh\xi}d\xi$$

Using the integral from (102) and changing the order of integration in (117) we have Apelblat and Kravitsky (1981) [41]

$$G(t,\nu) = \int_\nu^\infty \mathrm{J}_\xi(t)d\xi = I(t,\nu) + \frac{1}{\pi}\int_\nu^\infty \frac{e^{-\nu\zeta - t\sinh\zeta}[\zeta\sin(\pi\nu) + \pi\cos(\pi\nu)]}{\pi^2+\zeta^2}d\zeta =$$

$$\frac{1}{2} + \frac{1}{\pi}\int_0^\pi \sin(t\sin\zeta - \nu\zeta)d\zeta \tag{118}$$

$$\frac{\partial \mathrm{J}_\nu(t)}{\partial \nu} = \frac{1}{\pi}\int_0^\pi \xi\sin(t\sin\xi - \nu\xi)d\xi$$

For $\nu = 0$, by introducing (103), the integral in (118) together with the asymptotic expressions are

$$G(t,0)=\int_0^\infty \mathrm{J}_\xi(t)\,d\xi=\frac{1}{2}+\frac{1}{\pi}\int_0^\pi \sin(t\sin\zeta)\,d\zeta$$

$$G(t\to\infty,0)\to 1+\frac{\left(\frac{2}{\pi}\right)^{3/2}}{\sqrt{t}}\sin\left(t-\frac{\pi}{4}\right) \tag{119}$$

$$G(t\to 0,0)\to\frac{1}{2}+\frac{Si(\pi)}{\pi}t \quad ; \quad Si(\pi)=1.85155...$$

The Laplace transform procedure, as applied above, is also suitable for the integral Bessel function $Ji_\nu(t)$ Apelblat and Kravitsky (1981) [41]

$$Ji_\nu(t)=-\int_t^\infty \frac{J_\nu(\xi)}{\xi}\,d\xi$$

$$L\{Ji_\nu(t)\}=\frac{1}{\nu s}\left[\frac{1}{(s+\sqrt{s^2+1})^\nu}-1\right] \quad ; \quad \mathrm{Re}\,\nu>0 \tag{120}$$

$$L\left\{\frac{\partial Ji_\nu(t)}{\partial\nu}\right\}=\frac{1}{\nu s}\left[\frac{1-(s+\sqrt{s^2+1})^{-\nu}}{\nu}-\frac{\ln(s+\sqrt{s^2+1})}{(s+\sqrt{s^2+1})^\nu}\right]$$

Performing the inversion of (120) by using the complex integral formula in (49), the integral representation of the integral Bessel function is

$$Ji_\nu(t)=-\frac{1}{2\nu}-\frac{1}{\pi\nu}\int_0^\pi \sin(t\sin\xi-\nu\xi)\cot\xi\,d\xi+ \frac{\sin(\pi\nu)}{\pi\nu}\int_0^\infty e^{-\nu\xi-t\sinh\xi}\coth\xi\,d\xi \tag{121}$$

and the first derivative with respect to the order is given by

$$\frac{\partial Ji_\nu(t)}{\partial\nu}=\frac{Ji_\nu(t)}{\nu}+\frac{1}{\pi\nu}\int_0^\pi \xi\cos(t\sin\xi-\nu\xi)\cot\xi\,d\xi- \frac{1}{\pi\nu}\int_0^\infty e^{-\nu\xi-t\sinh\xi}[\xi\sin(\pi\nu)-\pi\cos(\pi\nu)]\coth\xi\,d\xi \tag{122}$$

However, using the inverse Laplace transforms Oberhettinger and Badii (1973) [67]

$$L^{-1}\left\{\frac{\ln(s+\sqrt{s^2+1})}{s}\right\} = Ji_0(t)$$
$$L^{-1}\left\{\frac{1}{(s+\sqrt{s^2+1})}\right\} = \frac{\nu}{t}J_\nu(t) \tag{123}$$

The derivative in (122) can be expressed in a more compact form of the convolution integral

$$\frac{\partial Ji_\nu(t)}{\partial \nu} = \frac{Ji_\nu(t)}{\nu} - \int_0^t Ji_0(t-\xi)J_\nu(\xi)\frac{d\xi}{\xi} \quad ; \quad \mathrm{Re}\,\nu > 0 \tag{124}$$

Or by changing the variable of integration, it is reduced to

$$\frac{\partial Ji_\nu(t)}{\partial \nu} = \frac{Ji_\nu(t)}{\nu} - 2\int_0^{\pi/2} \tan\theta\, Ji_0[t(\sin\theta)^2]J_\nu[t(\cos\theta)^2]d\theta \tag{125}$$

The Laplace transform procedures allow the evaluation of the derivatives and integrals with respect to the order of the Struve functions $H_\nu(z)$, the modified Struve functions $L_\nu(z)$ and the Kelvin functions $ber_\nu(z)$ and $bei_\nu(z)$ Apelblat (1989, 1991) [43, 44]. These functions are defined in Abramowitz and Stegun (1965) [63].

Final results of derivations of the derivatives of the Struve functions are presented by Apelblat (1989) [43]

$$\frac{\partial H_\nu(t)}{\partial \nu} = \left[\ln\left(\frac{t}{2}\right)+\gamma\right]H_\nu(t) + t\int_0^1 \xi^\nu \ln(1-\xi^2)H_{\nu-1}(t\xi)d\xi \quad ; \quad \mathrm{Re}\,\nu > -\frac{1}{2} \tag{126}$$
$$\frac{\partial L_\nu(t)}{\partial \nu} = \left[\ln\left(\frac{t}{2}\right)+\gamma\right]L_\nu(t) + t\int_0^1 \xi^\nu \ln(1-\xi^2)L_{\nu-1}(t\xi)d\xi$$

and the Kelvin functions are Apelblat (1991) [44]

$$\frac{\partial ber_\nu(t)}{\partial \nu} = \ln\left(\frac{t}{2}\right) ber_\nu(t) - \frac{3\pi}{4} bei_\nu(t) -$$

$$\frac{t}{2^{3/2}} \int_0^1 \xi^{(\nu-1)/2} [\gamma + \ln(1-\xi)][ber_{\nu-1}(t\sqrt{\xi}) + bei_{\nu-1}(t\sqrt{\xi})] d\xi$$

$$\frac{\partial bei_\nu(t)}{\partial \nu} = \ln\left(\frac{t}{2}\right) bei(t) + \frac{3\pi}{4} ber(t) + \tag{127}$$

$$\frac{t}{2^{3/2}} \int_0^1 \xi^{(\nu-1)/2} [\gamma + \ln(1-\xi)][ber_{\nu-1}(t\sqrt{\xi}) - bei_{\nu-1}(t\sqrt{\xi})] d\xi$$

The expressions in (127) are valid for $\nu > 0$, for $\nu = 0$ (see Apelblat (1991) [44].)

The Asymptotic Limit of the Bessel Function Expressed as the Shifted Dirac Delta Function $\delta(t\text{-}1)$ and the Laplace Transforms.

The impulse functions which represent a sudden excitation of physical systems are a powerful tool in many fields of applied mathematics, physics and engineering. They were introduced into mathematics over a long period of time in various forms by Fourier, Cauchy, Poisson, Hermite, Kirchhoff, Kelvin and Heaviside. Dirac started in 1926 to use the delta function $\delta(t)$ in quantum mechanics, and this impulse function that is important in many areas, is also involved with integral transforms. The representation of the shifted Dirac delta function which is related to the order of the Bessel function of the first kind was derived by Lamborn in 1969 [70], in the following way

$$\delta(\xi - 1) = \lim_{\nu\to\infty} \{\nu J_\nu(\nu\xi)\} \tag{128}$$

Starting with this expression, Apelblat (1999, 2000, 2004, 2008) [47-50] derived many mathematical relations which included representations of elementary and special functions, series, limits, integrals and asymptotic and functional relations. The results, associated directly with the Laplace transforms, are briefly presented here. If the Lamborn delta expression in (128) is multiplied by a function $f(t\xi)$ and integrated from zero to infinity, and the sampling property of delta function is applied, we have

$$f(t) = \int_0^\infty f(t\xi)\,\delta(\xi-1)\,d\xi = \lim_{\nu\to\infty}\left[\nu\int_0^\infty f(t\xi)\,J_\nu(\nu\xi)\,d\xi\right] \tag{129}$$

Since many of the Bessel infinite integrals are known, it is possible to apply this and similar expressions to obtain various mathematical relations. If integral transforms of a function $f(t)$ with respect to various kernels $K(t,s)$ are defined by

$$T\{f(t)\} = \int_0^\infty K(s,t)\,f(t)\,dt = T(s) \tag{130}$$

then with $f(t) = J_\nu(\nu t)$ by using (129) it follows that

$$\lim_{\nu\to\infty}\left[\nu\int_0^\infty K(s,t)\,J_\nu(t\nu)\,dt\right] = \int_0^\infty \delta(\xi-1)\,K(s,\xi)\,d\xi = K(s,1) \tag{131}$$

In the case of the Laplace transform, by considering that the Laplace transform of the Bessel function is given in (100), we have the following limit from (131)

$$\lim_{\nu\to\infty}\left[\frac{\nu^{\nu+1}}{\sqrt{s^2+1}\;[s+\sqrt{s^2+1}]^\nu}\right] = e^{-s} \quad ; \quad \mathrm{Re}\,s > 0 \tag{132}$$

If in the expression (131) a function $f(t,\lambda)$ is also introduced, then it is possible to obtain integral transforms

$$\begin{aligned} T\{J_\nu(\nu t),\xi\} &= \int_0^\infty K(\xi,t)\,J_\nu(\nu t)\,dt \\ T\{f(\xi,\lambda),t\} &= \int_0^\infty K(\xi,t)\,f(\xi,\lambda)\,d\xi \end{aligned} \tag{133}$$

and in the limit

$$\begin{aligned} &\lim_{\nu\to\infty}\left[\nu\int_0^\infty f(\xi,\lambda)\,T\{J_\nu(\nu t),\xi\}\,d\xi\right] = \\ &\lim_{\nu\to\infty}\left[\nu\int_0^\infty J_\nu(\nu t)\,T\{f(\xi,\lambda),t\}\,dt\right] = T(1,\lambda) \end{aligned} \tag{134}$$

which for the Laplace transform $L\{f(x,\lambda),t\} = F(t,\lambda)$, using (100), takes the following form

$$\lim_{\nu\to\infty}\left[\nu^{\nu+1}\int_0^\infty \frac{f(\xi,\lambda)}{\sqrt{\nu^2+\xi^2}\,[\xi+\sqrt{\nu^2+\xi^2}\,]^\nu}d\xi\right]= \tag{135}$$

$$\lim_{\nu\to\infty}\left[\nu\int_0^\infty J_\nu(\nu t)F(t,\lambda)\,dt\right]=F(1,\lambda)$$

For example, in the simplest case, $f(t,\lambda) = t^\lambda$, we have from (135), the representation of the gamma function

$$\lim_{\nu\to\infty}\left[\nu^{\nu+1}\int_0^\infty \frac{\xi^\lambda}{\sqrt{\nu^2+\xi^2}\,[\xi+\sqrt{\nu^2+\xi^2}\,]^\nu}d\xi\right]=\Gamma(\lambda+1) \quad ; \quad \mathrm{Re}\,\lambda>-1$$

(136)

Using tables of the Laplace transforms [65-67], permits to derive integral representations for many elementary and special functions in the form of asymptotic limits (similar to that presented in (136)) and a long list of them is given by Apelblat (2012) [53].

Laplace Transforms and the Mittag-Leffler Functions.

In the last four decades enormous attention is devoted to the Mittag-Leffler functions coming from various scientific and engineering disciplines. At the beginning of the previous century, these functions were introduced by the Swedish mathematician G.M. Mittag-Leffler as the generalization of exponential function Gorenflo, Kilbas, Mainardi and Rogosin (2014) [71]. Nowadays, the Mittag-Leffler functions play an important role in the modeling of many physical phenomena, such as the kinetics of chemical reactions, time and space fractional diffusion, nonlinear waves, viscoelastic systems, neural networks, electric field relaxations and statistical distributions. In mathematics, these functions are applied in fractional calculus, in the solution of systems with fractional differential and integral equations. The one and two-parameter Mittag-Leffler functions are defined by the following power series

$$E_{\alpha}(t) = \sum_{k=0}^{\infty} \frac{t^k}{\Gamma(\alpha k+1)} \quad ; \quad \mathrm{Re}\,\alpha > 0$$

$$E_{\alpha,\beta}(t) = \sum_{k=0}^{\infty} \frac{t^k}{\Gamma(\alpha k+\beta)} \quad ; \quad \mathrm{Re}\,\alpha > 0 \quad ; \quad \mathrm{Re}\,\beta > 0 \tag{137}$$

Many properties of these functions can be deduced by considering the simple forms of their Laplace transforms

$$L\left\{E_{\alpha}(\pm t^{\alpha})\right\} = \frac{s^{\alpha-1}}{s^{\alpha} \mp 1} \quad ; \quad \mathrm{Re}\,s > 1$$

$$L\left\{t^{\beta-1}E_{\alpha,\beta}(\pm \lambda t^{\alpha})\right\} = \frac{s^{\alpha-\beta}}{s^{\alpha} \mp \lambda} \quad ; \quad \mathrm{Re}\,s > |\lambda|^{1/\alpha} \tag{138}$$

They allow for the derivation of the inverse transforms and for easily recognizing functions for particular values of the parameters α and β as either elementary or special functions. If α is a positive integer, for example $\alpha = 1$ and $\alpha = 2$, then from (138) and tables of inverse transforms [65-67] it follows that

$$E_1(t) = L^{-1}\left\{L\left\{E_1(t)\right\}\right\} = L^{-1}\left\{\frac{1}{s-1}\right\} = e^{t}$$

$$E_1(-t) = L^{-1}\left\{L\left\{E_1(-t)\right\}\right\} = L^{-1}\left\{\frac{1}{s+1}\right\} = e^{-t}$$

$$E_1(t^2) = L^{-1}\left\{L\left\{E_1(t^2)\right\}\right\} = L^{-1}\left\{\frac{s}{s^2-1}\right\} = \cosh t \tag{139}$$

$$E_1(-t^2) = L^{-1}\left\{L\left\{E_1(-t^2)\right\}\right\} = L^{-1}\left\{\frac{s}{s^2+1}\right\} = \cos t$$

If $\alpha = 1/2$

$$E_{1/2}(\pm\sqrt{t}) = L^{-1}\left\{\frac{1}{\sqrt{s}\,(\sqrt{s} \mp 1)}\right\} = L^{-1}\left\{\mp\frac{1}{\sqrt{s}} \pm \frac{1}{\sqrt{s} \mp 1}\right\}$$

$$L^{-1}\left\{\frac{1}{\sqrt{s}}\right\} = \frac{1}{\sqrt{\pi t}} \quad ; \quad L^{-1}\left\{\frac{1}{\sqrt{s} \mp 1}\right\} = \frac{1}{\sqrt{\pi t}} + e^{t}\,erfc(\mp\sqrt{t}) \tag{140}$$

$$erfc(-\sqrt{t}) = erf(\sqrt{t}) - 1$$

we have

$$E_{1/2}(\pm\sqrt{t}) = e^{t}[1 \pm erf(\sqrt{t})] \tag{141}$$

By performing algebraic manipulations with the Laplace transforms it is possible to deduce some interrelation between the Mittag-Leffler functions. By introducing $\beta = \alpha + 1$ and $\lambda = 1$, in the second equation of (138)

$$L\left\{t^{\alpha} E_{\alpha,\alpha+1}(t^{\alpha})\right\} = \frac{s^{\alpha-1}}{s(s^{\alpha}-1)} = \frac{s^{\alpha-1}}{s^{\alpha}-1} - \frac{1}{s} \tag{142}$$

the inverse of (142) is

$$t^{\alpha} E_{\alpha,\alpha+1}(t^{\alpha}) = E_{\alpha}(t^{\alpha}) - 1 \tag{143}$$

Similarly, for the two-parameter Mittag-Leffler function, we have

$$L\left\{t^{\alpha+\beta-1} E_{\alpha,\alpha+\beta}(t^{\alpha})\right\} = \frac{s^{\alpha-\beta}}{s^{\beta}(s^{\alpha}-1)}$$

$$L\left\{t^{\beta-1} E_{\alpha,\beta}(t^{\alpha}) - \frac{t^{\beta-1}}{\Gamma(\beta)}\right\} = \frac{s^{\alpha-\beta}}{s^{\alpha}-1} - \frac{1}{s^{\beta}} = \frac{s^{\alpha-\beta}}{s^{\beta}(s^{\alpha}-1)} \tag{144}$$

$$t^{\alpha} E_{\alpha,\alpha+\beta}(t^{\alpha}) = E_{\alpha,\beta}(t^{\alpha}) - \frac{1}{\Gamma(\beta)}$$

If the Laplace transforms of the Mittag-Leffler function are written in the following form

$$L\left\{E_{\alpha}(t^{\alpha})\right\} = \frac{s^{\alpha-1}}{s^{\alpha}-1} = \frac{(\sqrt{s})^{2\alpha-1}}{\sqrt{s}\left[(\sqrt{s})^{2\alpha}-1\right]}$$

$$L\left\{E_{\alpha}(t^{\alpha})\right\} = \frac{(s^{1/3})^{3\alpha-1}}{s^{2/3}\left[(s^{1/3})^{3\alpha}-1\right]} \tag{145}$$

$$L\left\{1 - E_{\alpha}(t^{\alpha})\right\} = \frac{1}{s} - \frac{s^{\alpha-1}}{s^{\alpha}-1} = \frac{\left(\frac{1}{s}\right)^{\alpha-1}}{s^{2}\left[\left(\frac{1}{s}\right)^{\alpha}-1\right]}$$

then by using the substitution formulas of the Laplace transform Apelblat (2012) [53]

$$L\{f(t)\} = F(s)$$

$$L^{-1}\left\{\frac{1}{\sqrt{s}}F(\sqrt{s})\right\} = \frac{1}{\sqrt{\pi t}}\int_0^\infty e^{-\xi^2/4t} f(\xi)\,d\xi$$

$$L^{-1}\left\{\frac{1}{s^{2/3}}F(s^{1/3})\right\} = \frac{1}{\pi}\int_0^\infty \sqrt{\frac{\xi}{t}}\,K_{1/3}\left(\frac{2\xi^{3/2}}{\sqrt{27t}}\right) f(\xi)\,d\xi \qquad (146)$$

$$L^{-1}\left\{\frac{1}{s^2}F\left(\frac{1}{s}\right)\right\} = \int_0^\infty \sqrt{\frac{t}{\xi}}\,J_1(2\sqrt{t\xi})\,f(\xi)\,d\xi$$

it is possible to derive the self-representation of the Mittag-Leffler functions

$$E_\alpha(t^\alpha) = \frac{1}{\sqrt{\pi t}}\int_0^\infty e^{-\xi^2/4t} E_{2\alpha}(\xi^{2\alpha})\,d\xi$$

$$E_\alpha(t^\alpha) = \frac{1}{\pi}\int_0^\infty \sqrt{\frac{\xi}{t}}\,K_{1/3}\left(\frac{2\xi^{3/2}}{\sqrt{27t}}\right) E_{3\alpha}(\xi^{3\alpha})\,d\xi \qquad (147)$$

$$\frac{E_\alpha(t^\alpha)-1}{\sqrt{t}} = \int_0^\infty J_1(2\sqrt{t\xi})\,E_\alpha(\xi^\alpha)\frac{d\xi}{\sqrt{\xi}}$$

Laplace transforms manipulations of the Mittag-Leffler functions allow for the presentation of them as convolution integrals

$$L\left\{E_\alpha(t^\alpha)\right\} = \frac{s^{\alpha-1}}{s^\alpha-1} = \frac{s^{2\alpha-1}}{s^{2\alpha}-1} + \frac{s^{2\alpha-1}}{s^{2\alpha}-1}\cdot\frac{1}{s^\alpha}$$

$$L\left\{t^{\beta-1}E_{\alpha,\beta}(t^\alpha)\right\} = \frac{s^{\alpha-\beta}}{s^\alpha-1} = \frac{s^{2\alpha-\beta}}{s^{2\alpha}-1} + \frac{s^{2\alpha-\beta}}{s^{2\alpha}-1}\cdot\frac{1}{s^\alpha} \qquad (148)$$

From the product theorem it follows that

$$E_{\alpha}(t^{\alpha}) = E_{2\alpha}(t^{2\alpha}) + \int_{0}^{t} \frac{(t-\xi)^{\alpha-1}}{\Gamma(\alpha)} E_{2\alpha}(\xi^{2\alpha})\, d\xi$$

$$E_{\alpha,\beta}(t^{\alpha}) = E_{2\alpha,\beta}(t^{2\alpha}) + \int_{0}^{t} \left(\frac{\xi}{t}\right)^{\beta-1} \frac{(t-\xi)^{\alpha-1}}{\Gamma(\alpha)} E_{2\alpha}(\xi^{2\alpha})\, d\xi \qquad (149)$$

Other different forms of (138) and the corresponding convolution integrals are given by Apelblat (2020) [54].

The operational rules of the Laplace transform easily permit the differentiation and integration operations with the Mittag-Leffler functions by considering that these operations are equivalent to the multiplication or the division of the Laplace transforms by the variable s. For example, in the case of first derivative of $E_{\alpha}(t^{\alpha})$, by considering that the function at the origin has value of unity we have Apelblat (2020) [54]

$$L\left\{\frac{dE_{\alpha}(t^{\alpha})}{dt}\right\} = s\left(\frac{s^{\alpha-1}}{s^{\alpha}-1}\right) - 1 = \frac{s^{\alpha-1}}{s^{\alpha}-1} \cdot \frac{1}{s^{\alpha-1}} \quad ; \quad \mathrm{Re}\,\alpha > 0 \quad ; \quad \mathrm{Re}\,s > 1$$

$$\frac{dE_{\alpha}(t^{\alpha})}{dt} = \int_{0}^{t} \frac{(t-\xi)^{\alpha-2}}{\Gamma(\alpha-1)} E_{\alpha}(\xi^{\alpha})\, d\xi = t^{\alpha-1} E_{\alpha,\alpha}(t^{\alpha}) \qquad (150)$$

The Laplace transform is also suitable to determine sums of infinite and finite series of the Mittag-Leffler functions by writing the series of transforms in the following form

$$L\left\{\sum_{k=1}^{\infty} t^{k-1} E_{\alpha,k}(t^{\alpha})\right\} = \frac{s^{\alpha-1}}{s^{\alpha}-1} + \frac{s^{\alpha-2}}{s^{\alpha}-1} + \frac{s^{\alpha-3}}{s^{\alpha}-1} + \ldots + \frac{s^{\alpha-k}}{s^{\alpha}-1} + \ldots =$$

$$\frac{s^{\alpha-1}}{s^{\alpha}-1} + \frac{s^{\alpha-1}}{s^{\alpha}-1}\left\{\frac{1}{s} + \frac{1}{s^{2}} + \frac{1}{s^{3}} + \ldots + \frac{1}{s^{k}} + \ldots\right\} = \frac{s^{\alpha-1}}{s^{\alpha}-1} + \frac{s^{\alpha-1}}{s^{\alpha}-1} \cdot \frac{1}{s-1} \qquad (151)$$

The inverse of (151) leads to the sum of infinite series

$$\sum_{k=1}^{\infty} t^{k-1} E_{\alpha,k}(t^{\alpha}) = E_{\alpha,1}(t^{\alpha}) + tE_{\alpha,2}(t^{\alpha}) + t^{2}E_{\alpha,3}(t^{\alpha}) + \ldots + t^{k-1}E_{\alpha,k}(t^{\alpha}) + \ldots =$$

$$E_{\alpha}(t^{\alpha}) + \int_{0}^{t} e^{(t-\xi)} E_{\alpha}(\xi^{\alpha})\, d\xi \qquad (152)$$

$$E_{\alpha,1}(t^{\alpha}) = E_{\alpha}(t^{\alpha})$$

In the case of finite series having N terms, we have by a similar procedure

$$\sum_{k=1}^{N} t^{k-1} E_{\alpha,k}(t^{\alpha}) = E_{\alpha,1}(t^{\alpha}) + tE_{\alpha,2}(t^{\alpha}) + t^2 E_{\alpha,3}(t^{\alpha}) + ... + t^{N-1} E_{\alpha,N}(t^{\alpha}) =$$

$$E_{\alpha}(t^{\alpha}) + \int_0^t e^{(t-\xi)} \left[E_{\alpha}(\xi^{\alpha}) - \xi^N E_{\alpha,N+1}(\xi^{\alpha}) \right] d\xi \qquad (153)$$

Various infinite series of the Mittag-Leffler functions are presented by Apelblat (2020) [54], and it was established that if a function $f(t)$ is expanded into the Taylor series, and if it is possible to invert the Laplace transform $L^{-1}[F(1/s)-1]$, then the sum of series can be determined.

Operational rules of the Laplace transform are also suitable in the evaluation of derivatives and integrals of the Mittag-Leffler functions with respect to parameters α and β. The Laplace transform of the derivative of the Mittag-Leffler function $E_{\alpha}(t^{\alpha})$ with respect to α is

$$L\left\{\frac{\partial E_{\alpha}(t^{\alpha})}{\partial \alpha}\right\} = \frac{\partial}{\partial \alpha}\left(\frac{s^{\alpha-1}}{s^{\alpha}-1}\right) = \left[\frac{s^{\alpha-1}\ln s}{s^{\alpha}-1} - \frac{s^{2\alpha-1}\ln s}{(s^{\alpha}-1)^2}\right] = -\frac{s^{\alpha-1}\ln s}{(s^{\alpha}-1)(s^{\alpha}-1)} = \qquad (154)$$

$$-\frac{s^{\alpha-1}}{s^{\alpha}-1}\cdot\frac{s^{\alpha-1}}{s^{\alpha}-1}\cdot\frac{\ln s}{s^{\alpha-1}}$$

The convolution theorem can be applied to (154) because the inverse of the third term in the product of (154) is known Erdélyi, Magnus, Oberhettinger and Tricomi. (1954) [65]

$$L^{-1}\left\{\frac{\ln s}{s^{\alpha-1}}\right\} = \frac{t^{\alpha-2}}{\Gamma(\alpha-1)}[\psi(\alpha-1) - \ln t] \quad ; \quad \mathrm{Re}\,\alpha > 1 \qquad (155)$$

Therefore using (138), the derivative of the Mittag-Leffler function with respect to parameter α is given by

$$\frac{\partial E_{\alpha}(t^{\alpha})}{\partial \alpha} = E_{\alpha}(t^{\alpha}) * E_{\alpha}(t^{\alpha}) * \left\{ \ln t - \frac{t^{\alpha-2}}{\Gamma(\alpha-1)}[\psi(\alpha-1) \right\} \qquad (156)$$

Similarly, it is possible to obtain Apelblat (2020) [54]

$$\frac{\partial [t^{\beta-1} E_{\alpha,\beta}(t^{\alpha})]}{\partial \alpha} = L^{-1}\left\{ -\frac{s^{\alpha-1}}{s^{\alpha}-1}\cdot\frac{s^{\alpha-\beta}}{s^{\alpha}-1}\cdot\frac{\ln s}{s^{\alpha-1}} \right\} =$$

$$E_{\alpha}(t^{\alpha}) * \left[t^{\beta-1} E_{\alpha,\beta}(t^{\alpha}) \right] * \left\{ \ln t - \frac{t^{\alpha-2}}{\Gamma(\alpha-1)}[\psi(\alpha-1) \right\} \qquad (157)$$

If parameter β is considered as a variable, the Laplace transform of derivative of

$t^{\beta-1}E_{\alpha,\beta}(t^{\alpha})$ is

$$L\left\{\frac{\partial[t^{\beta-1}E_{\alpha,\beta}(t^{\alpha})]}{\partial\beta}\right\}=\frac{\partial}{\partial\beta}\left(\frac{s^{\alpha-\beta}}{s^{\alpha}-1}\right)=-\frac{s^{\alpha-\beta}\ln s}{s^{\alpha}-1}=-\frac{s^{\alpha-(\beta-\lambda)}}{s^{\alpha}-1}\cdot\frac{\ln s}{s^{\lambda}} \quad (158)$$

$\mathrm{Re}(\beta-\lambda)>0$

and the inverse of (158) is

$$\frac{\partial[t^{\beta-1}E_{\alpha,\beta}(t^{\alpha})]}{\partial\beta}=t^{\beta-1}\ln t\,E_{\alpha,\beta}(t^{\alpha})+t^{\beta-1}\frac{\partial E_{\alpha,\beta}(t^{\alpha})}{\partial\beta}=$$
$$t^{\beta-\lambda-1}E_{\alpha,\beta-\lambda}(t^{\alpha})*\left[\frac{t^{\lambda-1}}{\Gamma(\lambda)}[\ln t-\psi(\lambda)]\right] \quad (159)$$

Since the Laplace transform of the Mittag-Leffler function $t^{\beta-1}E_{\alpha,\beta}(t^{\alpha})$ has a simple form it is possible to derive the integral of it with respect to the parameter β

$$L\left\{\int_0^{\lambda}t^{\beta-1}E_{\alpha,\beta}(t^{\alpha})\,d\beta\right\}=\int_0^{\infty}e^{-st}\left[\int_0^{\lambda}t^{\beta-1}E_{\alpha,\beta}(t^{\alpha})\,d\beta\right]dt=$$
$$\int_0^{\lambda}\left[\int_0^{\infty}e^{-st}t^{\beta-1}E_{\alpha,\beta}(t^{\alpha})\,dt\right]d\beta=\int_0^{\lambda}\frac{s^{\alpha-\beta}}{s^{\alpha}-1}d\beta=\frac{s^{\alpha}}{s^{\alpha}-1}\cdot\frac{1}{\ln s}-\frac{s^{\alpha-\lambda}}{s^{\alpha}-1}\cdot\frac{1}{\ln s} \quad (160)$$

but

$$L^{-1}\left\{\frac{s^{\alpha}}{s^{\alpha}-1}\right\}=1+t^{\alpha-1}E_{\alpha,\alpha}(t^{\alpha})$$
$$L^{-1}\left\{\frac{1}{\ln s}\right\}=\int_0^{\infty}\frac{t^{\xi-1}}{\Gamma(\xi)}d\xi \quad (161)$$

and therefore

$$\int_0^{\lambda}t^{\beta-1}E_{\alpha,\beta}(t^{\alpha})\,d\beta=\left[1+t^{\alpha-1}E_{\alpha,\alpha}(t^{\alpha})-t^{\lambda-1}E_{\alpha,\lambda}(t^{\alpha})\right]*\int_0^{\infty}\frac{t^{\xi-1}}{\Gamma(\xi)}d\xi \quad (162)$$

The integrals in (161) and (162) are closely linked with the Volterra functions Apelblat (2008, 2009) [51, 52] and the convolutions integrals in (150) - (153), (156), (157), (159) and (162), by changing the variable of integration, can be written as finite integrals (see (109)).

Similar to the way the integral Bessel functions are defined by Abramowitz and Stegun (1965) [63], it is possible to introduce the integral Mittag-Leffler functions Apelblat and. Gonzáles-Santander (2021) [59]

$$Ei_{\alpha}(t)=\int_{0}^{t}\frac{[Ei_{\alpha}(\xi)-1]}{\xi}d\xi=\sum_{k=1}^{\infty}\frac{t^{k}}{k\,\Gamma(\alpha k+1)}$$

$$Ei_{\alpha,\beta}(t)=\int_{0}^{t}\frac{\left[Ei_{\alpha,\beta}(\xi)-\dfrac{1}{\Gamma(\beta)}\right]}{\xi}d\xi=\sum_{k=1}^{\infty}\frac{t^{k}}{k\,\Gamma(\alpha k+\beta)} \tag{163}$$

and their Laplace transforms are

$$L\{Ei_{\alpha}(t)\}=\sum_{k=0}^{\infty}\frac{(k+1)!}{(k+1)\Gamma[\alpha(k+1)+1]}\left(\frac{1}{s}\right)^{k+2} \quad ; \quad \mathrm{Re}\, s>1$$

$$L\{Ei_{\alpha,\beta}(t)\}=\sum_{k=0}^{\infty}\frac{(k+1)!}{(k+1)\Gamma[\alpha(k+1)+\beta]}\left(\frac{1}{s}\right)^{k+2} \tag{164}$$

For particular values of parameters α and β, these infinite series in (163) and (164) can be expressed in a closed form and results are tabulated by Apelblat and Gonzáles-Santander (2021) [59]. For example, in the case given in (139), $\alpha = \beta = 1$, we have

$$E_{1,1}(t)=E_{1}(t)=e^{t}$$

$$Ei_{1}(t)=-\gamma-\ln t+Shi(t)+Chi(t) \tag{165}$$

$$L\{Ei_{1}(t)\}=-\frac{1}{s}\ln\left(1-\frac{1}{s}\right)$$

where the integral hyperbolic trigonometric functions are defined by

$$Shi(t)=\int_{0}^{t}\frac{\sinh\xi}{\xi}d\xi$$

$$Chi(t)=\gamma+\ln t-\int_{0}^{t}\frac{1-\cosh\xi}{\xi}d\xi \tag{166}$$

Laplace Transforms and the Wright Functions.

Initially, regarded as some kind of a generalization of the Bessel functions, the Wright functions are defined in the following way Mainardi (2010) [72]

$$\mathrm{W}_{\lambda,\mu}(t)=\sum_{k=0}^{\infty}\frac{t^k}{k!\Gamma(\lambda k+\mu)} \tag{167}$$

Nowadays, they are considered as independent special functions used in the solution of the linear partial fractional differential equations. These differential equations are applied to describe many physical phenomena which take place in the natural sciences and in the engineering disciplines. It is worthwhile to mention that they occur in statistical mechanics, in geophysics, in meteorology, in condensed and soft matter physics, in fractional diffusion and kinetics, in socio-economic models and in many other areas. Depending on the value of parameter λ, these functions are called the Wright functions of the first kind when $\lambda \geq 0$, and with $-1 < \lambda < 0$, the Wright functions of the second kind

$$\mathrm{W}_{-\nu,\mu}(t)=\sum_{k=0}^{\infty}\frac{t^k}{k!\Gamma(-\nu k+\mu)} \quad ; \quad 0<\nu<1 \quad ; \quad \mu\geq 0 \tag{168}$$

The Laplace transforms of the Wright functions of the second kind are Gorenflo, Luchko and Mainardi (1999) [73], Mainardi and Consiglio (2020) [74]

$$\begin{aligned} &L\{\mathrm{W}_{-\nu,\mu}(-t)\}=E_{\nu,\mu+\nu}(-s) \quad ; \quad 0<\nu<1 \quad ; \quad \mu\geq 0 \\ &L\left\{t^{\mu-1}\mathrm{W}_{-\nu,\mu}\left(-\frac{\lambda}{t^{\nu}}\right)\right\}=\frac{1}{s^{\mu}}e^{-\lambda s^{\nu}} \quad ; \quad \lambda>0 \end{aligned} \tag{169}$$

In the cases of $\mu = 0$ and $\mu = 1 - \nu$, these functions are called as the Mainardi auxiliary functions

$$\begin{aligned} &F_{\nu}(t)=\mathrm{W}_{-\nu,0}(-t)=\sum_{k=0}^{\infty}\frac{(-t)^k}{k!\Gamma(-\nu k)} \quad ; \quad 0<\nu<1 \\ &M_{\nu}(t)=\mathrm{W}_{-\nu,1-\nu}(-t)=\sum_{k=0}^{\infty}\frac{(-t)^k}{k!\Gamma[-\nu(k+1)+1]} \\ &F_{\nu}(t)=\nu t M_{\nu}(t) \end{aligned} \tag{170}$$

and these functions have the following Laplace transforms

$$
\begin{aligned}
&L\{F_\nu(t)\} = E_{\nu,\nu}(-s) \quad ; \quad 0<\nu<1 \quad ; \quad \lambda>0 \\
&L\left\{\frac{1}{t}F_\nu\left(\frac{\lambda}{t^\nu}\right)\right\} = e^{-\lambda s^\nu} \\
&L\{M_\nu(t)\} = E_{\nu,1}(-s) \\
&L\left\{\frac{1}{t^\nu}M_\nu\left(\frac{\lambda}{t^\nu}\right)\right\} = s^{\nu-1}e^{-\lambda s^\nu}
\end{aligned}
\tag{171}
$$

The inversion transform of the exponential functions which appear in (169) and (171) is not easy to obtain and therefore this problem attracted many well-known mathematicians over the 1945-1970 period (Humbert (1945) [75], Pollard (1946) [76], Wlodarski (1952) [77], Mikusinski (1951, 1953, 1958, 1959) [78-81], Wintner (1956) [82], Ragab (1958) [83] and Stankovič (1970) [84]) In particular cases where ν is either 1/4, 1/3, 1/2, 2/3 and $\nu = k/m, > 0$, $k < m$, the inverse Laplace transforms can be expressed in terms of various special functions (for details see Apelblat and Mainardi (2020) [55]). In this problem, it is important to mention the product theorem that was generalized by Efros in the following form (1935) [85]

$$
\begin{aligned}
&L\{f(t)\} = F(s) \quad ; \quad G(s,\lambda) = H(s)e^{-\lambda q(s)} \\
&L^{-1}\{H(s)\,F[q(s)]\} = \int_0^\infty f(\xi)\,g(\xi,\lambda)\,d\xi
\end{aligned}
\tag{172}
$$

A special case with $H(s) = 1/s^\beta$ and $q(s) = s^\nu$, with $0 < \nu < 1$, was discussed by Wlodarski (1952) [77]. By introducing $\beta = 1 - \nu$, the Wlodarski formula can be applied to the Wright functions of the second kind Efros (1935) [55]

$$
\begin{aligned}
&L\{g(t)\} = G(s) \\
&L^{-1}\left\{\frac{G(s^\nu)}{s^{1-\nu}}\right\} = \int_0^\infty g(\xi)L^{-1}\left\{\frac{e^{-\xi s^\nu})}{s^{1-\nu}}\right\}d\xi \quad ; \quad 0<\nu<1
\end{aligned}
\tag{173}
$$

Since

$$
L\left\{\frac{1}{\nu}F_\nu\left(\frac{\lambda}{t^\nu}\right)\right\} = L\left\{\frac{\lambda}{t^\nu}M_\nu\left(\frac{\lambda}{t^\nu}\right)\right\} = \frac{e^{-\lambda s^\nu}}{s^{1-\nu}}e^{-\lambda s^\nu} \quad ; \quad 0<\nu<1 \quad ; \quad \lambda>0
\tag{174}
$$

it follows from (173) that

$$\nu L^{-1}\left\{\frac{G(s^{\nu})}{s^{1-\nu}}\right\}=\int_0^{\infty} g(\xi)F_{\nu}\left(\frac{\xi}{t^{\nu}}\right)\frac{d\xi}{\xi} \quad ; \quad 0<\nu<1$$

$$t^{\nu}L^{-1}\left\{\frac{G(s^{\nu})}{s^{1-\nu}}\right\}=\int_0^{\infty} g(\xi)M_{\nu}\left(\frac{\xi}{t^{\nu}}\right)d\xi$$

(175)

Thus, if the left-hand side Laplace transforms in (175) can be inverted with the help of tables and mathematical manipulations, the infinite integrals with the Mainardi functions $F_{\nu}(t)$ and $M_{\nu}(t)$ can be evaluated. The integrands of these integrals can include elementary functions (power, exponential, logarithmic, trigonometric and hyperbolic functions) and special functions (the error functions, the Mittag-Leffler, the modified Bessel and the Volterra functions). In the case of the modified Bessel functions of the second kind and order one-third they are Mainardi (2010) [72], Gorenflo, Luchko and Mainardi (1999) [73]

$$F_{1/3}\left(\frac{\xi}{t^{1/3}}\right)=\frac{\xi^{3/2}}{3\pi t^{1/3}}K_{1/3}(\phi) \quad ; \quad \phi=\frac{2\xi^{3/2}}{\sqrt{27t}}$$

$$M_{1/3}\left(\frac{\xi}{t^{1/3}}\right)=\frac{\xi^{1/2}}{\pi t^{1/6}}K_{1/3}(\phi) \qquad (176)$$

Derived using (175), integral identities are expressed mainly in the form of convolution integrals. These and other integrals are presented in Table 4.

Table 4. The integrals of the Mainardi functions $F_{\nu}(t)$ and $M_{\nu}(t)$.

The inverse Laplace transforms are based on (175) formulas.
$\int_0^{\infty} \xi^{\lambda-1} F_{\nu}\left(\frac{\xi}{t^{\nu}}\right)d\xi=\frac{\Gamma(\lambda)t^{\lambda\nu}}{\Gamma(\lambda\nu)} \quad ; \quad 0<\nu<1 \quad ; \quad \lambda>0$
$\int_0^{\infty} \xi^{\lambda} M_{\nu}\left(\frac{\xi}{t^{\nu}}\right)d\xi=\frac{\Gamma(\lambda)t^{(\lambda+1)\nu}}{\nu\Gamma(\lambda\nu)}$

Table 4. (Continued)

The inverse Laplace transforms are based on (175) formulas.
$\int_0^\lambda F_\nu\left(\frac{\xi}{t^\nu}\right)\frac{d\xi}{\xi}=\nu\left[1-\mathrm{W}_{-\nu,1}\left(-\frac{\lambda}{t^\nu}\right)\right]$; $\mathrm{W}_{-\nu,1}\left(-\frac{\lambda}{t^\nu}\right)=L^{-1}\left\{\frac{e^{-\lambda s^\nu}}{s}\right\}$
$\int_0^\lambda M_\nu\left(\frac{\xi}{t^\nu}\right)d\xi=t^\nu\left[1-\mathrm{W}_{-\nu,1}\left(-\frac{\lambda}{t^\nu}\right)\right]$
$\int_0^1(1-\xi)F_\nu\left(\frac{\xi}{t^\nu}\right)\frac{d\xi}{\xi}=\nu\left[1-\frac{t^\nu}{\Gamma(\nu+1)}+\frac{t^{\nu-1}}{\Gamma(\nu)}*\mathrm{W}_{-\nu,1}\left(-\frac{1}{t^\nu}\right)\right]$
$\int_0^1(1-\xi)M_\nu\left(\frac{\xi}{t^\nu}\right)d\xi=t^\nu\left[1-\frac{t^\nu}{\Gamma(\nu+1)}+\frac{t^{\nu-1}}{\Gamma(\nu)}*\mathrm{W}_{-\nu,1}\left(-\frac{1}{t^\nu}\right)\right]$
$\int_\lambda^\infty(\xi-\lambda)^\mu F_\nu\left(\frac{\xi}{t^\nu}\right)\frac{d\xi}{\xi}=\frac{\nu\Gamma(\mu+1)}{\Gamma(\mu\nu)}t^{\nu\mu-1}*\mathrm{W}_{-\nu,1}\left(-\frac{\lambda}{t^\nu}\right)$; $\mu>0$
$\int_\lambda^\infty(\xi-\lambda)^\mu M_\nu\left(\frac{\xi}{t^\nu}\right)d\xi=\frac{\Gamma(\mu+1)}{\Gamma(\mu\nu)}t^{\nu(\mu+1)-1}*\mathrm{W}_{-\nu,1}\left(-\frac{\lambda}{t^\nu}\right)$
$\int_0^\infty e^{-\lambda\xi}F_\nu\left(\frac{\xi}{t^\nu}\right)\frac{d\xi}{\xi}=\nu\,E_\nu(-\lambda t^\nu)$
$\int_0^\infty e^{-\lambda\xi}M_\nu\left(\frac{\xi}{t^\nu}\right)d\xi=t^\nu\,E_\nu(-\lambda t^\nu)$
$\int_\lambda^\infty[1-e^{-(\xi-\lambda)}]F_\nu\left(\frac{\xi}{t^\nu}\right)\frac{d\xi}{\xi}=\frac{\nu}{\Gamma(\nu-1)}\left[[t^{\nu-2}*E_\nu(-t^\nu)]*\mathrm{W}_{-\nu,1}\left(-\frac{\lambda}{t^\nu}\right)\right]$
$\int_\lambda^\infty[1-e^{-(\xi-\lambda)}]M_\nu\left(\frac{\xi}{t^\nu}\right)d\xi=\frac{t^\nu}{\Gamma(\nu-1)}\left[[t^{\nu-2}*E_\nu(-t^\nu)]*\mathrm{W}_{-\nu,1}\left(-\frac{\lambda}{t^\nu}\right)\right]$

The inverse Laplace transforms are based on (175) formulas.
$\int_0^\infty \ln\xi\, F_\nu\left(\frac{\xi}{t^\nu}\right)\frac{d\xi}{\xi} = \nu[(\nu-1)\gamma + \nu\ln t]$
$\int_0^\infty \ln\xi\, M_\nu\left(\frac{\xi}{t^\nu}\right)d\xi = t^\nu[(\nu-1)\gamma + \nu\ln t]$
$\int_0^\infty \xi^{\lambda-1}\ln\xi\, F_\nu\left(\frac{\xi}{t^\nu}\right)\frac{d\xi}{\xi} = \frac{\nu\Gamma(\lambda)[\psi(\lambda)+\gamma\nu]t^{\nu(\lambda-1)}}{\Gamma[\nu(\lambda-1)+1]} + \frac{\nu\Gamma(\lambda)}{\Gamma[\nu(\lambda-1)]}t^{\nu(\lambda-1)-1} * \ln(t^\nu)$
$\int_0^\infty \xi^{\lambda-1}\ln\xi\, M_\nu\left(\frac{\xi}{t^\nu}\right)d\xi = \frac{\Gamma(\lambda)[\psi(\lambda)+\gamma\nu]t^{\nu\lambda}}{\Gamma[\nu(\lambda-1)+1]} + \frac{\Gamma(\lambda)t^\nu}{\Gamma[\nu(\lambda-1)]}t^{\nu(\lambda-1)-1} * \ln(t^\nu)$
$\int_0^\infty \sin(\lambda\xi)\, F_\nu\left(\frac{\xi}{t^\nu}\right)\frac{d\xi}{\xi} = \frac{\nu\lambda}{\Gamma(\nu)}[t^{\nu-1} * E_{2\nu}(-\lambda^2 t^{2\nu})]$
$\int_0^\infty \sin(\lambda\xi)\, M_\nu\left(\frac{\xi}{t^\nu}\right)d\xi = \frac{\lambda t^\nu}{\Gamma(\nu)}[t^{\nu-1} * E_{2\nu}(-\lambda^2 t^{2\nu})]$
$\int_0^\infty \cos(\lambda\xi)\, F_\nu\left(\frac{\xi}{t^\nu}\right)\frac{d\xi}{\xi} = \nu E_{2\nu}(-\lambda^2 t^{2\nu})$
$\int_0^\infty \cos(\lambda\xi)\, M_\nu\left(\frac{\xi}{t^\nu}\right)d\xi = t^\nu E_{2\nu}(-\lambda^2 t^{2\nu})$
$\int_0^\infty \sinh(\lambda\xi)\, F_\nu\left(\frac{\xi}{t^\nu}\right)\frac{d\xi}{\xi} = \frac{\nu\lambda}{\Gamma(\nu)}[t^{\nu-1} * E_{2\nu}(\lambda^2 t^{2\nu})]$
$\int_0^\infty \sinh(\lambda\xi)\, M_\nu\left(\frac{\xi}{t^\nu}\right)d\xi = \frac{\lambda t^\nu}{\Gamma(\nu)}[t^{\nu-1} * E_{2\nu}(\lambda^2 t^{2\nu})]$
$\int_0^\infty \cosh(\lambda\xi)\, F_\nu\left(\frac{\xi}{t^\nu}\right)\frac{d\xi}{\xi} = \nu E_{2\nu}(\lambda^2 t^{2\nu})$

Table 4. (Continued)

The inverse Laplace transforms are based on (175) formulas.
$\int_0^\infty \cosh(\lambda\xi) M_\nu\left(\frac{\xi}{t^\nu}\right) d\xi = t^\nu E_{2\nu}(\lambda^2 t^{2\nu})$
$\int_0^\infty \sin(\lambda\xi)\sinh(\lambda\xi) F_\nu\left(\frac{\xi}{t^\nu}\right)\frac{d\xi}{\xi} = \frac{2\nu\lambda^2 t^{2\nu-1}}{\Gamma(2\nu)}[t^{2\nu-1} * E_{4\nu}(-4\lambda^4 t^{4\nu})]$
$\int_0^\infty \sin(\lambda\xi)\sinh(\lambda\xi) M_\nu\left(\frac{\xi}{t^\nu}\right) d\xi = \frac{2\nu\lambda^2 t^\nu}{\Gamma(2\nu)}[t^{2\nu-1} * E_{4\nu}(-4\lambda^4 t^{4\nu})]$
$\int_0^\infty \cos(\lambda\xi)\cosh(\lambda\xi) F_\nu\left(\frac{\xi}{t^\nu}\right)\frac{d\xi}{\xi} = \nu E_{4\nu}(-4\lambda^4 t^{4\nu})$
$\int_0^\infty \cos(\lambda\xi)\cosh(\lambda\xi) M_\nu\left(\frac{\xi}{t^\nu}\right) d\xi = t^\nu E_{4\nu}(-4\lambda^4 t^{4\nu})$
$\int_0^\infty E_\alpha(\pm\lambda\xi^\alpha) F_\nu\left(\frac{\xi}{t^\nu}\right)\frac{d\xi}{\xi} = \nu E_{\alpha\nu}(\pm\lambda t^\alpha) \quad ; \quad \alpha > 0$
$\int_0^\infty E_\alpha(\pm\lambda\xi^\alpha) M_\nu\left(\frac{\xi}{t^\nu}\right) d\xi = t^\nu E_{\alpha\nu}(\pm\lambda t^\alpha)$
$\int_0^\infty \xi^{\beta-1} E_{\alpha,\beta}(\pm\lambda\xi^\alpha) F_\nu\left(\frac{\xi}{t^\nu}\right)\frac{d\xi}{\xi} = \nu t^{(\beta-1)\nu} E_{\alpha\nu,(\beta-1)\nu+1}(\pm\lambda t^\alpha) \quad ; \quad \alpha, \beta > 0$
$\int_0^\infty \xi^{\beta-1} E_{\alpha,\beta}(\pm\lambda\xi^\alpha) M_\nu\left(\frac{\xi}{t^\nu}\right) d\xi = t^{\beta\nu} E_{\alpha\nu,(\beta-1)\nu+1}(\pm\lambda t^\alpha)$
$\int_0^\infty erf\left(\frac{\lambda}{2\sqrt{\xi}}\right) F_\nu\left(\frac{\xi}{t^\nu}\right)\frac{d\xi}{\xi} = \nu\left[1 - \mathrm{W}_{-\nu/2,1}\left(-\frac{\lambda}{t^{\nu/2}}\right)\right] \quad ; \quad \mathrm{W}_{-\nu/2,1}\left(-\frac{\lambda}{t^{\nu/2}}\right) = L^{-1}\left\{\frac{e^{-\lambda s^{\nu/2}}}{s}\right\}$

The inverse Laplace transforms are based on (175) formulas.
$\int_0^\infty erf\left(\frac{\lambda}{2\sqrt{\xi}}\right) M_\nu\left(\frac{\xi}{t^\nu}\right) d\xi = t^\nu \left[1 - \mathrm{W}_{-\nu/2,1}\left(-\frac{\lambda}{t^{\nu/2}}\right)\right]$
$\int_0^\infty \nu(\lambda\xi) F_\nu\left(\frac{\xi}{t^\nu}\right) \frac{d\xi}{\xi} = \nu(\lambda^{1/\nu} t) \quad ; \quad \nu(t) = \int_0^\infty \frac{t^\zeta}{\Gamma(\zeta+1)} d\zeta$
$\int_0^\infty \nu(\lambda\xi) M_\nu\left(\frac{\xi}{t^\nu}\right) d\xi = \frac{t^\nu}{\nu} \nu(\lambda^{1/\nu} t)$
$\int_0^\infty \nu(\lambda\xi, \alpha) F_\nu\left(\frac{\xi}{t^\nu}\right) \frac{d\xi}{\xi} = \lambda^\alpha \nu(\lambda^{1/\nu} t, \alpha\nu) \quad ; \quad \nu(t,\alpha) = \int_0^\infty \frac{t^{\zeta+\alpha}}{\Gamma(\zeta+\alpha+1)} d\zeta$
$\int_0^\infty \nu(\lambda\xi, \alpha) M_\nu\left(\frac{\xi}{t^\nu}\right) d\xi = \frac{\lambda^\alpha t^\nu}{\nu} \nu(\lambda^{1/\nu} t, \alpha\nu)$
$\int_0^\infty \nu(\xi, \alpha) F_\nu\left(\frac{\xi}{t^\nu}\right) d\xi = (\alpha+1)\nu[t, (\alpha+1)\nu] + \frac{1}{\nu}\mu[t, 1, (\alpha+1)\nu]$
$\int_0^\infty \xi\, \nu(\xi, \alpha) M_\nu\left(\frac{\xi}{t^\nu}\right) d\xi = t^\nu\, \nu[t, (\alpha+1)\nu] + \frac{1}{\nu}\mu[t, 1, (\alpha+1)\nu]$
$\int_0^\infty \mu(\lambda\xi, \beta, \alpha) F_\nu\left(\frac{\xi}{t^\nu}\right) \frac{d\xi}{\xi} = \frac{\lambda^\alpha}{\nu^\beta} \mu(\lambda^{1/\nu} t, \beta, \alpha\nu) \quad ; \quad \mu(t, \beta, \alpha) = \int_0^\infty \frac{\zeta^\beta t^{\zeta+\alpha}}{\Gamma(\beta+1)\Gamma(\zeta+\alpha+1)} d\zeta$
$\int_0^\infty \mu(\lambda\xi, \beta, \alpha) M_\nu\left(\frac{\xi}{t^\nu}\right) d\xi = \frac{\lambda^\alpha t^\nu}{\nu^{\beta+1}} \mu(\lambda^{1/\nu} t, \beta, \alpha\nu)$
$\int_0^\infty \sqrt{\xi}\, K_{1/3}(\phi) d\xi = \pi\sqrt{t} \quad ; \quad K_{1/3}(\phi) = \frac{3\pi t^{1/3}}{\xi^{3/2}} F_{1/3}\left(\frac{\xi}{t^{1/3}}\right) \quad ; \quad \phi = \frac{2\xi^{3/2}}{\sqrt{27t}}$

Table 4. (Continued)

The inverse Laplace transforms are based on (175) formulas.
$\int_0^\infty \xi^{3/2} K_{1/3}(\phi)\,d\xi = \dfrac{3\pi t^{5/6}}{\Gamma\left(\dfrac{1}{3}\right)}$; $K_{1/3}(\phi) = \dfrac{\pi t^{1/6}}{\xi^{1/2}} M_{1/3}\left(\dfrac{\xi}{t^{1/3}}\right)$
$\int_0^\infty \xi^{\lambda+1/2} K_{1/3}(\phi)\,d\xi = \dfrac{3\pi\,\Gamma(\lambda)\,t^{\lambda/3+1/2}}{\Gamma\left(\dfrac{\lambda}{3}\right)}$
$\int_0^\infty \xi^{2n+1/2} K_{1/3}(\phi)\,d\xi = \dfrac{3\pi\,\Gamma(2n)\,t^{2n/3+1/2}}{\Gamma\left(\dfrac{2n}{3}\right)}$
$\int_0^\infty \xi^{2n+3/2} K_{1/3}(\phi)\,d\xi = \dfrac{3\pi\Gamma(2n+1)\,t^{2n/3+5/6}}{\Gamma\left(\dfrac{2n+1}{3}\right)}$

By performing complex integration and the help of (49), the inversion of the exponential function, presented in (169), is Apelblat and Mainardi (2021) [57], Stankovič (1970) [84]

$$f_{\nu,\mu}(t) = L^{-1}\left\{\frac{e^{-s^\nu}}{s^\mu}\right\} = \frac{1}{\pi}\int_0^\infty \frac{e^{-t\xi - \xi^\nu \cos(\pi\nu)}}{\xi^\xi} \sin[\xi^\nu \sin(\pi\nu) + \pi\mu]\,d\xi \tag{177}$$

$$f_{\nu,\mu}(t) = t^{\mu-1} \mathrm{W}_{\nu,\mu}\left(-\frac{1}{t^\nu}\right) \quad ; \quad 0 < \nu < 1 \quad ; \quad \mu < 1$$

The case of $\mu = 0$, was already derived by Mikusinski (1951, 1953, 1958, 1959) [78-81]. For $\mu =$

$1 - \nu$, the Wright functions in (177) are reduced to the Mainardi functions

$$f_{\nu,1-\nu}(t) = \frac{1}{\nu} F_\nu\left(\frac{1}{t^\nu}\right) = \frac{1}{t^\nu} M_\nu\left(\frac{1}{t^\nu}\right) \quad ; \quad 0 < \nu < 1 \tag{178}$$

Applying simple operational rules of the Laplace transform Apelblat (2012) [53], it is possible to obtain that Apelblat and Mainardi (2021) [57]

$$f_{\nu,\mu}(t \to 0) \to 0$$

$$f_{\nu,\mu}(t \to \infty) \to \frac{1}{t^{\mu-1}} \left\{ \frac{1}{\Gamma(\mu)} - \frac{1}{\Gamma(\mu-\nu)t^{\nu}} + \frac{1}{2\Gamma(\mu-2\nu)t^{2\nu}} - \ldots \right\}$$

$$\frac{d f_{\nu,\mu}(t)}{dt} = f_{\nu,\mu-1}(t) = \frac{1}{t}[(\mu-1) f_{\nu,\mu}(t) + \nu\, f_{\nu,\mu-\nu}(t)] \tag{179}$$

$$\int_0^t f_{\nu,\mu}(\xi)\, d\xi = f_{\nu,\mu+1}(t)$$

and

$$f_{\nu,\mu+\lambda}(t) = \frac{t^{\lambda-1}}{\Gamma(\lambda)} * f_{\nu,\mu}(t)$$

$$f_{\nu,\mu}(t) * f_{\nu,\mu}(t) = \int_0^t f_{\nu,\mu}(t-\xi) f_{\nu,\mu}(\xi) d\xi = 2^{(2\mu-1)/\nu} f_{\nu,2\mu}\left(\frac{t}{2^{1/\nu}}\right)$$

$$f_{\nu,\mu}(t) * f_{\nu,\mu+\lambda}(t) = \int_0^t f_{\nu,\mu}(t-\xi) f_{\nu,\mu+\lambda}(\xi) d\xi = 2^{(2\mu+\lambda-1)/\nu} f_{\nu,\mu+\lambda}\left(\frac{t}{2^{1/\nu}}\right) \tag{180}$$

$$\frac{\partial f_{\nu,\mu}(t)}{\partial \nu} = \ln t * f_{\nu,\mu-\nu-1}(t) + \gamma f_{\nu,\mu-\nu}(t)$$

$$\frac{\partial f_{\nu,\mu}(t)}{\partial \mu} = -\mu\, f_{\nu,\mu+1}(t)$$

Similarly, by applying formulas in (175), various integrals with $f_{\nu,\mu}(t)$ functions can be derived by using the Wlodarski expression in (173). These infinite and convolution integrals are presented in Table 5

Table 5. The integrals with $f_{\nu,\mu}(t)$ functions.

The inverse Laplace transforms are based on (175) formulas.
$\int_0^\infty \xi^{\lambda} f_{\nu,\mu}(\xi)\, d\xi = \frac{\Gamma(\lambda+1) t^{(\lambda+1)\nu+\mu-1}}{\Gamma[(\lambda+1)\nu+\mu]} \quad ; \quad 0<\nu<1 \quad ; \quad \mu<1 \quad ; \quad \lambda>0$

Table 5. (Continued)

The inverse Laplace transforms are based on (175) formulas.
$\int_0^{\lambda} f_{\nu,\mu}(\xi)\,d\xi = \frac{t^{\nu+\mu-1}}{\Gamma(\nu+\mu)} - \lambda^{[(\mu-1)/\nu+1]} f_{\nu,\nu+\mu}\left(\frac{t}{\lambda^{1/\nu}}\right)$
$\int_0^{1} (1-\xi)\, f_{\nu,\mu}(\xi)\,d\xi = \frac{t^{2\nu+\mu-1}}{\Gamma(2\nu+\mu)} - \frac{t^{\nu+\mu-1}}{\Gamma(\nu+\mu)} + f_{\nu,2\nu+\mu}(t)$
$\int_{\lambda}^{\infty} (t-\lambda)^{\xi}\, f_{\nu,\mu}(\xi)\,d\xi = \lambda^{[(1/\nu+1)\mu-1/\nu+1]} f_{\nu,\nu+\nu\mu+\mu}\left(\frac{t}{\lambda^{1/\nu}}\right)$
$\int_0^{\infty} e^{-\alpha\xi} f_{\nu,\mu}(\xi)\,d\xi = \frac{t^{\nu+\mu-2}}{\Gamma(\nu+\mu-1)} * E_{\nu}(-\alpha t^{\alpha}) \quad ; \quad \alpha > 0$
$\int_{\lambda}^{\infty} \left[1-e^{-(\xi-\lambda)}\right] f_{\nu,\mu}(\xi)\,d\xi = \lambda^{(\nu+\mu-1)/\nu} E_{\nu}(-t^{\nu}) * f_{\nu,2\nu+\mu-1}\left(\frac{t}{\lambda^{1/\nu}}\right)$
$\int_{\lambda}^{\infty} e^{-(\xi-\lambda)} f_{\nu,\mu}(\xi)\,d\xi = \lambda^{(\nu+\mu-2)/\nu} E_{\nu}(-t^{\nu}) * f_{\nu,\nu+\mu-1}\left(\frac{t}{\lambda^{1/\nu}}\right)$
$\int_0^{\infty} \ln\xi\, f_{\nu,\mu}(\xi)\,d\xi = \frac{\gamma(\nu-1)t^{\nu+\mu-1}}{\Gamma(\nu+\mu)} + \frac{\nu t^{\nu+\mu-2}}{\Gamma(\nu+\mu-1)} * \ln t$
$\int_0^{\infty} \xi^{\lambda-1} \ln\xi\, f_{\nu,\mu}(\xi)\,d\xi = \frac{\Gamma(\lambda)[\psi(\lambda)+\gamma\nu]t^{\lambda\nu+\mu-1}}{\Gamma(\lambda\nu+\mu)} + \frac{\nu\Gamma(\lambda)t^{\lambda\nu+\mu-2}}{\Gamma(\lambda\nu+\mu-1)} * \ln(t^{\nu})$
$\int_0^{\infty} \sin(\lambda\xi)\, f_{\nu,\mu}(\xi)\,d\xi = \frac{\lambda t^{2\nu+\mu-2}}{\Gamma(2\nu+\mu-1)} * E_{2\nu}(-\lambda^2 t^{2\nu})$
$\int_0^{\infty} \cos(\lambda\xi)\, f_{\nu,\mu}(\xi)\,d\xi = \frac{t^{\nu+\mu-2}}{\Gamma(\nu+\mu-1)} * E_{2\nu}(-\lambda^2 t^{2\nu})$

The inverse Laplace transforms are based on (175) formulas.
$\int_0^\infty \sinh(\lambda\xi) f_{\nu,\mu}(\xi)\, d\xi = \frac{\lambda t^{2\nu+\mu-2}}{\Gamma(2\nu+\mu-1)} * E_{2\nu}(\lambda^2 t^{2\nu})$
$\int_0^\infty \cosh(\lambda\xi) f_{\nu,\mu}(\xi)\, d\xi = \frac{t^{\nu+\mu-2}}{\Gamma(\nu+\mu-1)} * E_{2\nu}(\lambda^2 t^{2\nu})$
$\int_0^\infty \sin(\lambda\xi)\sinh(\lambda\xi) f_{\nu,\mu}(\xi)\, d\xi = \frac{2\lambda^2 t^{3\nu+\mu-2}}{\Gamma(3\nu+\mu-1)} * E_{4\nu}(-4\lambda^4 t^{4\nu})$
$\int_0^\infty \cos(\lambda\xi)\cosh(\lambda\xi) f_{\nu,\mu}(\xi)\, d\xi = \frac{t^{\nu+\mu-2}}{\Gamma(\nu+\mu-1)} * E_{4\nu}(-4\lambda^4 t^{4\nu})$
$\int_0^\infty E_\alpha(\pm\lambda\xi^\alpha) f_{\nu,\mu}(\xi)\, d\xi = t^{\nu+\mu-1} E_{\alpha\nu,\nu+\mu}(\pm\lambda t^\alpha)$
$\int_0^\infty \xi^{\beta-1} E_{\alpha,\beta}(\pm\lambda\xi^\alpha) f_{\nu,\mu}(\xi)\, d\xi = t^{\beta\nu+\mu-1} E_{\alpha\nu,\beta\nu+\mu}(\pm\lambda t^\alpha)$
$\int_0^\infty erf\left(\frac{\lambda}{2\sqrt{\xi}}\right) f_{\nu,\mu}(\xi)\, d\xi = \frac{t^{\mu-1}}{\Gamma(\mu)} - \lambda^{2(\mu-1)/\nu} f_{\nu/2,\mu}\left(\frac{t^{\nu/2}}{\lambda^{2/\nu}}\right)$
$\int_0^\infty \nu(\lambda\xi) f_{\nu,\mu}(\xi)\, d\xi = \frac{t^{\nu+\mu-2}}{\nu\Gamma(\nu+\mu-1)} * \nu(\lambda^{1/\nu} t)$
$\int_0^\infty \nu(\lambda\xi,\alpha) f_{\nu,\mu}(\xi)\, d\xi = \frac{t^{\mu-2}}{\nu\lambda\Gamma(\mu-1)} * \nu[\lambda^{1/\nu} t,(\alpha+1)\nu]$
$\int_0^\infty \mu(\lambda\xi,\beta,\alpha) f_{\nu,\mu}(\xi)\, d\xi = \frac{t^{\mu-2}}{\lambda\Gamma(\mu-1)\nu^{\beta+1}} * \mu[\lambda^{1/\nu} t,\beta,(\alpha+1)\nu]$

Generally, the Laplace transforms of the Wright functions of the first kind are expressed in terms of the two-parameter Mittag-Leffler functions Gorenflo, Luchko and Mainardi (1999) [73], Mainardi and Consiglio (2020) [74]

$$L\left\{W_{\alpha,\beta}(\pm\lambda t)\right\} = \frac{1}{s} E_{\alpha,\beta}\left(\pm\frac{\lambda}{s}\right) \quad ; \quad \alpha > 0 \quad ; \quad \lambda > 0$$

(181)

By applying operational rules of the Laplace transform, we have Apelblat and Mainardi (2022) [60]

$$L\left\{t\, W_{\alpha,\beta}(\lambda t)\right\} = \frac{1}{\alpha\lambda s^2}\left[(\alpha\lambda - \beta + 1) E_{\alpha,\beta}\left(\frac{\lambda}{s}\right) + E_{\alpha,\beta-1}\left(\frac{\lambda}{s}\right)\right]$$

$$L\left\{e^{\pm\rho t}\, W_{\alpha,\beta}(\lambda t)\right\} = \frac{1}{s\mp\rho} E_{\alpha,\beta}\left(\frac{\lambda}{s\mp\rho}\right) \quad ; \quad \lambda, \rho > 0$$

$$L\left\{\sinh(\rho t)\, W_{\alpha,\beta}(\lambda t)\right\} = \frac{1}{2}\left[\frac{1}{s-\rho} E_{\alpha,\beta}\left(\frac{\lambda}{s-\rho}\right) - \frac{1}{s+\rho} E_{\alpha,\beta}\left(\frac{\lambda}{s+\rho}\right)\right]$$

$$L\left\{\cosh(\rho t)\, W_{\alpha,\beta}(\lambda t)\right\} = \frac{1}{2}\left[\frac{1}{s-\rho} E_{\alpha,\beta}\left(\frac{\lambda}{s-\rho}\right) + \frac{1}{s+\rho} E_{\alpha,\beta}\left(\frac{\lambda}{s+\rho}\right)\right]$$

(182)

A large group of the Laplace transforms of the Wright functions of the first kind is available when they are reduced to the Bessel functions (for details see Apelblat and Mainardi (2022) [60]).

Laplace Transforms and the Bateman and Havelock Functions

In the solution of some turbulent flow problems, Bateman introduced a new function (which nowadays is called the Bateman function) of argument x and order ν Bateman (1931) [86]

$$k_\nu(x) = \frac{\pi}{2} \int_0^{\pi/2} \cos(x \tan\theta - \nu\theta)\, d\theta \tag{183}$$

These functions can be linked to the confluent hypergeometric and Whittaker functions Apelblat, Consiglio and Mainardi (2021), Erdélyi, Magnus, Oberhettinger and Tricomi (1955) [58, 69]. Basic properties of the Bateman functions and the allied Havelock functions which are associated with surface waves problems, are described by Chaudhur (1968) [87]

$$h_\nu(x) = \frac{\pi}{2} \int_0^{\pi/2} \sin(x \tan\theta - \nu\theta)\, d\theta \tag{184}$$

They were recently reviewed by Apelblat, Consiglio and Mainardi (2021) [58].

The Laplace transforms of the Bateman functions are Erdélyi, Magnus, Oberhettinger and Tricomi (1954) [65], Roberts and Kaufman (1968) [66]

$$\begin{aligned}
&L\{k_0(t)\} = \frac{1}{s+1} \quad ; \quad \operatorname{Re} s > -1 \\
&L\{k_{2n+2}(t)\} = \frac{2(1-s)^n}{(s+1)^{n+2}} \quad ; \quad m,n = 0,1,2,\ldots \\
&L\{k_{2\nu}(t)\} = \frac{\sin(\pi\nu)}{2\pi\nu(1-\nu)}\, {}_2F_1\left(1,2;2-\nu;-\frac{s-1}{2}\right) \quad ; \quad \operatorname{Re} s > 0 \\
&L\left\{\frac{1}{t} k_{2m+2}\left(\frac{t}{2}\right) k_{2n+2}\left(\frac{t}{2}\right)\right\} = \frac{(-1)^{m+n}}{(s+1)^{m+n+2}}\, {}_2F_1\left(-m,-n;2;\frac{1}{s^2}\right)
\end{aligned} \tag{185}$$

By using the initial and final values theorems of operational calculus it is possible to show that Apelblat, Consiglio and Mainardi (2021) [58]

$$\begin{aligned}
&k_0(t \to 0) \to 1 \\
&k_0(t \to \infty) \to 0 \\
&k_{2n+2}(t \to 0) \to 0 \\
&k_{2n+2}(t \to \infty) \to 0
\end{aligned} \tag{186}$$

Since the Bateman function is a particular case of the Whittaker function $W_{\nu,1/2}(t)$ Abramowitz and Stegun (1965) [63], Erdélyi, Magnus, Oberhettinger and Tricomi (1955) [69]

$$k_{2\nu}\left(\frac{t}{2}\right) = \frac{1}{\Gamma(\nu+1)} W_{\nu,1/2}(t) \tag{187}$$

it is possible to derive the Laplace transforms by using particular cases of the Whittaker function

$$L\left\{t^{\alpha-1}k_{2\nu}\left(\frac{t}{2}\right)\right\}=$$

$$\frac{\Gamma(\alpha)\Gamma(\alpha+1)}{\Gamma(\nu+1)\Gamma(\alpha-\nu+1)}\left(\frac{2}{1+2s}\right)^{\alpha+1}{}_2F_1\left(\alpha+1,-\nu;\alpha-\nu+1;\frac{2s-1}{2s+1}\right) \tag{188}$$

$$L\left\{t^{\nu}\,e^{1/2t}k_{2\nu}\left(\frac{2}{t}\right)\right\}=\frac{2^{1-2\nu}}{\Gamma(\nu+1)\,s^{\nu+1/2}}S_{2\nu,1}(2\sqrt{s}) \quad ; \quad \mathrm{Re}(\nu\pm\frac{1}{2})>-\frac{1}{2}$$

$$L\left\{t^{-\nu}\,e^{-1/2t}k_{2\nu}\left(\frac{2}{t}\right)\right\}=\frac{2s^{\nu-1/2}}{\Gamma(\nu+1)}K_1(2\sqrt{s})$$

where $S_{\nu,\mu}(z)$ denotes the Lommel function Abramowitz and Stegun (1965) [63].

In analogy to the integral Bessel functions, Chaudhur (1968) [87] introduced the integral Bateman function

$$ki_{2n}(t)=-\int_t^{\infty}\frac{ki_{2n}(\xi)}{\xi} \quad ; \quad t>0$$

(189)

that is related to the Laguerre polynomials.

Using the operational rules, their asymptotic properties at the origin are

$$ki_{2n}(t\to 0)\to\begin{cases}\dfrac{2}{n} & ; \quad n=2k+1 \quad ; \quad k=0,1,2,3,\ldots\\ 0 & ; \quad n=2k\end{cases} \tag{190}$$

The corresponding Laplace transforms are

$$L\{ki_0(t)\}=-\frac{\ln s}{s}$$

$$L\{ki_2(t)\}=-\frac{2}{s+1} \tag{191}$$

$$L\{ki_{2n}(t)\}=\frac{1}{ns}\left[\left(\frac{1-s}{s+1}\right)^n-1\right]$$

For the Havelock functions (184) we have Apelblat, Consiglio and Mainardi (2021) [58]

$$h_0(t) = \frac{1}{2}\left[e^t li(e^{-t}) - e^{-t} li(e^t)\right] \quad ; \quad li(t) = \int_0^t \frac{d\xi}{\ln \xi}$$

$$h_0(t \to 0) \to 0 \qquad (192)$$

$$h_0(t \to \infty) \to 0$$

and

$$L\{h_0(t)\} = \frac{2 \ln s}{\pi (s^2 - 1)}$$

$$L\{h_1(t)\} = \frac{2}{\pi (s+1)}\left[\frac{\sec^{-1}(s)}{\sqrt{s^2 - 1}} - 1\right] \qquad (193)$$

$$L\{h_2(t)\} = \frac{2(s + 1 + \ln s)}{\pi (s^2 + 1)^2}$$

Laplace Transforms and the Volterra Functions.

The Volterra functions appeared in mathematics in the context of solutions of integral equations of convolution type with the logarithmic kernels and their interrelations with various special functions [51, 52, 69, 88-93]. These functions were previously considered in several places, mainly associated with derivation of various integrals. These functions and their Laplace transforms are presented in (82) and (83). Here the Volterra functions will be discussed by considering the form of their Laplace transforms and how this can be used to derive their functional representations in the form of direct and inverse transforms and integral equalities.

Starting with the complex integration of (83) by applying (49), the integral representation of the Volterra function $\nu(t,\alpha)$ is Apelblat (2008) [51]

$$\nu(t,\alpha) = \frac{1}{2\pi i}\int_{c-i\infty}^{c+i\infty} \frac{e^{st}}{s^{\alpha+1} \ln s} ds = e^t + \frac{1}{\pi}\int_0^\infty \frac{e^{-t\xi} \sin\left[\pi(\alpha+1) + \tan^{-1}\left(\frac{\pi}{\xi}\right)\right]}{\xi^{\alpha+1}\sqrt{\pi^2 + (\ln \xi)^2}} d\xi \qquad (194)$$

For $\alpha = 0$, the integral representation in (194) reduces to the Ramanujan result presented in (79). The asymptotic behaviour of this functions which is based on the initial and final values theorems of operational calculus is

$$\nu(t \to 0,\alpha) \to \frac{t^{\alpha}}{\Gamma(\alpha+1)}\left\{\frac{1}{\ln(1/t)} - \frac{\psi(\alpha+1)}{[\ln(1/t)]^2} + ...\right\} \tag{195}$$

$$\nu(t \to \infty,\alpha) \to e^{t}$$

If the Laplace transform of $\nu(t,\alpha)$ is presented in the following form

$$L\{\nu(t,\alpha)\} = \frac{1}{s^{\alpha+1}\ln s} = \frac{1}{s^{\alpha_1}}\cdot\frac{1}{s^{\alpha_2+1}\ln s} \tag{196}$$

$$\alpha = \alpha_1 + \alpha_2$$

then by using the product theorem of the Laplace transform we have

$$\nu(t,\alpha) = \frac{t^{\alpha_1-1}}{\Gamma(\alpha_1)} * \nu(t,\alpha_2) = \frac{1}{\Gamma(\alpha_1)}\int_0^t (t-\xi)^{\alpha_1-1}\nu(\xi,\alpha_2)\,d\xi \tag{197}$$

$$\alpha = \alpha_1 + \alpha_2$$

Similarly

$$\mu(t,\beta,\alpha) = \frac{t^{\alpha_1-1}}{\Gamma(\alpha_1)} * \mu(t,\beta,\alpha_2) = \mu(t,\beta_1,\alpha_1) * \mu(t,\beta_2-1,\alpha_2-1)$$

$$\alpha = \alpha_1 + \alpha_2 \quad ; \quad \beta = \beta_1 + \beta_2$$

(198)

From the multiplication of the original function by the variable t it follows that

$$L\{t\,\nu(t,\alpha)\} = -\frac{d}{ds}\left[\frac{1}{s^{\alpha+1}\ln s}\right] = \frac{\alpha+1}{s^{\alpha+2}\ln s} + \frac{1}{s^{\alpha+2}(\ln s)^2} \tag{199}$$

and the inversion of (199) gives the following recurrence relation

$$\mu(t,1,\alpha) = t\,\nu(t,\alpha) - (\alpha+1)\nu(t,\alpha+1) \tag{200}$$

Similarly, from

$$t\,\mu(t,\beta,\alpha) = -L^{-1}\left\{\frac{d}{ds}\left[\frac{1}{s^{\alpha+1}(\ln s)^{\beta+1}}\right]\right\} \tag{201}$$

we have the interrelation between the generalized Volterra functions

$$t\,\mu(t,\beta,\alpha) = (\alpha+1)\,\mu(t,\beta,\alpha+1) + (\beta+1)\,\mu(t,\beta+1,\alpha+1) \tag{202}$$

The expressions in (200) and (202) can be written in the alternative form for terms of convolution integrals

$$t\nu(t,\alpha)=\nu(t,\alpha)*[(\alpha+1)+\nu(t)]$$
$$t\mu(t,\beta,\alpha)=\mu(t,\beta,\alpha)*[[(\alpha+1)+(\beta+1)\nu(t)] \tag{203}$$

Considering the substitution formula

$$L\{f(t)\}=F(s)$$

$$L^{-1}\left\{\frac{1}{\sqrt{s}}F(\sqrt{s})\right\}=\frac{1}{\sqrt{\pi t}}\int_0^\infty e^{-\xi^2/4t}\, f(\xi)\,d\xi \tag{204}$$

and by presenting the Laplace inverse transform of the Volterra function in the following form

$$2\nu(t)=L^{-1}\left\{\frac{2}{s\ln s}\right\}=L^{-1}\left\{\frac{1}{\sqrt{s}\,\ln(\sqrt{s})}\right\} \tag{205}$$

we have from (204) and (205) that

$$\nu(t)=\frac{1}{2\sqrt{\pi t}}\int_0^\infty e^{-\xi^2/4t}\,\nu(\xi)\,d\xi \tag{206}$$

Similarly it is possible to obtain

$$\mu(t,\beta,\alpha)=\frac{1}{2^{\beta+1}\sqrt{\pi t}}\int_0^\infty e^{-\xi^2/4t}\,\mu(\xi,\beta,2\alpha)\,d\xi \tag{207}$$

Using the operational rule Apelblat (2012) [53]

$$L^{-1}\left\{\frac{1}{s^{2\lambda+1}}F\left(\frac{1}{s}\right)\right\}=\int_0^\infty\left(\frac{t}{\xi}\right)^{\lambda} J_{2\lambda}(2\sqrt{t\xi})\,f(\xi)\,d\xi \tag{208}$$

and representing the Laplace transform in the following form

$$L\{\nu(t,\alpha)\}=\frac{1}{s^{\alpha+1}\ln s}=-\frac{1}{s^{\alpha+2}\left[\frac{1}{s}\ln\left(\frac{1}{s}\right)\right]} \tag{209}$$

we have

$$\nu(t,\alpha)=-\int_0^\infty\left(\frac{t}{\xi}\right)^{(\alpha+1)/2} J_{\alpha+1}(2\sqrt{t\xi})\,\nu(\xi)\,d\xi \tag{210}$$

There are many Laplace transforms of various functions which can be expressed in terms of the Volterra functions [51,52,65,67]. If a function $f(t,\lambda)$ has the Laplace transform in the form

$$L\{f(t,\lambda)\} = F_1(s)[F_2(s)]^{\lambda} \tag{211}$$

then

$$L\left\{\int_{\alpha}^{\infty} \frac{f(t,\xi)}{\Gamma(\xi+1)}\, d\xi\right\} = F_1(s)\nu[F_2(s),\alpha] \quad ; \quad \operatorname{Re}\alpha > -1$$

(212)

In this way, the Volterra functions with composite arguments are the Laplace transforms of infinite integrals presented in (212). For example, in the case of incomplete gamma function Oberhettinger and Badii (1973) [67]

$$L\{\gamma(\lambda,bt\} = \frac{\Gamma(\lambda)}{s}\left(\frac{b}{s+b}\right)^{\lambda} \quad ; \quad \operatorname{Re}\lambda > -1 \quad ; \quad \operatorname{Re}b > 0 \tag{213}$$

it follows from (212) that

$$L\left\{\int_{\alpha}^{\infty} \frac{\gamma(\xi,at)}{\Gamma(\xi)\Gamma(\xi+1)}\, d\xi\right\} = \frac{1}{s}\nu\left(\frac{b}{s+b},\alpha\right) \quad ; \quad \operatorname{Re}\alpha > -1 \quad ; \quad \operatorname{Re}(s-b) > 0 \tag{214}$$

and similarly in the case of the Bessel function of the first kind we have

$$L\{J_{\lambda}(bt)\} = \frac{1}{\sqrt{s^2+b^2}}\left[\frac{b}{s+\sqrt{s^2+b^2}}\right]^{\lambda} \quad ; \quad \operatorname{Re}\lambda > -1 \quad ; \quad \operatorname{Re}b > 0$$

$$L\left\{\int_{\alpha}^{\infty} \frac{J_{\xi}(bt)}{\Gamma(\xi+1)}\, d\xi\right\} = \frac{1}{\sqrt{s^2+b^2}}\nu\left(\frac{b}{s+\sqrt{s^2+b^2}},\alpha\right) \quad ; \quad \operatorname{Re}\alpha > -1 \tag{215}$$

The Volterra functions are also associated with the reciprocal gamma functions

$$L\left\{\frac{1}{\Gamma(t+1)}\right\} = \int_0^{\infty} \frac{e^{-st}}{\Gamma(t+1)}\, dt = \int_0^{\infty} \frac{(e^{-s})^t}{\Gamma(t+1)}\, dt = \nu(e^{-s}) \tag{216}$$

And, therefore, for all three Volterra functions we have the following Laplace transforms

$$L\left\{\frac{1}{\Gamma(t+1)}\right\}=\nu(e^{-s})$$

$$L\left\{\frac{1}{\Gamma(t+\alpha+1)}\right\}=e^{\alpha s}\,\nu(e^{-s},\alpha) \tag{217}$$

$$L\left\{\frac{t^{\beta}}{\Gamma(t+\alpha+1)}\right\}=\Gamma(\beta+1)\,e^{\alpha s}\,\mu(e^{-s},\beta,\alpha)$$

It is worthwhile mentioning the substitution formulas of the inverse Laplace transform that include ln$s$ as variables

$$L\{f(t)\}=F(s)$$

$$L\left\{\int_0^\infty \frac{t^{\xi-1}}{\Gamma(\xi)}f(\xi)\,d\xi\right\}=F(\ln s) \quad ; \quad \mathrm{Re}\,s>1 \tag{218}$$

$$L\left\{\int_0^\infty \frac{t^{\xi}}{\Gamma(\xi+1)}f(\xi)\,d\xi\right\}=\frac{F(\ln s)}{s}$$

These formulas and those presented in (83) permit establishing many transforms and integral relations between the Volterra functions and various special functions Apelblat (2008) [51]. In the case where variable t is replaced by the square root of t in the Volterra function, this leads to

$$L\left\{\frac{1}{\sqrt{t}}\nu(b\sqrt{t})\right\}=\frac{1}{\sqrt{s}}\int_0^\infty \frac{\left(\frac{b^2}{4s}\right)^{\xi/2}}{\Gamma\left(\frac{\xi}{2}+1\right)}d\xi=2\sqrt{\frac{\pi}{s}}\,\nu\left(\frac{b^2}{4s}\right) \quad ; \quad \mathrm{Re}\,b>0 \tag{219}$$

and similarly

$$L\left\{\nu(b\sqrt{t},\alpha)\right\} = \frac{\sqrt{\pi}\,b}{s^{3/2}}\nu\left(\frac{b^2}{4s},\frac{\alpha-1}{2}\right) \quad ; \quad \mathrm{Re}\,\alpha > -1$$

$$L\left\{\frac{1}{\sqrt{t}}\nu(b\sqrt{t},2\alpha)\right\} = 2\sqrt{\frac{\pi}{s}}\nu\left(\frac{b^2}{4s},\alpha\right) \tag{220}$$

$$L\left\{\mu(b\sqrt{t},\beta,\alpha\right\} = \frac{2^{\beta}\sqrt{\pi}\,b}{s^{3/2}}\mu\left(\frac{b^2}{4s},\beta,\frac{\alpha-1}{2}\right) \quad ; \quad \mathrm{Re}\,\beta > -1$$

$$L\left\{\frac{1}{\sqrt{t}}\mu(b\sqrt{t},\beta,2\alpha\right\} = 2^{\beta+1}\sqrt{\frac{\pi}{s}}\mu\left(\frac{b^2}{4s},\beta,\alpha\right)$$

If the Laplace transforms in the following form

$$L\left\{t^{\lambda} f(t)\right\} = F(\lambda,s) \quad ; \quad \mathrm{Re}\,\lambda > -1 \tag{221}$$

are known, then the Laplace transforms of the products of the Volterra function and *f*(*t*) are given by

$$\int_0^\infty e^{-st} f(t)\nu(bt,\alpha)\,dt = \int_0^\infty \frac{b^{\xi+\alpha}}{\Gamma(\xi+\alpha+1)}F(\xi+\alpha,s)\,d\xi \tag{222}$$

$\mathrm{Re}\,\alpha > -1 \quad ; \quad \mathrm{Re}\,b > 0 \quad ; \quad \mathrm{Re}\,s > 0$

For example, since Oberhettinger and Badii (1973) [67]

$$L\left\{t^{\lambda}e^{-c/t}\right\} = 2\left(\frac{c}{s}\right)^{(\lambda+1)/2} K_{\lambda+1}(2\sqrt{cs}) \quad ; \quad \mathrm{Re}\,c > 0 \tag{223}$$

it follows from (222) that

$$\int_0^\infty e^{-st-c/t}\,\nu(bt,\alpha)\,dt = \frac{2}{b}\int_0^\infty \frac{\left(b\sqrt{\frac{c}{s}}\right)^{\xi+\alpha+1}}{\Gamma(\xi+\alpha+1)} K_{\xi+\alpha+1}(2\sqrt{sc})\,d\xi \tag{224}$$

In this integral equality, the integral with respect to the order of the modified Bessel function of the second kind is expressed as the infinite integral of the Volterra function.

Similarly, if the Laplace transforms

$$L\left\{[f(t)]^{\lambda}\right\} = F(\lambda,s) \quad ; \quad \mathrm{Re}\,\lambda > -1$$
$$L\left\{g(t)[f(t)]^{\lambda}\right\} = H(\lambda,s) \tag{225}$$

are available, then we have

$$\int_0^\infty e^{-st}\,\nu[bf(t),\alpha]\,dt = \int_0^\infty \frac{b^{\xi+\alpha}}{\Gamma(\xi+\alpha+1)} F(\xi+\alpha,s)\,d\xi$$
$$\int_0^\infty e^{-st}\,g(t)\nu[bf(t),\alpha]\,dt = \int_0^\infty \frac{b^{\xi+\alpha}}{\Gamma(\xi+\alpha+1)} H(\xi+\alpha,s)\,d\xi \tag{226}$$

For example, using the Laplace transform

$$L\left\{\frac{1}{\sqrt{2t+t^2}}(2t+t^2)^{\lambda}\right\} = \frac{\Gamma\left(\lambda+\frac{1}{2}\right)e^{s}}{\sqrt{\pi}}\left(\frac{2}{s}\right)^{\lambda} K_{\lambda}(s) \tag{227}$$

it is possible to obtain the following integral equality

$$\int_0^\infty \frac{e^{-st}}{\sqrt{2t+t^2}}\nu(2t+t^2,\alpha)\,dt = \frac{e^{s}}{\sqrt{\pi}}\int_0^\infty \frac{\Gamma\left(\xi+\alpha+\frac{1}{2}\right)}{\Gamma(\xi+\alpha+1)}\left(\frac{2}{s}\right)^{\xi+\alpha} K_{\xi+\alpha}(s)\,d\xi \tag{228}$$

As in many other cases, e.g. (224) and (228), the Volterra functions are associated with the Bessel and related functions in mathematical operations with respect to the order (for see more details Apelblat (2008, 2020) [51, 56]).

Conclusion

The Laplace transform itself or in a combination with other integral transforms, and frequently together with the use of special functions, is a well-established part of applied mathematics that has significant relevance in many scientific and technical disciplines. On a graduate level education in science and engineering, the Laplace transform is traditionally directed to solving linear ordinary and partial differential equations. Usually, these equations describe simple situations in dynamic control systems, kinetics of the first order chemical reactions and elementary cases of heat and mass transfer. However, as illustrated in this chapter, there is no doubt that methods of operational calculus can also be applied to solving a broad spectrum of complex linear and nonlinear physical and technological problems. For more than half century, the present author successfully applied the Laplace transform techniques in two areas, initially solving ordinary and partial differential equations associated with the boundary layer flows and with mass transfer with or without a chemical reaction, and later to derive many functional relations and integrals of special functions. This Chapter demonstrates how various properties of the Mittag-Leffler, Volterra, Whittaker, Wright, Bessel, Kelvin, Struve, Bateman and Havelock functions can be easily established by applying Laplace transform methods, in a skillful way.

References

[1] McLachlan, N.W., and Humbert, P. 1950 Formulaire pour Calcul Symbolique. Paris: Gauthier-Villars.
[2] McLachlan, N.W. 1962. Laplace Transforms and Their Applications to Differential Equations. New York:Dover Publ. Inc.
[3] McLachlan, N.W. 1963. Complex Variable and Transform Calculus. Cambridge: Cambridge University Press.
[4] Carson, J.R. 1926. Electric Circuit Theory and Operational Calculus. New York: McGraw-Hill Book Co.
[5] Efros, A.M. and Danilevsky, A.M. 1937. Operational Calculus and Contour Integrals. Kharkov:ONTI Press.
[6] Gardner, M.F. and Barnes, J.L. 1942. Transients in Linear Systems Studied by the Laplace Transform. Vol. 1. Lumped-Constant Systems. New York:J. Wiley & Sons.
[7] Ghizzetti, A. 1943. Calcolo Simbolico. Bologna:N. Zanichelli.
[8] Widder, D.V. 1948. The Laplace Transform. Princeton:Princeton University Press.
[9] Parodi, M. 1948. Applications Physique de la Transform de la Laplace. Paris:CNRS.
[10] Goldman, S. 1949. Transform Calculus and Electrical Transients. Englewood Cliff: Printice Hall Inc.
[11] Doetsch, G. 1950-1956. Handbuch der Laplace Transform. Vols. 1, 2 and 3. Bassel:Birkhäuser-Verlag.
[12] Bronwell, A. 1953. Advanced Mathematics in Physics and Engineering. New York: McGraw-Hill Book Co.

[13] Parodi, M. 1957. Introduction a l'Etude de l'Analyse Symbolique. Paris:Gauthier- Villars.
[14] Churchill, R.V. 1958. Operational Mathematics. New York:McGraw-Hill Book Co.
[15] Denis-Papin, M.,. Kaufman, A. and Faure, 1961. R. Exercises de Calcul Opérationnel Avec Leurs Solutions. Paris: Éditions Eyrolles.
[16] Lepage, W.R. 1961. Complex Variables and the Laplace Transform for Engineers. New York:McGraw-Hill Book Co..
[17] Denis-Papin, M. and Kaufman, A. 1963. Cours de Calcul Opérationnel Appliqué. Paris:Éditions Albin Michel.
[18] Carslaw, H.S. and Jaeger, J.C. 1963. Operational Methods in Applied Mathematics. New York:Dover Publ. Inc.
[19] Fodor, G. 1963. Laplace Transforms in Engineering. Budapest:Akadémiai Kiadó.
[20] Brown, W.M. 1963. Analysis of Linear Time-Invariant Systems. New York:McGraw-Hill Book Co.
[21] van der Pol, B. and Bremmer, H. 1964. Operational Calculus Based on the Two-Sided Laplace Integral. Cambridge:Cambridge University Press.
[22] Spiegel, M.R. 1965. Theory and Problems of Laplace Transforms. New York:Schaum Publ. Co.
[23] Humbert, P. and Colombo, S. 1965. Calcul Symbolique et Ses Applications a la Physique Mathématique. Paris:Gauthier-Villars
[24] Ditkin, V.A. and Prudnikov, A.P. 1966. Operational Calculus. Moscow:Vysh'aya Shkola Publ.
[25] Berg, L. 1967. Introduction to the Operational Calculus. Amsterdam:North-Holland Publ. Co.
[26] Hladik, J. 1969. La Transform de Laplace a Plusieurs Variables. Paris:Masson et Cie Éditeurs.
[27] Hladik, J. 1969. Les Transforms Fonctionnelles. Paris:Dunod.
[28] Sneddon, I. 1971. The Use of Integral Transforms. New York:McGraw-Hill Book Co.
[29] Ghizzetti, A. and Ossicini, A. 1971. A Transformate di Laplace e Calcolo Simbolico. Torino:Unione Tipografico-Editrice Torinese.
[30] Colombo, S. and Lavoine. 1972. Transforms de Laplace et de Mellin. Formulaires, Mode D'Utilisation. Paris:Gauthier-Villars.
[31] Davies, B. 1978. Integral Transforms and Their Applications. Berlin:Springer-Verlag.
[32] Osiowski, J. 1981. Operational Calculus. Warsaw:WNT Press.

[33] Debnath, L. 2000. Integral Transforms and Their Applications. Boca Raton:CRC Press.

[34] Apelblat, A. 1967. Application of the Laplace transform to the solution of the boundary layer equations. I. J. Phys. Soc. Japan, 23:1157-1162.

[35] Apelblat, A. 1968. Application of the Laplace transform to the solution of the boundary layer equations. II. Magneto-hydrodynamic Blasius problem. J. Phys. Soc. Japan, 25:888-891.

[36] Apelblat, A. 1969. Application of the Laplace transform to the solution of the boundary layer equations. III. Magnetohydrodynamic Falkner-Skan problem. J. Phys. Soc. Japan, 27:235-239.

[37] Apelblat, A. 1980. Mass transfer with a chemical reaction of the first order. Analytical solutions. Chem. Eng. Journal 19:19-37.

[38] Apelblat, A. 1981. Repeating use of integral transforms - A new method for evaluation of some infinite integrals. IMA J. Appl. Math., 27:481-496.

[39] Apelblat, A. 1982. Mass transfer with a chemical reaction of the first order. Effect of axial diffusion. Chem. Eng. Journal 23:193-203.

[40] Apelblat, A. 1985. Some integrals of gamma, polygamma and Volterra functions. IMA J. Appl. Math., 34:173-186.

[41] Apelblat, A. and Kravitsky, N. 1985. Integral representations of derivatives and integrals with respect to the order of the Bessel functions J□(t) and I□(t), the Anger function J□(t) and the integral Bessel function Ji□(t) IMA J. Appl. Math., 34:187-210.

[42] Apelblat, A. and Kravitsky, N. 1988. Improper integrals associated with the inverse Laplace transforms of [ln(s2+a2)]-n. Z. Angew. Math. Mech., ZAMM, 78:565-571.

[43] Apelblat, A. 1989. Derivatives and integrals with respect to the order of the Struve functions H□(x) and L□(x). J. Math. Anal. Appl., 137:17-36.

[44] Apelblat, A. 1991. Integral representation of Kelvin functions and their derivatives with respect to the order. J. Appl. Math. Phys., ZAMP, 42:708-714.

[45] Apelblat, A. 1994. Application of Laplace transform to evaluation of integrals. J. Math. Anal. Appl., 186:237-253.

[46] Apelblat, A. 1998. Application of the Laplace, Mellin and Stieltjes transforms in the evaluation of integrals. In Proceedings of the Eighth International Colloquium on Differential Equations. Plovdiv, Bulgaria, 18-23 August 1997. Editor. Bainov, D. Utrecht:VSP VB, 25-34.

[47] Apelblat, A. 1999. The integral representation of functions as the asymptotic limit of infinite integral of the Bessel function J□(□x). Intern. J. Appl. Math., 1:19-27.

[48] Apelblat, A. 2000. The asymptotic limit of integral of the Bessel function J□(□x) as integral representation of elementary and special functions. Intern. J. Appl. Math., 2:743-762..

[49] Apelblat, A. 2004. The application of the Dirac delta function □(x-1) to the evaluation of limits and integrals of elementary and special functions. Intern. J. Appl. Math., 16: 323 - 339.

[50] Apelblat, A. 2008. The evaluation of the asymptotic relations, limits of series, integrals and integral representations of elementary and special functions using the shifted Dirac delta function □(x-1). Computing Letters (CoLe) 4:11-19.

[51] Apelblat, A. 2008. Volterra Functions, New York: Nova Publishers.

[52] Apelblat, A. 2009. Integral Transforms and Volterra Functions, New York: Nova Publishers.

[53] Apelblat, A. 2012. Laplace Transforms and Their Applications. New York:Nova Publishers.

[54] Apelblat, A. 2020. Differentiation of the Mittag-Leffler functions with respect to parameters in the Laplace transform approach. Mathematics, Art. No 657, 8:1-22.

[55] Apelblat, A. and Mainardi, F. 2020. Application of the Efros theorem to the Wright functions of the second kind and other results. Lecture Notes TICMT 21:9-28.

[56] A. Apelblat, A. 2020. Bessel and Related Functions. Mathematical Operations with Respect to the Order. Volume 1: Theoretical Aspects. Berlin:De Gruyter.

[57] Apelblat, A. and Mainardi, F. 2021. Application of the Efros theorem to the function represented by the inverse of Laplace transform of s□exp(-s□). MDPI Symmetry, Art. No 354. 13:1-15.

[58] Apelblat, A., Consiglio, A. and Mainardi, F. 2021. The Bateman functions revisited after 90 years. A survey of old and new results. MDPI Mathematics, Art. No 1273, 9-27.

[59] Apelblat, A. and. Gonzáles-Santander, J.L. 2021. The integral Mittag-Leffler, Whittaker and Wright functions. MDPI Mathematics, 9 :3255

[60] Apelblat, A. and Mainardi, F. 2022. Differentiation of the Wright functions with respect to parameters and other results. MDPI Appl. Sci., Art. No 12805, 12:1-17.

[61] Meksyn, D. 1961. New Methods in Laminar Boundary Layer Theory. New York: Pergamon Press.

[62] Whittaker, E. and Watson, G.A. 1963. A Course of Modern Analysis. Cambridge:Cambridge Univ. Press.

[63] Abramowitz, A. and Stegun, I.E. 1965. Handbook of Mathematical Functions with Formulas, Graphs, and Mathematical Tables. Washington DC: U.S. National Bureau of Standards.

[64] Apelblat, A. 1996 Table of Integrals and Series. Thun und Frankfurt am Main:Verlag Harri Deutsch.

[65] Erdélyi, A., Magnus, W., Oberhettinger, F. and Tricomi, F.G. 1954 Tables of Integral Transforms. New York:McGraw-Hill Book Co.

[66] Roberts, G.E. and Kaufman, H. 1968 Table of Laplace Transforms. Philadelphia:Saunders.

[67] Oberhettinger, F. and Badii, L. 1973 Tables of Laplace Transforms Berlin:Springer- Verlag.

[68] Hardy, G.H. 1959 Ramanujan-Twelve Lectures on Subjects Suggested by His Life and Work. New York:Chelsea Publ. Co.

[69] Erdélyi, A., Magnus, W., Oberhettinger, F. and Tricomi, F.G. 1955 Higher Transcendental Functions. Vol. 3. New York:McGraw-Hill Book Co.

[70] Lamborn, B.N.A. 1969. An expression of the Dirac delta function. SIAM Review11:603.

[71] Gorenflo, R., Kilbas, A.A., Mainardi, F. and Rogosin, S.V. 2014 Mittag-Leffler Functions. Related Topics and Applications. Berlin:Springer.

[72] Mainardi, F. 2010 Fractional Calculus and Waves in Linear Viscoelesticity. London:Imperial College Press.

[73] Gorenflo, R., Luchko, Yu. and Mainardi, F. 1999 Analytical properties and applications of the Wright functions. Frac. Calc. Appl. Anal., 2:383-414.

[74] Mainardi, F. and Consiglio, A. 2020 The Wright functions of the second kind in mathematical physics. Mathematics, Art. No 884, 8:1-26.

[75] Humbert, P. 1945 Nouvelle correspondances symboliques. Bull. Soc. Math., France, 69:121- 129.

[76] Pollard, H. 1946 The representation of exp (-x□) as a Laplace integral. Bull. Amer. Math. Soc., 52:908-910.

[77] Wlodarski, L. 1952 Sur une formule de Eftros. Studia Math., 13:183-187.

[78] Mikusinski, 1951 J. Sur les fonctions exponentielles du calcul opératoire. Studia Math., 12:208-224..

[79] Mikusinski, J. 1953 Sur la croissance de la function opérationelle exp(-s□□). Bull. Acad. Polon. 4:423-425.

[80] Mikusinski, J. 1958 Sur la function dont la transformée de Laplace est exp(-s□□). Bull. Acad. Polon. 6:691-693.

[81] Mikusinski, J. 1959 On the function whose Laplace transform is exp(-s□). Studia Math., 18:195-198.

[82] A. Wintner, A. 1956 Cauchy's stable distributions and an "explicit formula" of Mellin. Amer. J. Math., 78:819-861, 1956.

[83] Ragab, F.M. 1958 The inverse Laplace transform of a exponential function. Comm. Pure Appl. Math., 11:115-127.

[84] Stankovič, B 1970 On the function of E.M. Wright. Publications de L'Institute Mathématique. Nouvelle série, 10:113-124.

[85] A.M. Efros. 1935 Some applications of operational calculus in analysis. Mat. Sbornik, 42:699-705.

[86] Bateman, H. 1931 The k-function, a particular case of the confluent hypergeometric function. Trans. Amer. Math. Soc., 33:817-831.

[87] Chaudhur, J. 1968 On Bateman-integer function. Math. Zeitschr. 78:25-32.

[88] Volterra, V. 1916 Theoria delle potenze, dei logaritmi e delle funzioni di composizione. A. Acc. dei Lincei, Mem. Ser. 5a vol.11.

[89] Volterra, V. and Péres, J. 1924 Leçons sur la Composition et les Functions Permutables. Paris:Gauthier-Villars.

[90] Colombo, S. 1948 Sur la function □(t,n) C.R. Acad. Sci., 226:1235-1236.

[91] Colombo, S. 1952 Sur les équations intégrales de Volterra a noyaux logarithmiques. C.R. Acad. Sci., 235:928-929.

[91] Colombo, S. 1953 Sur quelques transcendantes introduites par la résolution de équations intégrales de Volterra a noyaux logarithmiques. Bull. Sci. Math., 77:89-104.

[92] Colombo, S. 1955 Sur la function □(x,n) et □(x,m,n). Bull. Sci Math. 79:72-78.

[93] Colombo, S. 1964 Étude des Transcendantes Intervenant Dans la Résolution des Équations Intégrales de Volterra a Noyaux Logarithmiques. Studia Univ. Lovanium No 18, Paris:Gauthier-Villars.

Chapter 6

Laplace Sum Rules in Quantum ChromoDynamics

Stephan Narison*

Laboratoire Univers et Particules de Montpellier (LUPM)
NRS-IN2P3, and University of Montpellier, France
Institute of High-Energy Physics of Madagascar (iHEPMAD)
University of Ankatso, Antananarivo, Madagascar

Abstract

We shortly review some applications of the (inverse) Laplace (LSR) transform sum rules in Quantum ChromoDynamics (QCD) for extracting the fundamental QCD parameters (coupling constant α_s, quark and gluon condensates) and the hadron properties (masses and decay constants). Links of LSR to some other forms of QCD spectral sum rules are also discussed. As prototype examples, we discuss in detail the ρ and π meson sum rules.

Keywords: QCD spectral sum rules, QCD coupling α_s quark masses and condensates, Hadron masses and couplings

---

* Corresponding Author's Email: snarison@yahoo.fr

In: The Laplace Transform and Its Applications in Science
Editor: Victor Martinez-Luaces
ISBN: 979-8-89113-747-9

1. Introduction

Different spectral sum rules based on dispersion relations called current algebra sum rules have been used before QCD [1]. Among these, there are the Weinberg [2] and Das-Mathur-Okubo (DMO) [3] sum rules which assume the asymptotic realizations of chiral $SU(2)_L \otimes SU(2)_R$ and $SU(2)_V$ flavour symmetries to derive constraints on the vector and axial-vector mesons masses and couplings. One of the famous predictions from the Weinberg sum rules is the mass relation:

$$M_{A_1} \simeq \sqrt{2} M_\rho, \tag{1}$$

which is reproduced within the errors by the data.

Within the advent of QCD, corrections to these sum rules have been studied [4, 5, 6] in 1979-80. At about the same time, Shifman-Vainshtein and Zakharov (hereafter referred as SVZ) [7, 8] [†], have introduced the QCD spectral sum rules which are the improvement of the usual dispersion relation obeyed by e.g. the hadron two-point function:

$$\Pi_H(q^2) = i\int d^4x\, e^{iqx} \langle 0|\mathcal{T} \mathcal{O}_H(x)\, (\mathcal{O}_H(0))^\dagger\, |0\rangle = \int_{t_>}^{\infty} \frac{dt}{t-q^2-i\epsilon} \frac{1}{\pi} \mathrm{Im}\Pi(t) + \cdots , \tag{2}$$

where $\cdots$ mean subtraction constants polynomial in the momentum $q^2 \equiv -Q^2 < 0$.

$\mathcal{O}_H(x)$ is a generic notation for a hadronic current:

– Quark bilinear local current $\bar\psi_1 \Gamma_{12} \psi_2$ for mesons. Γ_{12} is any Dirac matrices which specify the quantum numbers of the corresponding meson state (and its radial excitations),

– Quark trilinear local current $\psi_1\psi_2\psi_3$ for baryons,

– Four-quark $(\bar\psi_1\Gamma_{12}\psi_2)(\bar\psi_3\Gamma_{34}\psi_4)$ or diquark anti-diquark $(\bar\psi_1\Gamma_{12}\bar\psi_2)(\psi_3\Gamma_{34}\psi_4)$ local current for molecules or tetraquark states,

– Pentaquark $\psi_1\psi_2\psi_3\psi_4\psi_5$ states,

– Gluon local currents $G^2, G^3 \cdots$ for gluonia/glueball states,

– Quark-gluon local currents $\bar\psi_1 G \psi_2, \bar\psi_1 G^2 \psi_2 \cdots$ for hybrid states.

In the case of a $\bar\psi_1\psi_2$ meson bound state built from a quark u and the anti-quark $\bar d$, the current reads:

$$\mathcal{O}_H(x) = \bar u \Gamma d(x), \tag{3}$$

where Γ denotes any Dirac matrices which specify the quantum number of the associated hadron. $\Gamma = \gamma_u, \gamma_\mu\gamma_5, \gamma_5, 1$ for the vector, axial-vector, pseudoscalar and scalar currents which correspond respectively to the ρ, A_1, π and scalar quarkonium mesons.

[†] For reviews, see e.g. the books [9, 10], the reviews in [11, 12, 13, 14] and the recent ones in Refs.[15, 16, 17].

2. The SVZ Borel/Laplace Sum Rules (LSR)

Besides the Shifman-Vainshtein-Zakharov (SVZ) (see Fig. 1) theoretical improvement of the perturbative QCD expression within the Operator Product Expansion (OPE) where quark and gluon condensates of higher and higher dimensions are supposed to approximate the non-perturbative QCD contributions which we shall discuss in the next section. The phenomenological success of QCD (spectral) sum rules or SVZ sum rules also comes from the improvement of the usual dispersion relation which is the bridge between the high-energy QCD expression and the measurable spectral function at low energy. In the example of the light isovector ($I = 1$) vector current:

$$J_H^\mu(x) = \frac{1}{2}[: \bar{\psi}_u \gamma^\mu \psi_u - \bar{\psi}_d \gamma^\mu \psi_d :]. \tag{4}$$

the spectral function is related to the $e^+e^- \to$ Hadrons total cross-section via the optical theorem:

$$R_{ee} \equiv \frac{\sigma(e^+e^- \to \text{Hadrons})}{\sigma(e^+e^- \to \mu^+\mu^-)} = \left(\frac{3}{2}\right) 8\pi \, \text{Im}\Pi_H(t). \tag{5}$$

Figure 1. Left to right : Arkady Vainshtein, SN, Valya Zakharov, Mikhael Shifman at the Munich conference (2006).

2.1. The Form of the Laplace Sum Rules

This improvement has been achieved by working with large number n of derivatives and large value of the momentum transfer Q^2 but taking their ratio $\tau \equiv n/Q^2$ finite leading to the so-called SVZ Borel/Laplace or Exponential sum rules (LSR) and their ratios,[1] :

$$\mathcal{L}_0^c(\tau,\mu) \equiv \lim_{\substack{Q^2, n\to\infty \\ n/Q^2\equiv\tau}} \frac{(-Q^2)^n}{(n-1)!}\frac{\partial^n \Pi_H}{(\partial Q^2)^n} = \int_{t>}^{t_c} dt\, e^{-t\tau}\frac{1}{\pi}\mathrm{Im}\Pi_H(t,\mu)\,,$$

$$\mathcal{R}_{10}^c(\tau) \equiv \frac{\mathcal{L}_1^c}{\mathcal{L}_0^c} = \frac{\int_{t>}^{t_c} dt\, e^{-t\tau} t\,\mathrm{Im}\Pi_H(t,\mu)}{\int_{t>}^{t_c} dt\, e^{-t\tau}\mathrm{Im}\Pi_H(t,\mu)}, \tag{6}$$

where τ is the LSR variable, $t>$ is the hadronic threshold. Here t_c is the threshold of the "QCD continuum" which parametrizes, from the discontinuity of the Feynman diagrams, the spectral function $\mathrm{Im}\,\Pi_H(t, m_Q^2, \mu^2)$.

◇ Thanks to the exponential weight and for moderate values of τ, the previous sum rule improvements enhance the low energy contribution to the spectral integral which is accessible experimentally.

2.2. The Minimal Duality Ansatz (MDA)

In the often case where the data on the spectral function are not available, one usually parametrizes it via the minimal duality ansatz :

$$\frac{1}{\pi}\,\mathrm{Im}\Pi_H(t) \simeq f_H^2 M_H^{2d}\delta(t-M_H^2) + \text{"QCD continuum"}\,\theta(t-t_c), \tag{7}$$

in order to predict the masses and couplings of the lowest ground state and in some case the ones of its radial excitations. d depends on the dimension of the current, f_H is the hadron decay constant normalized as $f_\pi = 132$ MeV. Its accuracy has been tested in various light and heavy quark channels $e^+e^- \to \rho, J/\psi, \Upsilon, \ldots$ where complete data are available [9, 10] and in the π-pseudoscalar channel where an improved parametrization of the 3π channel within chiral perturbation theory has been used [20].

[1]Non-relativistic version of this sum rule has been discussed by [18, 13], while the inclusion of the PT α_s correction to the QCD expression has shown that it has the property of an inverse Laplace transform in SNR [19], then the name LSR.

2.3. Hadron Masses from the Ratio of Moments

Within the MDA parametrization, the ratio of sum rules $\mathcal{R}_n^c(\tau)$ is used to extract the mass of the lowest ground state as it is equal to its square:

$$M_H^2 \simeq \mathcal{R}_{10}^c(\tau). \tag{8}$$

However, this analysis cannot be done blindly without studying / checking the moments $\mathcal{L}_{0,1}$ which in some cases can violate positivity of the specral integral for some values of the sum rule variables (τ) though their ratio can lead to a positive number identified with the hadron mass squared.

2.4. Hadron Mass-splittings from the Double Ratio of Sum Rule (DRSR) [21]

The hadron H and H' mass-splitting, like the one due to $SU(3)$ breakings, can be derived from the quantity:

$$r_{H'/H} \equiv \frac{\mathcal{R}_{10}^c(\tau')|_{H'}}{\mathcal{R}_{10}^c(\tau)|_H} \simeq \frac{M_{H'}^2}{M_H^2}, \tag{9}$$

provided that the optimal value: $\tau_0' \simeq \tau_0$. A similar quantity can be used for the heavy quark moments.

3. Link of LSR to Some Other Sum Rules

3.1. Local Duality Finite Energy Sum Rules (FESR)

It has the form:

$$\mathcal{F}_n^c(t_c,\mu) = \int_{t>}^{t_c} dt\, t^n \frac{1}{\pi}\text{Im}\, \Pi_H(t,\mu)\ , \tag{10}$$

where $n \geq 0$ is the degree of the sum rule. This sum rule has been used before QCD in Refs. [22, 23] and within QCD in Refs. [24, 25, 26].

This sum rule can be derived from LSR by using a small τ-expansion and by matching for a given τ its QCD and experimental side.

FESR can be used to fix the value of the QCD continuum threshold t_c from its dual constraint with the mass and decay constant of the ground state [27, 28]. Contrary to LSR, FESR emphasizes the role of higher masses radial excitations to the integral. Therefore, it requires a good control of the high-energy part of the spectral function.

3.2. Gaussian Sum Rules

It has the form [27, 28]:

$$G_n^c(s,\sigma,\mu) = \frac{1}{\sqrt{4\pi\sigma}}\int_{t>}^{t_c} dt\, e^{-\frac{(t+s)^2}{4\pi}} \frac{1}{\pi}\mathrm{Im}\Pi_H(t,\mu)\;, \tag{11}$$

for a Gaussian centered at s with a finite width resolution $\sqrt{4\pi\sigma}$. It has been also shown in Ref. [27, 28] that the LSR can be derived from the Gaussian sum rule using the ζ-prescription.

4. The SVZ - Operator Product Expansion (OPE)

◇ According to SVZ, the Right Hand Side (RHS) of the two-point function can be evaluated in QCD within the Operator Product Expansion (OPE) provided that $\Lambda^2 \ll Q^2 \equiv -q^2, m_Q^2$ where $\Lambda \simeq (342 \pm 8)$ MeV is the QCD scale for 3 flavours and m_Q is the heavy quark mass. In this way, it reads:

$$4\pi^2\Pi_H(q^2,m_Q^2,\mu) = \sum_{D=0,1,..} \frac{C_{2D}(q^2,m_Q^2,\mu)}{Q^{2D}}\langle O_{2D}(\mu)\rangle\;, \tag{12}$$

where, in addition to the usual perturbative QCD contribution (unit operator), one has added the ones due to non-perturbative gauge invariant quark and gluon condensates $\langle O_{2D}(\mu)\rangle$ having a dimension $2D$ which have been assumed to parametrize approximately the not yet under good control QCD confinement. C_{2D} are separated calculable Wilson coefficients in Perturbative (PT) QCD:

4.1. The Usual Perturbative (PT) Contribution

It corresponds to $D = 0$ while the quadratic quark mass corrections enter via $D = 1$.

4.2. The Quark and Gluon Condensates

They contribute through the OPE. Up to $2D = 6$, they are successively the :

– $2D = 4$ quark and gluon $m\langle\bar\psi\psi\rangle$ and $\langle\alpha_s G^2\rangle$,

– $2D = 5$ mixed quark-gluon: $\langle\bar\psi\sigma^{\mu\nu}\frac{\lambda_a}{2}G^a_{\mu\nu}\psi\rangle$

– $2D = 6$ four-quark and three-gluon: $\langle\bar\psi\Gamma_1\psi\bar\psi\Gamma_2\psi\rangle$ and $\langle g^3 f_{abc}G^a_{\mu\rho}G^{b,\rho}_{\nu}G^{c,\nu}_{\rho}\rangle$.

– and so on.

4.3. The $2D = 4$ Condensates

The quark condensate $m\langle\bar{\psi}\psi\rangle$ and the part of the trace of the energy-momentum transfer: $\theta^{\mu}_{\mu}|_g \equiv m\gamma\langle\bar{\psi}\psi\rangle + (1/4)\beta\langle G^a_{\mu\nu}G^{\mu\nu}_a\rangle$ are known to be subtraction μ-independent where γ, β are the quark mass anomalous dimension and Callan-Symanzik β-function.

– The $\langle\bar{\psi}\psi\rangle$ condensate can be deduced from the well-known Gell-Mann, Oakes, Renner relation [29]:

$$(m_u + m_d)\langle\bar{\psi}_u\psi_u + \bar{\psi}_d\psi_d\rangle = -m_\pi^2 f_\pi^2, \tag{13}$$

once the running light quark mass is known (m_π, $f_\pi = 132$ MeV are the pion mass and decay constant) or directly from light baryon sum rules [30]. With the value of the running u, d quark masses given in Table 2 in Ref. [15], one can deduce the value of $\langle\bar{\psi}\psi\rangle$ in this Table 2.

– The original value of the $2D = 4$ gluon condensate $\langle\alpha_s G^2\rangle = 0.04$ GeV^4 of SVZ [7, 12] has been claimed [18, 13] from charmonium sum rules and Finite Energy Sum Rule (FESR) for $e^+e^- \to I = 1$ Hadrons [27, 28] to be underestimated. Recent analysis from $e^+e^- \to$ Hadrons, τ-decay and charmonium confirm these claims with the present updated average (see the different determinations in Table 1 of [31]):

$$\langle\alpha_s G^2\rangle = (6.49 \pm 0.35) \times 10^{-2}\ \text{GeV}^4. \tag{14}$$

4.4. The $2D = 5$ Quark-Gluon Mixed Condensate

It is usually parametrized as $g\langle\bar{\psi}G\psi\rangle = M_0^2\langle\bar{\psi}\psi\rangle$. It mixes under renormalization and runs as $(\alpha_s)^{1/(6\beta_1)}$ in the chiral limit $m = 0$ [32]. The scale :

$$M_0^2 = 0.8(2)\ \text{GeV}^2, \tag{15}$$

has been phenomenologically estimated from light baryons [33, 34, 35] and heavy-light mesons [36] sum rules.

4.5. The $2D = 6$ Four-Quark Condensates

The renormalization of dimension-six condensates has been studied in [32] where it has been shown that these condensates mix under renormalization and then cannot be compatible with the vacuum saturation assumption used by SVZ. Its phenomenological estimate from τ-like decays [37], $e^+e^- \to$ Hadrons data [38], τ-decay [39], FESR [27, 28] and baryon [34] sum rules, leads to the average :

$$\rho\alpha_s\langle\bar{\psi}\psi\rangle^2 \simeq 5.8(9)10^{-4}\,\text{GeV}^6, \tag{16}$$

indicating a huge violation of the vacuum saturation or factorization:

$$\alpha_s\langle\bar\psi\psi\rangle^2|_{fac} = 1.0(9)\times 10^{-4}\,\text{GeV}^4, \tag{17}$$

by a factor ρ about 5.8.

◇ Fixing the ratio $\langle\alpha_s G^2\rangle/\rho\alpha_s\langle\bar\psi\psi\rangle^2 = 106(12)$ GeV^{-2} as quoted in Ref. [40] which reduces the analysis to a one-parameter fit, one deduces from LSR [37]:

$$\langle\alpha_s G^2\rangle = (6.1\pm 0.7)10^{-2}\,\text{GeV}^6, \tag{18}$$

which shows the self-consistency of the previous numbers. Some other consistency tests can be found in [37].

4.6. The $2D = 6\ g^3 f_{abc}\langle G^a G^b G^c\rangle$ Condensate

It does not mix under renormalization and behaves as $(\alpha_s)^{23/(6\beta_1)}$ [32], where $\beta_1 = -(1/2)(11-2n_f/3)$ is the first coefficient of the β-function and n_f is number of quark flavours. The first improvement of the estimate of the $g^3 f_{abc}\langle G^a G^b G^c\rangle$ condensate was the recent direct determination of the ratio of the dimension-six gluon condensate $\langle g^3 f_{abc} G^3\rangle$ over the dimension-four one $\langle\alpha_s G^2\rangle$ using heavy quark sum rules with the value [41]:

$$\rho_G \equiv \langle g^3 f_{abc} G^3\rangle/\langle\alpha_s G^2\rangle = (8.2\pm 1.0)\,\text{GeV}^2, \tag{19}$$

which differs significantly from the instanton liquid model estimate [42, 43, 44]. This result may question the validity of the instanton result. Earlier lattice results in pure Yang-Mills found: $\rho_G \approx 1.2$ GeV^2 [45] such that it is important to have new lattice results for this quantity. Note however, that the value given in Eq. 19 might also be an effective value of all unknown high-dimension condensates not taken into account in the analysis of [41] when requiring the fit of the data by the truncated OPE. We have seen in some examples [16, 15] that the effect of $\langle g^3 f_{abc} G^3\rangle$ is a small correction at the stability region where the optimal results are extracted.

4.7. Higher Dimension Condensates

Usually, the truncation of the OPE up to $2D = 6$ provides enough information for extracting the masses and couplings of the ground state hadrons with a good accuracy. In many papers, some classes of higher dimension terms up to 2D=12 ! are added in the OPE. However, it is not clear if such term gives the dominant contributions compared to the omitted ones having the same dimension. The size of these high-dimension condensates is not also under a good control due to the eventual violation of the vacuum saturation used for their estimate.

4.8. Beyond the SVZ-OPE

Different contributions beyond the standard SVZ-OPE (tachyonic gluon mass, small size instantons, duality violation) can be consulted in the recent reviews [16, 15].

5. The Different Terminologies of the SVZ Sum Rules

5.1. The Origin of the Name: Borel Sum Rule from the OPE

Applying the operator in Eq. 6 to the QCD expression in Eq. 12, the OPE of the 1st moment becomes:

$$\mathcal{L}_0(\tau, m_q^2, \mu) = \sum_{D=0,1,..} C_{2D}(\tau, m_q^2, \mu) \frac{\tau^D}{(D)!} \langle O_{2D}(\mu) \rangle \ , \tag{20}$$

where the appearance of the $(D)!$ factor in the OPE indicates the property of the Borel transform for the moment sum rule thus the name *Borel sum rule*. For this purpose, we recall that if we consider a function $f(x)$, its Borel transform is:

$$\tilde{f}(\lambda) = \frac{1}{2\pi\, i} \int_{c-i\infty}^{c+i\infty} e^{\lambda/x} f(x)\, x\, d(1/x), \tag{21}$$

where the integration contour runs to the right of all the singularities of the function $f(x)$. Then, the inverse Borel transform reads:

$$x f(x) = \int_0^\infty f(x)\, e^{-\lambda/x} d\lambda. \tag{22}$$

Therefore, if

$$f(x) = a_0 + a_1 x + a_2 x^2 + \cdots + a_k x^k + \cdots \ , \tag{23}$$

then :

$$\tilde{f}(\lambda) = \frac{a_0}{0!} + \frac{a_1}{1!}\lambda + \frac{a_2}{2!}\lambda^2 + \cdots + \frac{a_k}{k!}\lambda^k + \cdots \ . \tag{24}$$

The great advantage of this property is the improved convergence of the OPE for the Borel transform $\tilde{f}(\lambda)$ compared to the one of the original two-point function $f(x)$.

Figure 2. Left to right : SN, Eduardo de Rafael, Christine Kattner (soprano singer) @ QCD 2000 - Montpellier.

5.2. The Origin of the Name: Laplace Sum Rule (LSR) Including α_s^n Corrections

In Ref. [19], S. Narison and E. de Rafael (referred hereafter as SNR) [2] have remarked that working with the renormalization group resummed QCD expression including radiative corrections, the sum rule has instead a Laplace transform property from which all QCD expressions can be derived.

The Laplace transform sum rule (LSR) can be derived from the dispersion relation (Hilbert transform) in Eq. 2 from the useful formulae (see e.g. the Appendix of [9, 10]):

$$
\begin{aligned}
\mathcal{L}\left[\frac{1}{(Q^2+m^2)^\alpha}\right] &= \frac{\tau^\alpha}{\Gamma(\alpha)}e^{-m^2\tau},\\
\mathcal{L}\left[\frac{1}{(Q^2)^\alpha}\ln\frac{Q^2}{\nu^2}\right] &= \frac{\tau^\alpha}{\Gamma(\alpha)}\Big[-\ln\tau\nu^2+\psi(\alpha)\Big],\\
\mathcal{L}\left[\frac{1}{(Q^2)^\alpha}\ln^2\frac{Q^2}{\nu^2}\right] &= \frac{\tau^\alpha}{\Gamma(\alpha)}\Big[\ln^2\tau\nu^2-2\psi(\alpha)\ln\tau\nu^2+\psi^2(\alpha)-\psi'(\alpha)\Big],\\
\mathcal{L}\left[\frac{1}{(x)^\alpha}\frac{1}{(\ln x)^\beta}\right] &= y\,\mu(y,\beta-1,\alpha-1),\\
&\overset{y\to 0}{\simeq} \frac{y^{\alpha+1}}{\Gamma(\alpha)}\frac{1}{(-\ln y)^\beta}\left[1+\beta\,\psi(\alpha)\frac{1}{\ln y}+\mathcal{O}\left(\frac{1}{\ln^2 y}\right)\right]\quad\text{for}\quad y\equiv\tau\Lambda^2\\
\mathcal{L}\left[\frac{\ln\ln x}{x^\alpha(\ln x)^\beta}\right] &\overset{y\to 0}{\simeq} \frac{y^{\alpha+1}}{\Gamma(\alpha)}\frac{\ln\ln y}{(-\ln y)^\beta}\left[1+\beta\,\psi(\alpha)\frac{1}{\ln y}+\mathcal{O}\left(\frac{1}{\ln^2 y}\right)\right]
\end{aligned}
\tag{25}
$$

[2]The main idea comes from Prof. E. de Rafael as I was a student in that time.

where $\mathcal{L}$ is the (inverse) Laplace transform operator and :

$$
\begin{aligned}
\psi(z) &\equiv \frac{d}{dz}\ln\Gamma(z) , \\
\mu(y,\beta,\alpha) &\equiv \int_0^\infty dx \, \frac{y^{\alpha+x}x^\beta}{\Gamma(\beta+1)\,\Gamma(\alpha+x+1)}, \\
\mu(y,-\beta,\alpha) &= (-1)^{\beta-1}\frac{d^{\beta-1}}{(dx)^{\beta-1}}\left[\frac{y^{\alpha-x}}{\Gamma(\alpha+x+1)}\right]_{x=0} . \qquad (26)
\end{aligned}
$$

with the useful properties:

$$
\begin{aligned}
\psi(1) &= -\gamma_e = -0.5772... \quad \text{(Euler constant)}, \\
\mu(y,-1,\alpha) &= \frac{y^\alpha}{\Gamma(\alpha+1)}, \\
\mu(y,-2,\alpha) &= \frac{y^\alpha}{\Gamma(\alpha+1)}[-\ln y + \psi(\alpha+1)], \\
\mu(y,-3,\alpha) &= \frac{y^\alpha}{\Gamma(\alpha+1)}[-\ln^2 y - 2\psi(\alpha+1)\ln y + \psi^2(\alpha+1) - \psi'(\alpha+1)]. \quad (27)
\end{aligned}
$$

In addition to previous formulae, the expression of the integral:

$$
\int_0^{t_c} dt\, t^n\, e^{-t\tau} = (n-1)!\,\tau^{-n}(1-\rho^n) \quad \text{with} \quad \rho^n = e^{-t_c\tau}\left(1 + t_c\tau + \cdots \frac{(t_c\tau)^n}{n!}\right) \quad (28)
$$

is useful for the treatment of the QCD continuum contribution to the LSR from the QCD spectral function as parametrized in Eq. 7.

6. The QCD Expressions of the Prototype ρ Meson and Pion LSR

We illustrate the derivation of the Laplace sum rule in the case of the vector and pseudoscalar (divergence of the axial-vector) currents:

$$
\begin{aligned}
V_\mu(x) &= \bar{u}(\gamma_\mu)d(x), \\
\partial_\mu A^\mu(x) &= (m_u + m_d)\bar{u}(i\gamma_5)d(x) \qquad (29)
\end{aligned}
$$

and the corresponding two-point functions $\Pi(q^2)$ and $\psi_5(q^2)$ used in Ref. [19] for demonstrating (for the first time) the Laplace transform properties of the SVZ sum rule [3] :

$$-(g_{\mu\nu}q^2 - q_\mu q_\nu)\Pi(q^2) = i\int d^4x\, e^{iqx}\langle 0|\mathcal{T}\, V_\mu(x)V_\nu^\dagger(0)|0\rangle,$$
$$\psi_5(q^2) = i\int d^4x\, e^{iqx}\langle 0|\mathcal{T}\, \partial^\mu A_\mu(x)\,(\partial^\mu A_\mu)^\dagger\,(0)|0\rangle \quad (30)$$

6.1. The ρ Meson Channel

As the corresponding two-point correlator is once substracted, it is convenient to work with its first derivative:

$$\chi^{(1)}(Q^2) \equiv \left(-\frac{d}{dQ^2}\right)\Pi(Q^2) = \frac{C_0}{Q^2} + \frac{C_2\langle O_2\rangle}{(Q^2)^2} + 2\frac{C_4\langle O_4\rangle}{(Q^2)^3} + 3\frac{C_6\langle O_6\rangle}{(Q^2)^4} + \cdots \quad (31)$$

with (see e.g. [9, 10]):

$$C_0 = \frac{1}{4\pi^2}\left[1 + a_s + \bar{a}_s^2 R_2 + \cdots\right],$$
$$C_2\langle O_2\rangle = 3(m_u^2 + m_d^2),$$
$$C_4\langle O_4\rangle = 4\pi^2\left[m_u\langle\bar{\psi}_u\psi_u\rangle + m_d\langle\bar{\psi}_d\psi_d\rangle\right] + \frac{\pi}{3}\langle\alpha_s G^2\rangle + \mathcal{O}(m_i^4\ln(m_i^2/q^2),$$
$$C_6\langle O_6\rangle = \frac{896}{81}\pi^3\rho\,\alpha_s\langle\bar{\psi}_q\psi_q\rangle^2, \quad (32)$$

where $R_2 = 1.986 - 0.115\,n_f$ for n_f flavours. $\chi^{(1)}(Q^2)$ is superconvergent and obeys the homogeneous Renormalization Group Equation (RGE):

$$\left(-Q^2\frac{\partial}{\partial Q^2} + \beta(\alpha_s)\alpha_s\frac{\partial}{\partial\alpha_s} - \sum_i[1+\gamma(\alpha_s)]x_i\frac{\partial}{\partial x_i}\right)\chi^{(1)}(Q^2), \quad (33)$$

where the running parameters to order α_s are (see e.g.[4, 46, 9, 10]):

$$a_s \equiv \frac{\bar{\alpha}_s(Q^2)}{\pi} = a_s^{(0)}\left[1 - a_s^{(0)}\frac{\beta_2}{\beta_1}\ln\ln(Q^2/\Lambda^2)\right]$$
$$\bar{m}_q(Q^2) \equiv x_i\sqrt{Q^2} \equiv \frac{\hat{m}_q}{(-\beta_1 a_s)^{\gamma_1/\beta_1)}}\left[1 + \frac{\beta_2}{\beta_1}\left(\frac{\gamma_1}{\beta_1} - \frac{\gamma_2}{\beta_2}\right)a_s + 2.707a_s^2\right], \quad (34)$$

with:

$$a_s^{(0)} = \frac{2}{-\beta_1\ln(Q^2/\Lambda^2)} \quad : \quad \Lambda = (342 \pm 8)\ \text{MeV for 3 flavours.} \quad (35)$$

[3] In these pedagogical examples, we shall limit to the perturbative QCD expression to order α_s^2. Analysis including higher order terms can be found in [48, 47].

$\beta_{1,2}$ and $\gamma_{1,2}$ are respectively the 1st and 2nd coefficients of the β function and mass anomalous dimension. For n_f quark flavours, they read:

$$\beta_1 = -\frac{11}{2}+\frac{1}{3}n_f, \quad \beta_2 = -\frac{51}{4}+\frac{19}{12}n_f, \quad \gamma_1 = 2, \quad \gamma_2 = \frac{101}{12}-\frac{5}{18}n_f. \quad (36)$$

Using the previous formulae of the Laplace transform into the QCD expressions of the corresponding two-point correlator, one can deduce the Laplace transform [19, 9, 10]:

$$\mathcal{L}_0^V \equiv \int_0^\infty e^{-t\tau}\frac{1}{\pi}\mathrm{Im}\Pi_V(t) = \tau^1\left(\frac{1}{4\pi^2}\right)\Big\{1 + a_s(\tau) + a_s^2(\tau)\left[R_2 - \frac{1}{2}\beta_1\gamma_E + (\beta_2/\beta_1)\ln\ln(\tau\Lambda^2)\right]$$
$$+4\pi^2\tau^2\left[m_u\langle\bar\psi_u\psi_u\rangle + m_d\langle\bar\psi_d\psi_d\rangle\right] + \frac{\pi}{3}\tau^2\langle\alpha_s G^2\rangle + \tau^3\left(\frac{1}{2}\right)\frac{896}{81}\pi^3\rho\,\alpha_s\langle\bar\psi_q\psi_q\rangle^2\Big\}, \quad (37)$$

where the factor $\rho = 1$ if one uses the vacuum saturation assumption for the four-quark condensate.

6.2. The Pion Channel

The corresponding two-point function behaves as $(q^2)^2\ln(-q^2/\nu^2)$and is then twice substracted. Thererefore, its 2nd derivative $\psi_5^{(2)}(q^2)$ is superficially convergent and obeys the RGE in Eq. 33. Taking the Laplace transform of $\psi_5^{(2)}$, one obtains: [19, 9, 10][4]:

$$\mathcal{L}_0^\pi = \frac{3}{8\pi^2}\frac{(\hat m_u+\hat m_d)^2}{\left(-\ln\sqrt{\tau}\Lambda\right)^{2\gamma_1/-\beta_1}}\tau^{-2}\Big\{1 - \left[\bar m_u^2+\bar m_d^2+(\bar m_u-\bar m_d)^2\right]\tau$$
$$+a_s\left[\frac{11}{3}+2\gamma_E+\frac{2}{-\beta_1}\left(\gamma_2-\gamma_1\frac{\beta_2}{\beta_1}\right)+2\gamma_1\frac{\beta_2}{\beta_1^2}\ln\left(-\ln\tau\Lambda^2\right)\right]$$
$$-\frac{8}{3}\pi^2\tau^2\left[\left(m_d-\frac{m_u}{2}\right)\langle\bar\psi_u\psi_u\rangle+\left(m_u-\frac{m_d}{2}\right)\langle\bar\psi_d\psi_d\rangle\right]+\frac{\pi}{3}\tau^2\langle\alpha_s G^2\rangle$$
$$-\frac{4\pi^2}{3}\tau^3\left[m_d\langle\bar\psi_u G\psi_u\rangle+\frac{32}{27}\pi\rho\alpha_s\left(\langle\bar\psi_u\psi_u\rangle^2+\langle\bar\psi_d\psi_d\rangle^2-9\langle\bar\psi_u\psi_u\rangle\langle\bar\psi_d\psi_d\rangle\right)\right]\Big\} \quad (38)$$

7. Optimization Procedure for the LSR

7.1. The SVZ Sum Rule Window

The second important step in the sum rule analysis is the extraction of the optimal result from the sum rule as, in principle, the Laplace sum rule (LSR) variable τ,

[4] For an updated expression including higher order contributions, see e.g. [47].

the degree of moments n and the QCD continuum threshold t_c are free external variables. In their original work, SVZ have adjusted the values of $M^2 \equiv 1/\tau$ and t_c using some guessed % contributions, for finding the sum rule window where the QCD continuum contribution is less than some input number while the ground state one is bigger than some input number and where, in this sum rule window, the OPE is expected to converge. The arbitrariness values of these numbers have created some doubts for non-experts on the results from the sum rules, in addition to the ones on the eventual non-correctness of the non-trivial QCD-OPE expressions. Unfortunately, many sum rules practitioners continue to use this inaccurate SVZ criterion.

7.2. τ-stability from Quantum Mechanics and J/Ψ

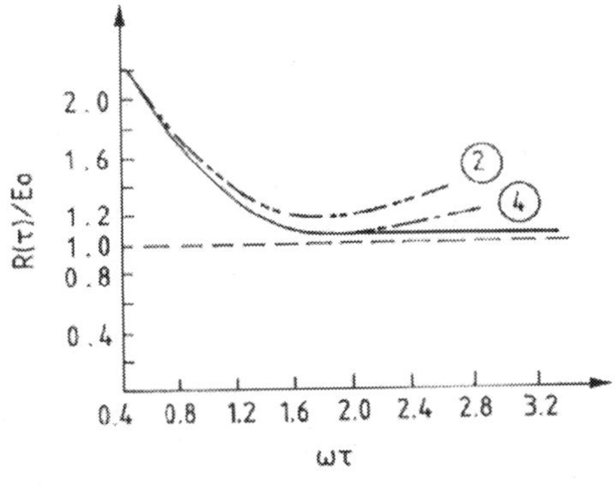

Figure 3. Behaviour of the ground state mass versus the time variable τ for different truncation of the series from Ref. [18, 13]. The horizontal line is the exact solution. ω is the harmonic oscillator frequency.

7.3. Examples of τ-stability from the Υ and B mesons

Hopefully, in the examples of harmonic oscillator in quantum mechanics and of the non-relativistic charmonium sum rules, Refs. [18, 13] have shown that the optimal information from the analysis for a truncated series is obtained at the minimum or inflexion point in τ of the approximate theoretical curves (see Figs 3 and 4). This

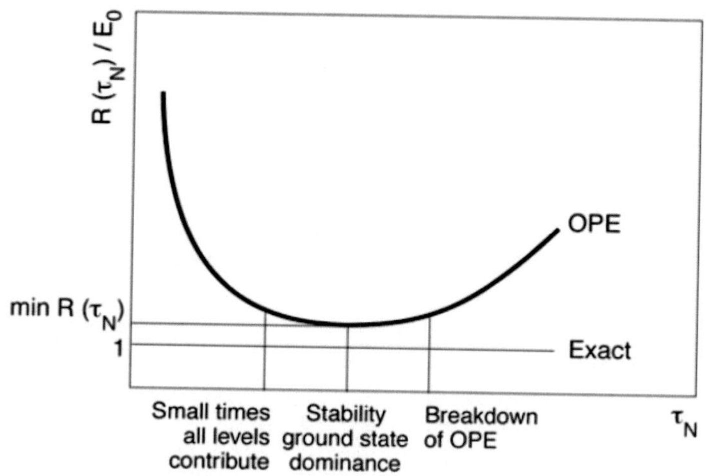

Figure 4. Schematic behaviour of the J/ψ mass versus the sum rule variable τ_N from Ref. [18, 13]. The horizontal line is the experimental mass.

optimal criterion corresponds to the *(principle of minimum sensitivity of the physical parameters on the external sum rule variable* τ*)* and where the exact solution is reached when one adds more and more terms in the approximate series.

In the case of hadrons, we illustrate in Figs. 5 and 6 the analysis for the Υ systems and for the B meson decay constant using relativistic sum rules.

7.4. t_c and μ-stabilities

Later on, we have extended this τ-stability criterion to the continuum threshold variable t_c [5] and to the arbitrary Perturbative (PT) subtraction point μ [6]. The lowest value of t_c corresponds to the beginning of τ-stability while the t_c-stability corresponds to a complete lowest ground state dominance in the spectral integral. In our analysis, we always consider the conservative optimal result inside this t_c-region [7].

[5] In many papers, the optimal value is extracted at the lowest value of t_c ! but the result still increases with the t_c changes.

[6] One can also use the RGE resummed solution which is equivalent to take $\mu^2 = 1/\tau$ but in some case the value of τ is relatively large such that the OPE is not well behaved. A such choice of μ value is often outside the μ-stability region.

[7] $\sqrt{t_c}$ is often identified to the mass of the 1st radial excitation which is a crude approximation as the QCD continuum smears all higher state contributions to the spectral function.

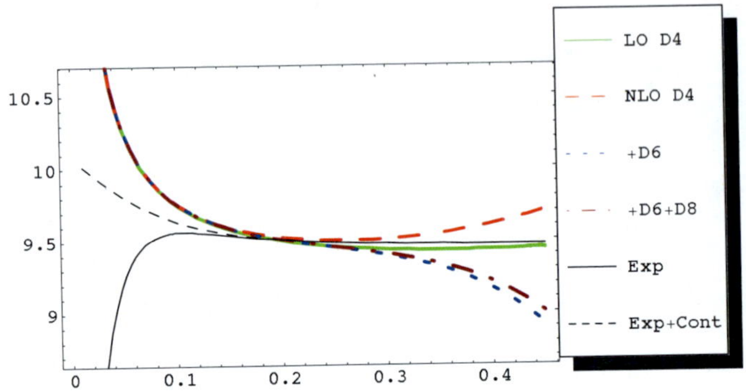

Figure 5. Behaviour of the ratio of moments $\sqrt{\mathcal{R}^b_{10}(\overline{m}_b^2)}$ in GeV versus τ in GeV^{-2} and for $\overline{m}_b(\overline{m}_b) = 4212$ MeV from [49]. The black continuous (resp. short dashed) curves are the experimental contribution including (resp. without) the QCD continuum (it is about the M_Υ). The green (thick continuous) is the PT contribution including the $D = 4$ condensate to LO. The long dashed (red) curve is the contribution including the α_s correction to the $D = 4$ contribution. The short dashed (blue) curve is the QCD expression including the $D = 6$ condensate and the dot-dashed (red-wine) is the QCD expression including the $D = 8$ contribution.

7.5. Ground State versus the QCD Continuum

For some low value of (t_c and large value of τ, one can have some flat curves or some (apparent) minimum in τ. Then, to check or/and to restrict the optimal region in this case, one also requests that the contribution of the ground state to e.g. the spectral integral (e.g. moment sum rule) is larger than the QCD continuum one. One can formulate this constraint in a more rigorous way as (see e.g. [51] for some examples of applications) :

$$R_{P/C} \equiv \frac{\int_{t>}^{t_c} dt\, e^{-t\tau}\, \mathrm{Im}\,\Pi(t)}{\int_{t_c}^{\infty} dt\, e^{-t\tau}\, \mathrm{Im}\,\Pi(t)} \geq 1. \tag{39}$$

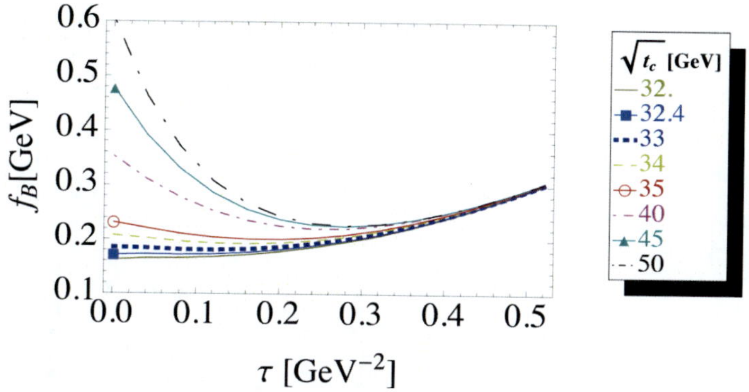

Figure 6. Behaviour of f_B from Ref. [50] versus the sum rule variable τ for different values of the QCD continuum threshold t_c and for a given $\mu = 3$ GeV and $\overline{m}_b(\overline{m}_b) =$ 4177 MeV.

8. Phenomenology of the ρ-meson LSR

8.1. Lower Bound on the ρ-meson Coupling from the Lowest Moment $\mathcal{L}_0$

We use the MDA parametrization of the spectral function and extract the ρ-meson coupling [8]. :

$$\frac{1}{\pi}\text{Im}\Pi_\rho(t) = \frac{M_\rho^2}{2\gamma_\rho^2} \text{ with } \Gamma_{\rho\to e^+e^-} = \frac{2}{3}\alpha^2\pi\frac{M_\rho}{2\gamma_\rho^2} \text{ and } \gamma_\rho^{exp} = 2.479(15). \quad (40)$$

◇ Using positivity of the spectral integral and evaluating the LSR at $\tau = 1/M_\rho^2$, SNR deduces the inequality to leading order of PT series [9]:

$$\gamma_\rho^2 \geq \frac{4\pi}{2\,e}\left[1 + \mathcal{O}(\alpha_s) - \frac{\pi}{3}\frac{\langle\alpha_s G^2\rangle}{M_\rho^4}\right] \Longrightarrow \gamma_\rho \geq 2.43. \quad (41)$$

We improve this bound by adding higher order terms in the PT series and including the contribution of the condensates up to dimension-six. We show the analysis in Fig.7 for different values of t_c and τ. The $R_{P/C}$ condition is shown by the cyan curve where the lowest ground state dominates in the delimited right region. Optimal results are obtained for the sets (τ, t_c) from (0.5,1.5) to (1.1,3.0) in units

[8] Analysis using the complete $e^+e^- \to I = 1$ Hadrons data can e.g. be found in Refs. [38, 48]
[9] Notice that this inequality has been considered as an estimate in the original SVZ paper [7].

of (GeV^{-2}, GeV^2). This leads to the improved lower bound at order α_s^2 :

$$\gamma_\rho \geq 1.83(13)_{t_c}(1)_{a_s}(1)_{G^2}(1)_{\bar\psi\psi^2}. \tag{42}$$

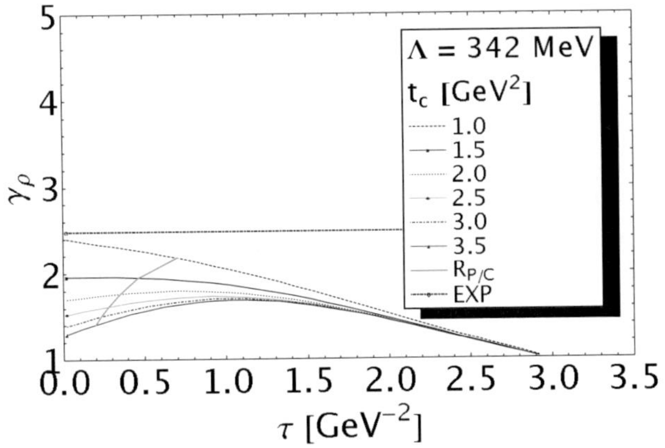

Figure 7. Lower bound of the γ_ρ coupling versus the sum rule variable τ for different values of the continuum threshold t_c.

8.2. The Lowest Moment $\mathcal{L}_0$ Using the $e^+e^- \to I = 1$ Hadrons Data

It is informative to compare the QCD prediction of $\mathcal{L}_0$ with the one where the complete $e^+e^- \to I = 1$ Hadrons data is used to estimate the spectral integral. This analysis has been done recently in Ref. [48] which we show in Fig. 8.

One can observe that there is a good agreement with the experimental e^+e^- data. and the theory prediction using the previous sets of QCD parameters up to $D = 6, 8$ condensates.

8.3. M_ρ from the Lowest Ratio of Moments $\mathcal{R}_{10}$

⋄ Within MDA, one expects that the ratio of moments provides the mass squared of the ρ-meson ground state. We show the analysis versus τ and for different values of t_c in Fig. 9. The optimal results are obtained for the sets (τ, t_c) from (0.5,1.2) to

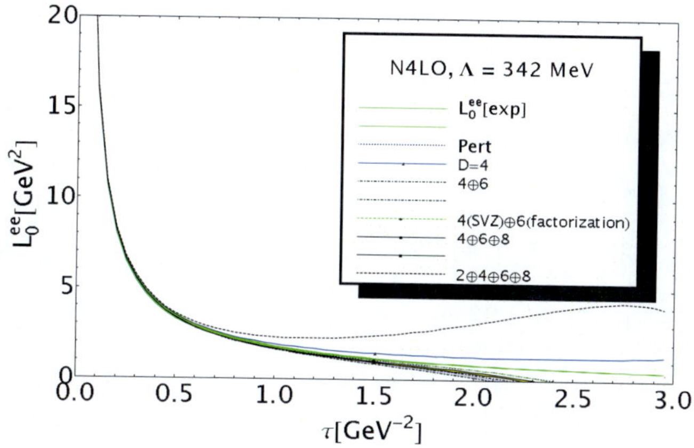

Figure 8. τ-behaviour of the lowest moment $\mathcal{L}_0$ for different truncation of the OPE.

(1.2,3.0) in units of (GeV^{-2}, GeV^2). We obtain:

$$M_\rho \simeq 848(109)_{t_c}(3)_{a_s}(3)_{G^2}(13)_{\bar\psi\psi^2}, \tag{43}$$

where the error comes mainly from the value of t_c which we have taken in the conservative range from the beginning of τ stability until the t_c-one.

⋄ One can restrict this range of t_c value using the constraint from the first moment of FESR. It reads [27, 28]:

$$\frac{M_\rho^2}{2\gamma_\rho^2} = \frac{t_c}{4\pi^2}\Big[1 + a_s(t_c) + a_s^2(t_c)\big[R_2 - \frac{\beta_1}{2} - (\beta_2/\beta_1)\ln\ln(tc/\Lambda^2)\big]\Big], \tag{44}$$

from which we deduce:

$$t_c \simeq 1.962(24)_{t_c}(2)_{\alpha_s}\ \mathrm{GeV}^2. \tag{45}$$

⋄ One should note that the low value of t_c often chosen by many "QCD sum rules practitioners" to reproduce the value of the experimental mass or to extract the mass of the ground state does not coi ncide with the duality constraint from FESR. At the value of t_c in Eq. 45, we obtain from the ratio of moments :

$$M_\rho \simeq 881\ \mathrm{MeV}. \tag{46}$$

⋄ From the previous analysis, the ratio of moments reproduces within 10% the experimental meson mass.

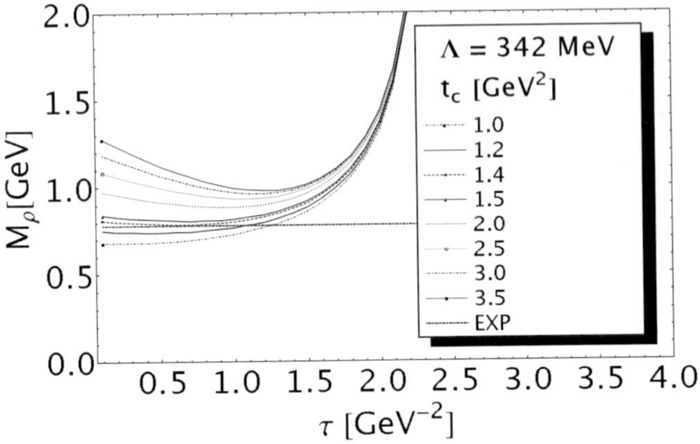

Figure 9. M_ρ versus the sum rule variable τ for different values of the continuum threshold t_c.

8.4. QCD Condensates from $\mathcal{R}_{10}$ Using $e^+e^- \to$ Hadrons Data

◇ The ratio of moments has the advantage to be less sensitive to α_s correction than the moment $\mathcal{L}_0$ as the PT correction starts to order α_s^2. It is then expected to be a good place for extracting the QCD condensates. This analysis has been initiated by Ref. [38] and revised recently in Ref. [48]. Using as input the value of the gluon condensate obtained from heavy quark sum rules [31, 15, 16] given in Eq. 14, one obtains:

$$\rho\alpha_s\langle\bar\psi\psi\rangle^2 \simeq 5.98(64)10^{-4}\,\text{GeV}^6, \quad C_8\langle O_8\rangle = (4.3\pm3.0)\times10^{-2}\,\text{GeV}^8, \quad (47)$$

where the value of the four-quark condensate is in good agreement with the one in Eq. 16.

◇ We show in Fig. 10 a comparison of the experimental expression of $\mathcal{R}_{10}$ with the QCD prediction for different truncations of the OPE. One can notice a good agreement when the condensates up to dimension 6,8 are included.

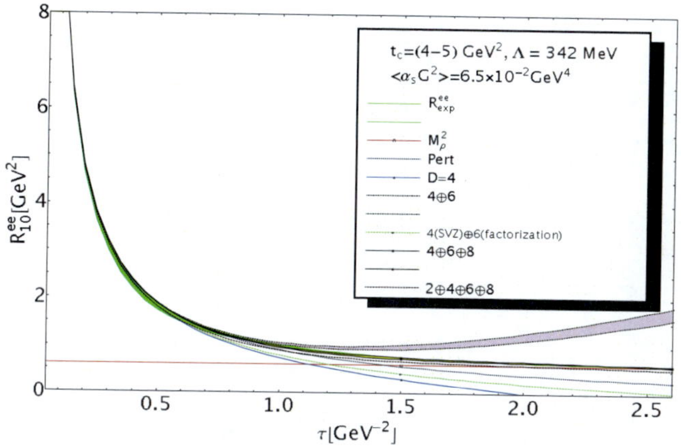

Figure 10. τ-behaviour of the ratio of moments $\mathcal{R}_{10}$ for different truncation of the OPE.

9. Sum of Light Quark Masses from the Pion LSR

9.1. Lower Bound on $(\hat{m}_u + \hat{m}_d)$

Using the pion sum rule, SNR has derived a lower bound on the RGI quark mass defined in Eq. 34 from the positivity of the spectral function and retaining the pion contribution. The results have been given for different values of the QCD scale Λ. Given the present progress on the determination of Λ, we show the new version of the figure given in Ref. [19] in Fig. 11 for different truncation of the PT series. We deduce the optimal lower bound at the maximum of the curves :

$$\begin{aligned}(\hat{m}_u + \hat{m}_d) &\geq 10.13(6)_{\alpha_s}(22)_{G^2}(37)_{\bar{\psi}\psi^2}\,\text{MeV} = 10.13(44)\,\text{MeV} \quad : \quad \text{NLO},\\ &\geq 7.92(26)_{\alpha_s}(1)_{G^2}(8)_{\bar{\psi}\psi^2}\,\text{MeV} = 7.92(27)\,\text{MeV} \quad : \quad \text{N4LO}.\end{aligned} \tag{48}$$

This bound can be improved by introducing the contribution of the next radial excitation $\pi(1300)$ [47] and/or some alternative approaches [52].

9.2. Estimate of $(\hat{m}_u + \hat{m}_d)$

One can transform the previous lower bound by introducing the $\pi(1300)\oplus$ QCD continuum for parametrizing the spectral function. Using the standard OPE, one

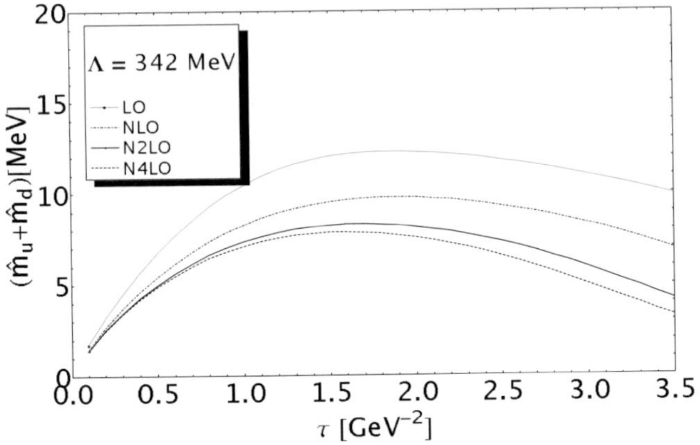

Figure 11. τ-behaviour of RGI $(\hat{m}_u + \hat{m}_d)$ lower bound for different truncation of the PT series.

obtains to order α_s^4 for the RGI and running masses $\overline{m}_q$ evaluated at 2 GeV [47] [10]:

$$(\hat{m}_u + \hat{m}_d) = 9.76(88)\ \text{MeV} \quad \rightarrow \quad (\overline{m}_u + \overline{m}_d)(2) = 8.08(61)\ \text{MeV}, \quad (49)$$

where one can notice that the main uncertainties and the size of the central value come from the parametrization of the spectral function (radial excitations $\oplus$ choice of the QCD continuum threshold t_c). The error due to the way for truncating the PT series only affects slightly the total error (see e.g. [53]). These results agree with lattice calculations [54] and from some other approaches quoted in PDG [55].

10. Some other Applications of LSR

These applications have been already summarized in the recent short reviews [16, 15] where orginal references and some dedicated reviews can be found. They concern the :

- Light baryons qqq' ligth quark states
- Heavy quarks $\bar{Q}Q$ sector.
- Heavy-light $\bar{Q}q$ states.
- So-called exotic states :

[10]Extension of the analysis to some other channels are discussed in this paper.

◇ Glueballs / gluonia gg, ggg states.

◇ Hybrids $\bar{q}gq, \cdots$ states.

◇ Light $(\bar{q}\bar{q}')(qq')$, heavy-light $(\bar{q}\bar{Q})(Qq) \cdots$, fully-heavy $(\bar{Q}\bar{Q})(QQ), \cdots$ four-quark states.

◇ Light $(\bar{q}q')(q\bar{q}')$, heavy-light $(\bar{q}Q)(Q\bar{q}) \cdots$, fully-heavy $(\bar{Q}Q)(\bar{Q}Q), \cdots$ molecule states.

We plan to develop these different parts in a more extended review on QCD spectral sum rules.

Concluding Remarks

LSR and (in general) QCD spectral sum rules (QSSR) are useful tools to tackle within a good approximation the properties of hadrons and for extracting the fundamental parameters of the QCD Lagrangian. If properly done, QSSR is a serious alternative or/and a supplement to the lattice QCD numerical simulations.

References

[1] De Alfaro V., Fubini S., Furlan G., Rossetti C., *Currents in hadron physics*, North Holland Pub. Co. Amsterdam-London (1973).

[2] Weinberg S.,*Phys. Rev. Lett.* **18** (1967) 507.

[3] Das T., Mathur V.S., Okubo S., *Phys. Rev. Lett.* **19** (1967) 470.

[4] Floratos E.G., Narison S. and de Rafael E., *Nucl. Phys.* **B115** (1979) 155 and references quoted therein.

[5] Narison S. and de Rafael E., *Nucl. Phys* **B169** (1980) 253 and references quoted therein..

[6] Narison S., *Z. Phys.* **C 22** (1984) 161.

[7] Shifman M.A., Vainshtein A.I., Zakharov V.I., *Nucl. Phys.* **B147** (1979) 385; *ibid*, *Nucl. Phys.* **B147** (1979) 448.

[8] Zakharov V.I., *Int. J. Mod .Phys.* **A14**, (1999) 4865.

[9] Narison S., *QCD spectral sum rules, World Sci. Lect. Notes Phys.* **26** (1989) 1.

[10] Narison S., *QCD as a theory of hadrons, Cambridge Monogr. Part. Phys. Nucl. Phys. Cosmol.* **17**. (2004) 1 [hep-ph/0205006].

[11] Narison S., *Phys. Rept.* **84 n.4** (1982) 263; *ibid*, *Riv. Nuovo Cim.* **10N2** (1987) 1.

[12] Reinders L.J., Rubinstein H.R., Yazaki S., *Phys. Reports* **127** (1985) 1.

[13] Bertlmann R.A., *Acta Phys. Austriaca* **53**, (1981) 305.

[14] Shifman M.A., *Prog. Theor. Phys. Suppl.* **131** (1998)1; Colangelo P. and Khodjamirian A., hep-ph/0010175; de Rafael E., hep-ph/9802448; Yndurain Y.J., *Monographs in Physics, Springer-Verlag, New York* (1999).

[15] Narison S. ,*Nucl.Part. Phys. Proc.***324-329** (2023) 94; ibid, **258-259** (2015) 189.

[16] Narison S., *Nucl. Part. Phys. Proc.***312-317** (2021) 87; ibid, **300-302** (2018) 153.

[17] Narison S., *Nucl. Part. Phys. Proc.* **309-311** (2020) 135; ibid **270-272** (2016) 143.

[18] Bell J.S., Bertlmann R.A., *ibid, Nucl. Phys.* **B187**, (1981) 285.

[19] Narison S., de Rafael E., *Phys. Lett.* **B103**, (1981) 57.

[20] Bijnens J., Prades J., de Rafael E., *Phys. Lett.* **B348** (1995) 226.

[21] Narison S., *Phys. Lett.* **B210** (1988) 238; *ibid, Phys. Lett.* **B337** (1994) 163.

[22] Logunov A.A., Soloviev L.D., Tavkhelidze A.N., *Phys. Lett.* **B24** (1967) 181.

[23] Sakurai J.J., *Phys. Lett.* **B46** (1973) 207.

[24] Chetyrkin K.G., Krasnikov N.V., Tavkhelidze A.N., *Phys. Lett.* **B76** (1978) 83.

[25] Kataev A.L., Krasnikov N.V. , Pivovarov A.A., *Phys. Lett.* **B123** (1983) 93.

[26] Gorishnii S.G., Kataev A.L., Larin S.A., *Phys. Lett.* **B135** (1984) 457.

[27] Bertlmann R.A., Launer G. and de Rafael E., *Nucl. Phys.* **B250**, (1985) 61.

[28] Bertlmann R.A., Dominguez C.A., Loewe M., Perrottet M. and de Rafael E., *Z. Phys.* **C39** (1988) 231.

[29] Gell-Mann M., Oakes R.J., Renner B., *Phys. Rev.* **175** (1968) 2195.

[30] Dosch H.G., Narison S., *Phys. Lett.* **B417** (1998) 173.

[31] Narison S., *Int. J. Mod. Phys.* **A33** (2018) no. 10, 185004; Addendum: *Int. J. Mod. Phys.* **A33** (2018) no.10, 1850045.

[32] Narison S., Tarrach R., *Phys. Lett.* **B125** (1983) 217.

[33] Ioffe B.L., *Prog. Part. Nucl. Phys.* **56** (2006) 232; ibid, **B188** (1981) 317; ibid, **B191** (1981) 591.

[34] Dosch H.G., *Non-Perturbative Methods* (Montpellier 1985); Chung Y. et al.*Z. Phys.* **C25** (1984) 151; Dosch H.G., Jamin M., Narison S., *Phys. Lett.* **B220** (1989) 251.

[35] Ovchinnikov A.A., Pivovarov A.A., *Yad. Fiz.* **48** (1988) 1135.

[36] Narison S., *Phys. Lett.* **B605** (2005) 319.

[37] Narison S., *Phys. Lett.* **B673** (2009) 30.

[38] Launer G., Narison S., Tarrach R., *Z. Phys.* **C26** (1984) 433.

[39] Dominguez C.A., Sola J., *Z. Phys.* **C40** (1988) 63.

[40] Narison S., *Phys. Lett.* **B300** (1993) 293; ibid **B361** (1995) 121.

[41] Narison S., *Phys. Lett.* **B693** (2010) 559, erratum *ibid*, **B705** (2011) 544; ibid, **B706** (2012) 412; ibid, **B707** (2012) 259.

[42] Nikolaev S.N. , Radyushkin A.V., *Phys. Lett.* **B124** (1983) 243.

[43] Schafer T., Shuryak E.V., *Rev. Mod. Phys.* **70** (1998) 323.

[44] Ioffe B.L., Samsonov A.V., *Phys. Atom. Nucl.* **63** (2000) 1448.

[45] Di Giacomo A., *Non-pertubative Methods*, ed. Narison,*World Scientific* (1985); Di Giacomo A., Rossi G.C., *Phys. Lett.* **B100** (1981) 481; D'Elia M., Di Giacomo A., Meggiolaro E., *Phys. Lett.* **B408** (1997) 315.

[46] Chetyrkin K.G., Kühn J.H. and M. Steinhauser M.,*Comput. Phys. Comm* **133** (2000) 43 and references therein.

[47] Narison S., *Phys. Lett.* **B738** (2014) 346 and references therein.

[48] Narison S., *Nucl.Phys.* **A 1039** (2023) 122744.

[49] Narison S., *Phys. Lett.* **B707** (2012) 259.

[50] Narison S., *Phys. Lett.* **B718** (2013) 1321; ibid, *Nucl. Phys. (Proc.Suppl.)* **234** (2013) 187.

[51] Albuquerque R.M., Narison S., Rabetiarivony D., *Nucl. Phys.* **A 1034** (2023) 122637.

[52] Lellouch L., de Rafael E., Taron J., *Phys.Lett.* **B414**(1997)195; Dominguez C.A., de Rafael E., *Annals Phys.* **174** (1987) 372.

[53] Alam Khan M.S.A., arXiv: 2306.10266 [hep-ph] (2023).

[54] Aoki S. et al., arXiv:1310.8555 [hep-lat] (2013).

[55] Workman R.L.et al. *(Particle Data Group)*, *Prog. Theor. Exp. Phys.* **2022** (2022) 083C01.

Chapter 7

Inverse Problems in Chemical Reactor Design Using the Laplace Transform

Victor Martinez-Luaces*
Chemical Engineering Institute, Faculty of Engineering, UdelaR, Montevideo, Uruguay
ProfeSTEAM Project, University of Granada, Granada, Spain

Abstract

In this work, a transfer function of a continuous stirred-tank reactor (CSTR) as well as the transfer functions of series and parallels of these devices are presented.
After the introduction, several inverse problems, which are the main parts of the chapter, arise. Specifically, the idea is to analyze whether or not any rational function can be the transfer function of a CSTRs system, and even more, if that CSTRs system is unique or not, and if small changes in the transfer function correspond to small changes in the system. In other words, the main purpose is to study the existence, uniqueness and stability of the inverse problem mentioned above.
The chapter shows that the three problems have a negative answer, that is, neither the existence, uniqueness, nor the stability can be assured for the mentioned inverse problem.

Keywords: Laplace transform, chemical reactor design, inverse problems, continuous stirred-tank reactors

AMS Subject Classification: 44A10, 34A55, 34A30

* Corresponding Author's Email: victorml@fing.edu.uy , victorml@correo.ugr.es.

In: The Laplace Transform and Its Applications in Science
Editor: Victor Martinez-Luaces
ISBN: 979-8-89113-747-9

Introduction

Let us consider a continuous stirred tank, like in Figure 1, where F represents the volumetric flow rate (for instance, in liters per second), $C_0(t), C(t)$ are the input and output concentrations of a certain solute like Sodium Chloride (for instance, in grams per liter) and finally t, m are time and solute mass, respectively [1, 2, 3, 4].

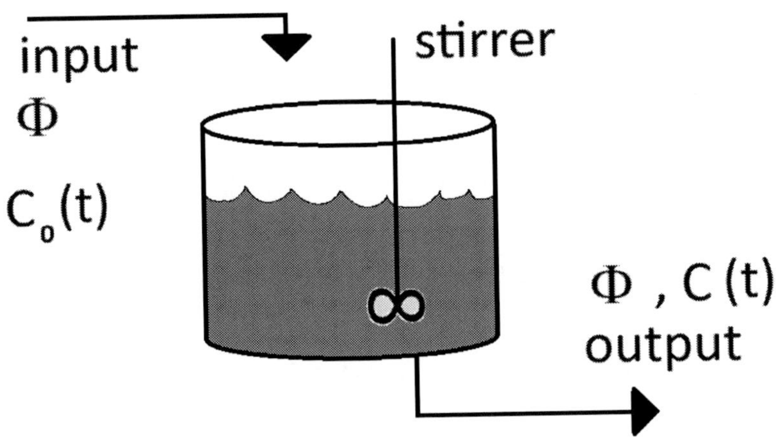

Figure 1. Scheme of a continuous stirred-tank reactor.

A mass balance gives:

$$\frac{dm}{dt} = F\,C_0(t) - F\,C(t) \qquad (1)$$

Which can be also written as:

$$\frac{d}{dt}\big(V\,C(t)\big) = F\,C_0(t) - F\,C(t) \qquad (2)$$

Or:

$$V\,C'(t) = F\,C_0(t) - F\,C(t) \qquad (3)$$

Applying Laplace Transform we obtain

$$V\mathscr{L}\{C'(t)\} = F\mathscr{L}\{C_0(t)\} - F\mathscr{L}\{C(t)\} \tag{4}$$

$$\mathscr{L}\{C'(t)\} = s\mathscr{L}\{C(t)\} - C(0) = s\mathscr{L}\{C(t)\} \tag{5}$$

Then, we have:

$$Vs\mathscr{L}\{C(t)\} = F\mathscr{L}\{C_0(t)\} - F\mathscr{L}\{C(t)\} \tag{6}$$

Or:

$$(F + Vs)\,\mathscr{L}\{C(t)\} = F\mathscr{L}\{C_0(t)\} \tag{7}$$

Then, the quotient of $\mathscr{L}\{C(t)\}$ and $\mathscr{L}\{C_0(t)\}$ is:

$$G(s) = \mathscr{L}\{C(t)\} \,/\, \mathscr{L}\{C_0(t)\} = \frac{F}{F+Vs} = \frac{1}{1+(V/F)\,s} = \frac{1}{1+t\,s} \tag{8}$$

Note that dimensionally $t = V/F$ is a time. In fact, $t = V/F$ is called the residence time in the reactor [5, 6, 7].

This quotient $G(s) = \mathscr{L}\{C(t)\} \,/\, \mathscr{L}\{C_0(t)\} = \frac{1}{1+\tau s}$ is the transfer function of the CSTR.

In the following sections, we consider the transfer function for a series of CSTRs and after that, we analyze the same situation for a CSTRs parallel system [8, 9].

Definition and Lemma

Before considering a series and parallels of CSTRs, we are going to introduce a definition and a previous lemma.

Definition: CSTR-polynomials

Let $A = \left\{\frac{1}{\tau_1}, \frac{1}{\tau_2}, \ldots, \frac{1}{\tau_n}\right\}$ be a set of reciprocals of residence times corresponding to several CSTRs and let $\{n_1, n_2, \ldots n_i\}$ be a subset of the index set $\{1, 2, \ldots, n\}$. Then, the associated CSTR-polynomial can be defined as:

$$P_{n_1 n_2 \ldots n_i}(s) \stackrel{\text{def}}{=} \left(s + \frac{1}{n_1}\right).\left(s + \frac{1}{n_2}\right)\ldots\left(s + \frac{1}{n_i}\right) \tag{9}$$

Lemma 1.

The CSTR-polynomial can be written as:

$$P_{n_1 n_2 \ldots n_i}(s) = s^i + \sigma_1 s^{i-1} + \sigma_2 s^{i-2} + \cdots + \sigma_{i-1} s + \sigma_i \tag{10}$$

Being σ_k the sum of products of numbers $\frac{1}{\tau_{n_1}}, \frac{1}{\tau_{n_2}}, \ldots, \frac{1}{\tau_{n_i}}$ taken k at a time.

Proof.

$$\begin{aligned} P_{n_1 n_2 \ldots n_i}(s) &= \left(s + \frac{1}{n_1}\right).\left(s + \frac{1}{n_2}\right)\ldots\left(s + \frac{1}{n_i}\right) \\ &= s^i + \left(\frac{1}{\tau_{n_1}} + \frac{1}{\tau_{n_2}} + \ldots + \frac{1}{\tau_{n_i}}\right) s^{i-1} \\ &+ \left(\frac{1}{\tau_{n_1}}\frac{1}{\tau_{n_2}} + \frac{1}{\tau_{n_1}}\frac{1}{\tau_{n_3}} + \cdots\right) s^{i-2} + \cdots + \left(\frac{1}{\tau_{n_1}}\frac{1}{\tau_{n_2}}\cdots\frac{1}{\tau_{n_i}}\right) \end{aligned}$$

Then:

$$P_{n_1 n_2 \ldots n_i}(s) = s^i + \sigma_1 s^{i-1} + \sigma_2 s^{i-2} + \cdots + \sigma_{i-1} s + \sigma_i \tag{10}$$

Where $\sigma_1 = \frac{1}{\tau_{n_1}} + \frac{1}{\tau_{n_2}} + \ldots + \frac{1}{\tau_{n_i}} = \sum_j \frac{1}{\tau_{n_j}}$,

$\sigma_2 = \frac{1}{\tau_{n_1}}\frac{1}{\tau_{n_2}} + \frac{1}{\tau_{n_1}}\frac{1}{\tau_{n_3}} + \cdots = \sum_{j \neq k} \frac{1}{\tau_{n_j}}\frac{1}{\tau_{n_k}}$ and, in general, σ_k is the sum of products of numbers $\frac{1}{\tau_{n_1}}, \frac{1}{\tau_{n_2}}, \ldots, \frac{1}{\tau_{n_i}}$ taken k at a time. So, the thesis is proved.

Series of CSTRs

Let us consider a series of CSTRs like in Figure 2. The transfer function for this system is given by the following theorem:

Theorem 1.

The transfer function for the CSTRs system of Figure 2 is given by:

$$G(s) = \frac{1}{(\prod_1^n \tau_i)\, P_{1\,2\ldots n}(s)} \tag{11}$$

Proof.

The transfer functions corresponding to the reactors are:

$$G_1(s) = \mathscr{L}\{C_1(t)\} \,/\, \mathscr{L}\{C_0(t)\}$$

$$G_2(s) = \mathscr{L}\{C_2(t)\} \,/\, \mathscr{L}\{C_1(t)\}$$

....

$$G_{n-1}(s) = \mathscr{L}\{C_{n-1}(t)\} \,/\, \mathscr{L}\{C_{n-2}(t)\}$$

$$G_n(s) = \mathscr{L}\{C_n(t)\} \,/\, \mathscr{L}\{C_{n-1}(t)\} \tag{12}$$

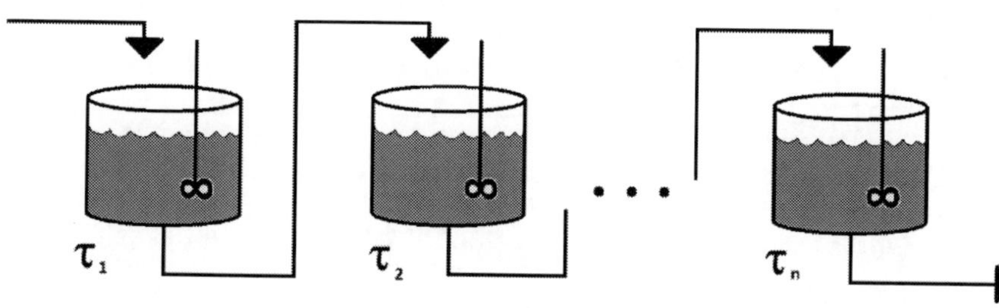

Figure 2. Series of two CSTRs.

Multiplying all these equalities we obtain:

$$\prod_{i=1}^n G_i(s) = G_1(s) \cdot G_2(s) \cdots G_n(s) = \mathscr{L}\{C_n(t)\} \,/\, \mathscr{L}\{C_0(t)\} = G(s) \tag{13}$$

If this general law is applied to continuous stirred tanks, the result is:

$$G(s) = \prod_{i=1}^{n} G_i(s) = \frac{1}{1+\tau_1\, s} \cdot \frac{1}{1+\tau_2\, s} \cdots \frac{1}{1+\tau_n\, s} = \frac{1}{(\prod_1^n \tau_i)\left(s+\frac{1}{\tau_1}\right)\left(s+\frac{1}{\tau_2}\right)\cdots\left(s+\frac{1}{\tau_n}\right)} =$$

$$= \frac{1}{(\prod_1^n \tau_i)\,(s^n+\sigma_1 s^{n-1}+\cdots+\sigma_{n-1}s+\sigma_n)} \quad (14)$$

Where σ_k is the sum of numbers $\frac{1}{\tau_1}, \frac{1}{\tau_2}, \ldots, \frac{1}{\tau_n}$ taken k at a time, and so we have:

$$G(s) = \frac{1}{(\prod_1^n \tau_i)\, P_{1\,2 \ldots n}(s)} \quad (11)$$

And the theorem is proved.

Some comments about a particular case.

If only two reactors are included in the CSTRs system, the transfer function is:

$$G(s) = \frac{1}{(\prod_1^2 \tau_i)\, P_{1\,2}(s)} = \frac{1}{(\tau_1\tau_2)\,(s^2+\sigma_1 s+\sigma_2)} \quad (15)$$

Moreover, $\sigma_1 = \sum_j \frac{1}{\tau_j} = \frac{1}{\tau_1} + \frac{1}{\tau_2}$, so $(\tau_1\tau_2)\sigma_1 = (\tau_1\tau_2)\left(\frac{1}{\tau_1} + \frac{1}{\tau_2}\right) = \tau_1 + \tau_2$

and

$\sigma_2 = \sum_{j \neq k} \frac{1}{\tau_j}\frac{1}{\tau_k} = \frac{1}{\tau_1}\frac{1}{\tau_2}$, so $(\tau_1\tau_2)\sigma_2 = (\tau_1\tau_2)\left(\frac{1}{\tau_1}\frac{1}{\tau_2}\right) = 1$

In other words:

$$G(s) = \frac{1}{(\tau_1\tau_2)\, s^2 + (\tau_1 + \tau_2)\, s + 1} \quad (16)$$

Three interesting observations can be made about this function:

1) the leading coefficient is the product of the residence times $\tau_1\tau_2$.
2) the constant term is one.
3) the polynomial $(\tau_1\tau_2)\, s^2 + (\tau_1 + \tau_2)\, s + 1$ has a non-negative discriminant, since $\Delta = (\tau_1 + \tau_2)^2 - 4\,(\tau_1\, \tau_2) = {\tau_1}^2 + {\tau_1}^2 + 2\,\tau_1\,\tau_2 - 4\,\tau_1\,\tau_2 = {\tau_1}^2 + {\tau_1}^2 - 2\,\tau_1\,\tau_2 = (\tau_1 - \tau_2)^2 \geq 0$.

The first two observations can be generalized, as it is shown in the following corollary:

Corollary 1.

In a series of n CSTRs, the transfer function is:

$$G(s) = \frac{1}{(\prod_i \tau_i)\, s^n + \cdots + 1} \tag{17}$$

That is, for the denominator, the leading coefficient is $(\prod_i \tau_i)$ and the constant term is one.

Proof.

The transfer function of a series of CSTRs can be written as:

$$G(s) = \frac{1}{(\prod_1^n \tau_i)\, P_{1\,2\ldots n}(s)} = \frac{1}{(\prod_1^n \tau_i)\,(s^n + \sigma_1 s^{n-1} + \cdots + \sigma_{n-1} s + \sigma_n)} = \frac{1}{(\prod_1^n \tau_i)\, s^n + \cdots + (\prod_1^n \tau_i)\sigma_n} \tag{18}$$

So, the leading coefficient is $(\prod_i \tau_i)$ and the constant term is $(\prod_1^n \tau_i)\sigma_n = \tau_1 \ldots \tau_n \frac{1}{\tau_1} \ldots \frac{1}{\tau_n} = 1$, then we have:

$$G(s) = \frac{1}{(\prod_1^n \tau_i)\, s^n + \cdots + 1} \tag{17}$$

Parallel of CSTRs

Let us consider a parallel of CSTRs like in Figure 3, where the big pipe is branched into n pipes, where $F_1, F_2, \ldots, F_n$ represent the flow rates in the branches and $C_1(t), C_2(t), \ldots, C_n(t)$ are the corresponding concentrations. The associated transfer function is obtained in the following theorem:

Theorem 2.

The transfer function for the CSTRs system of Figure 3 is given by:

$$G(s) = \frac{f_1}{\tau_1} \frac{P_{2\ldots n}(s)}{P_{1\,2\ldots n}(s)} + \cdots + \frac{f_n}{\tau_n} \frac{P_{1\ldots n-1}(s)}{P_{1\,2\ldots n}(s)} \tag{19}$$

Where $f_1 = \frac{F_1}{F}$, ... , $f_n = \frac{F_n}{F}$ are the flow fractions corresponding to the different branches.

Proof.

A flow balance at point A in Figure 3 gives:

$$F = F_1 + F_2 + \ldots + F_n \tag{20}$$

And a mass balance at point B, gives:

$$F\ C = F_1\ C_1 + F_2\ C_2 + \ldots + F_n\ C_n \tag{21}$$

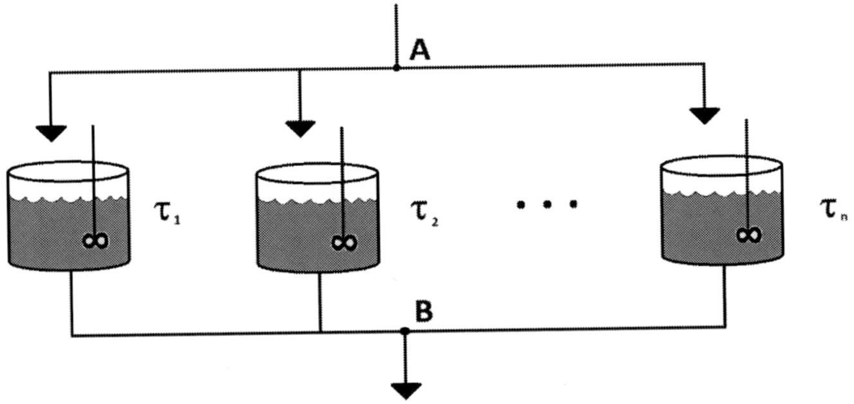

Figure 3. A parallel of n CSTRs.

Applying Laplace Transform to this last equality, we obtain:

$$F_1\ \mathscr{L}\{C_1(t)\} + F_2\ \mathscr{L}\{C_2(t)\} + \cdots + F_n\ \mathscr{L}\{C_n(t)\} = F\ \mathscr{L}\{C(t)\}$$

If this equality is divided into $\mathscr{L}\{C_0(t)\}$ then we have:

$$F_1\ \mathscr{L}\{C_1(t)\}\ /\ \mathscr{L}\{C_0(t)\} + F_2\ \mathscr{L}\{C_2(t)\}\ /\ \mathscr{L}\{C_0(t)\} + \cdots +$$

$$+ F_n\ \mathscr{L}\{C_n(t)\}\ /\ \mathscr{L}\{C_0(t)\} = F\ \mathscr{L}\{C(t)\}\ /\ \mathscr{L}\{C_0(t)\}$$

That is:

$$F_1\ G_1(s) + F_1\ G_1(s) + \ldots + F_n\ G_n(s) = F\ G(s) \tag{22}$$

Now, dividing into the volumetric flow F we obtain:

$$G(s) = \frac{F_1}{F}\, G_1(s) + \frac{F_2}{F}\, G_2(s) + \ ... + \frac{F_n}{F}\, G_n(s) = f_1\, G_1(s) + f_2\, G_2(s) + \ ... + f_n\, G_n(s) \qquad (23)$$

Since all the reactors are CSTRs, the result is:

$$G(s) = f_1 \frac{1}{1+\tau_1 s} + f_2 \frac{1}{1+\tau_2 s} + \ ... + f_n \frac{1}{1+\tau_n s} \qquad (24)$$

After some algebraic manipulations, we obtain:

$$\begin{aligned} G(s) &= f_1 \frac{(1+\tau_2 s) ... (1+\tau_n s)}{(1+\tau_1 s)(1+\tau_2 s) ... (1+\tau_n s)} + \cdots \\ &\quad + f_n \frac{(1+\tau_1 s) ... (1+\tau_{n-1} s)}{(1+\tau_1 s)(1+\tau_2 s) ... (1+\tau_n s)} \\ &= f_1 \frac{\tau_2 \ldots \tau_n (s+1/\tau_2) ... (s+1/\tau_n)}{\tau_1\tau_2 \ldots \tau_n (s+1/\tau_1)(s+1/\tau_2) ... (s+1/\tau_n)} \\ &\quad + \cdots f_n \frac{\tau_1 \ldots \tau_{n-1} (s+1/\tau_1) ... (s+1/\tau_{n-1})}{\tau_1\tau_2 \ldots \tau_n (s+1/\tau_1)(s+1/\tau_2) ... (s+1/\tau_n)} \end{aligned}$$

Finally, simplifying this expression results:

$$G(s) = \frac{f_1}{\tau_1} \frac{(s+1/\tau_2)...(s+1/\tau_n)}{(s+1/\tau_1)(s+1/\tau_2)...(s+1/\tau_n)} + \cdots + \frac{f_n}{\tau_n} \frac{(s+1/\tau_1)...(s+1/\tau_{n-1})}{(s+1/\tau_1)(s+1/\tau_2)...(s+1/\tau_n)} = \\ = \frac{f_1}{\tau_1} \frac{P_{2...n}(s)}{P_{1\,2...n}(s)} + \cdots + \frac{f_n}{\tau_n} \frac{P_{1...n-1}(s)}{P_{1\,2...n}(s)} \qquad (25)$$

And the thesis is proved.

Some comments about two particular cases.

Let us consider only two reactors in the CSTR parallel system. In this case the transfer function is:

$$G(s) = \frac{f_1}{\tau_1} \frac{P_2(s)}{P_{1\,2}(s)} + \frac{f_2}{\tau_2} \frac{P_1(s)}{P_{1\,2}(s)} = \frac{f_1\,\tau_2\,P_2(s) + f_2\,\tau_1\,P_1(s)}{\tau_1\tau_2 P_{1\,2}(s)} \qquad (26)$$

Then:

$$G(s) = \frac{f_1\,\tau_2 (s+1/\tau_2) + f_2\,\tau_1 (s+1/\tau_1)}{\tau_1\tau_2\,(s+1/\tau_1)(s+1/\tau_2)} = \frac{(f_1\,\tau_2 + f_2\,\tau_1)\, s + (f_1 + f_2)}{\tau_1\tau_2\, s^2 + (\tau_1 + \tau_2)\, s + 1} = \\ \frac{(f_1\,\tau_2 + f_2\,\tau_1)\, s + 1}{\tau_1\tau_2\, s^2 + (\tau_1 + \tau_2)\, s + 1} \qquad (27)$$

Three observations can be made regarding the denominator of the fraction:

1) the leading coefficient is the product of the residence times $\tau_1\tau_2$.

2) the constant term is 1.
3) the polynomial $(\tau_1\tau_2)\, s^2 + (\tau_1 + \tau_2)\, s + 1$ has a non-negative discriminant, as in the series case.

Regarding the numerator of the fraction, it should be noted that:

1) The constant term is one.
2) The leading coefficient is a convex linear combination of the residence times.

If three CSTRs in parallel are considered, the result is:

$$G(s) = \frac{f_1}{\tau_1}\frac{P_{2\,3}(s)}{P_{1\,2\,3}(s)} + \frac{f_2}{\tau_2}\frac{P_{1\,3}(s)}{P_{1\,2\,3}(s)} + \frac{f_3}{\tau_3}\frac{P_{1\,2}(s)}{P_{1\,2\,3}(s)}$$
$$= \frac{f_1\,\tau_2\tau_3\,P_{2\,3}(s) + f_2\,\tau_1\tau_3\,P_{1\,3}(s) + f_3\,\tau_1\tau_2\,P_{1\,2}(s)}{\tau_1\tau_2\tau_3\,P_{1\,2\,3}(s)}$$

Then:

$$G(s)$$
$$= \frac{f_1\,\tau_2\tau_3\,(s + 1/\tau_2)(s + 1/\tau_3)}{\tau_1\tau_2\tau_3\,(s + 1/\tau_1)(s + 1/\tau_2)(s + 1/\tau_3)}$$
$$+ \frac{f_2\,\tau_1\tau_3\,(s + 1/\tau_1)(s + 1/\tau_3)}{\tau_1\tau_2\tau_3\,(s + 1/\tau_1)(s + 1/\tau_2)(s + 1/\tau_3)}$$
$$+ \frac{f_3\,\tau_1\tau_2\,(s + 1/\tau_1)(s + 1/\tau_2)}{\tau_1\tau_2\tau_3\,(s + 1/\tau_1)(s + 1/\tau_2)(s + 1/\tau_3)}$$
$$= \frac{(f_1\,\tau_2\tau_3 + f_2\,\tau_1\tau_3 + f_3\,\tau_1\tau_2)\,s^2 + \cdots + 1}{\tau_1\tau_2\tau_3\,s^3 + (\tau_1\tau_2 + \tau_1\tau_3 + \tau_2\tau_3)\,s^2 + (\tau_1 + \tau_2 + \tau_3)\,s + 1}$$

So, once again the leading coefficient of the denominator is the product of the residence times, the constant term in both the numerator and denominator is one, and the leading coefficient of the numerator is a convex linear combination of the products $\{\tau_1\tau_2, \tau_1\tau_3, \tau_2\tau_3\}$.

Some of these results are generalized in the following corollary:

Corollary 2.

In a parallel of CSTRs, the transfer function is a rational one such that the constant term in both the numerator and denominator is 1, the denominator leading coefficient is the product of the residence times $\tau_1\tau_2 \dots \tau_n$ and the numerator leading coefficient is a convex linear combination of the products $\tau_2\tau_3 \dots \tau_n$, $\tau_1\tau_3 \dots \tau_n$ and $\tau_1\tau_2 \dots \tau_{n-1}$.

Proof.

Due to Theorem 3, the transfer function for a CSTRs parallel system is:

$$G(s) = \frac{f_1}{\tau_1}\frac{P_{2\dots n}(s)}{P_{1\,2\dots n}(s)} + \cdots + \frac{f_n}{\tau_n}\frac{P_{1\dots n-1}(s)}{P_{1\,2\dots n}(s)} \tag{19}$$

That is:

$$G(s) = \frac{f_1\,\tau_2 \dots \tau_n\,P_{2\dots n}(s) + \cdots + f_n\,\tau_1 \dots \tau_{n-1}\,P_{1\dots n-1}(s)}{\tau_1\tau_2 \dots \tau_n\,P_{1\,2\dots n}(s)} \tag{28}$$

Taking into account Lemma 1, the denominator can be written as:

$$\tau_1\tau_2 \dots \tau_n\,P_{1\,2\dots n}(s) = \tau_1\tau_2 \dots \tau_n\,(s^n + \sigma_1 s^{n-1} + \cdots + \sigma_{n-1}s + \sigma_n) =$$
$$\tau_1\tau_2 \dots \tau_n\;s^n + \cdots + \tau_1\tau_2 \dots \tau_n\,\frac{1}{\tau_1}\frac{1}{\tau_2}\dots\frac{1}{\tau_n} = \tau_1\tau_2 \dots \tau_n\;s^n + \cdots + 1$$

Similarly, the numerator is:

$$\begin{aligned} & f_1\,\tau_2 \dots \tau_n\,P_{2\dots n}(s) + \cdots + f_n\,\tau_1 \dots \tau_{n-1}\,P_{1\dots n-1}(s) \\ &= f_1\,\tau_2 \dots \tau_n\left(s^{n-1} + \cdots + \frac{1}{\tau_2}\dots\frac{1}{\tau_n}\right) + \cdots \\ &+ f_n\,\tau_1 \dots \tau_{n-1}\left(s^{n-1} + \cdots + \frac{1}{\tau_1}\dots\frac{1}{\tau_{n-1}}\right) \\ &= (f_1\,\tau_2 \dots \tau_n + \cdots + f_n\,\tau_1 \dots \tau_{n-1})s^{n-1} + \cdots \\ &+ (f_1 + \cdots + f_n) \\ &= (f_1\,\tau_2 \dots \tau_n + \cdots + f_n\,\tau_1 \dots \tau_{n-1})\,s^{n-1} + \cdots + 1 \end{aligned}$$

Then:

$$G(s) = \frac{(f_1\,\tau_2 \dots \tau_n + \cdots + f_n\,\tau_1 \dots \tau_{n-1})\,s^{n-1} + \cdots + 1}{\tau_1\tau_2 \dots \tau_n\;s^n + \cdots + 1} \tag{29}$$

So, both constant terms are 1, the numerator leading coefficient is $f_1\tau_2 \ldots \tau_n + \cdots + f_n\tau_1 \ldots \tau_{n-1}$ (a convex linear combination) and the leading coefficient is $\tau_1\tau_2 \ldots \tau_n = \prod_i \tau_i$ for the denominator. Then, the thesis is proved.

Several comments about the previous results.

In both cases –series and parallel– the transfer function is a quotient of polynomials such that the degree of the numerator is less than the degree of the denominator, so if both series and parallels are combined, the transfer function must be similar. However, in a combined system the degree of the denominator polynomial does not always coincide with the number of reactors.

For instance, in Figure 4 there is a series of two reactors and this sub-system in turn is in parallel with another reactor, resulting in a system of three CSTRs, however, this is not reflected in the transfer function.

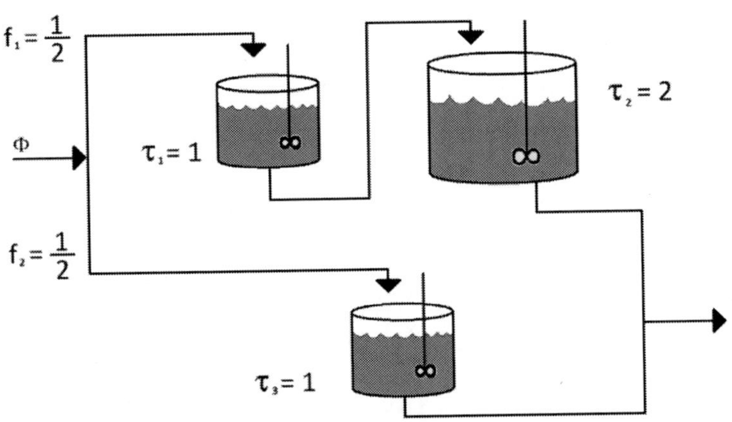

Figure 4. Series of two CSTRs in parallel with another reactor.

In this case the transfer function is:

$$G(s) = \frac{1}{2}\left(\frac{1}{1+s}\,\frac{1}{1+2\,s}\right) + \frac{1}{2}\,\frac{1}{1+s} = \frac{1}{2}\left(\frac{1}{1+2\,s} + 1\right)\frac{1}{1+s}$$
$$= \frac{1}{2}\,\frac{1+1+2s}{1+2\,s}\,\frac{1}{1+s} = \frac{1}{1+2\,s} \qquad (30)$$

That is, the CSTRs system has the same transfer function as a single CSTR with residence time $\tau = 2$, so the number of reactors cannot be determined from the transfer function.
Another example appears in Figure 5, where a sub-system consisting in a parallel of two CSTRs is connected in a series with another reactor.

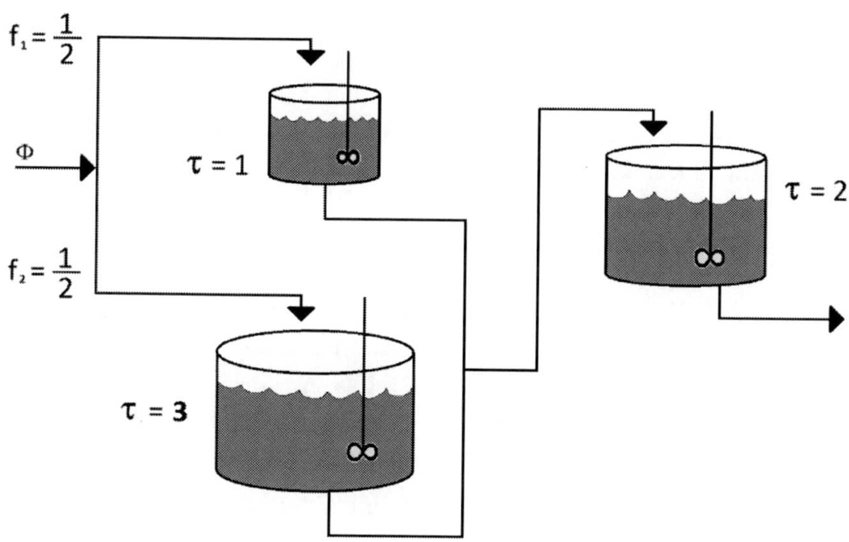

Figure 5. Two CSTRs in parallel followed by a series with another reactor.

Once again, results a system of three reactors, which transfer function is:

$$\begin{aligned} G(s) &= \left(\frac{1}{2}\,\frac{1}{1+s} + \frac{1}{2}\,\frac{1}{1+3\,s}\right)\frac{1}{1+2\,s} \\ &= \frac{1}{2}\left(\frac{1+3\,s+1+s}{(1+s)(1+3\,s)}\right)\frac{1}{1+2\,s} \\ &= \frac{1}{2}\,\frac{2+4\,s}{(1+s)(1+3\,s)}\,\frac{1}{1+2\,s} = \frac{1}{(1+s)(1+3\,s)} \end{aligned} \tag{31}$$

That is, the CSTRs system has the same transfer function as a series of two CSTRs with residence times $\tau = 1$ and $= 3$. Once again, the number of reactors cannot be determined from the transfer function.

Inverse problems about series and/or parallels of CSTRs

In a previous work, Martinez-Luaces, Fernandez-Plaza and Rico [10] described four groups of inverse problems: causation problems, specification problems, double (both causation and specification) and mixed (direct-inverse or inverse-direct).

In this chapter we focus on the first group, i.e., the causation inverse problems. Firstly, in order to explain them we consider direct problems. As Groetsch [11] mentioned "The type of direct problems we have in mind is that of determining the effect y of a given cause x when a definite mathematical model K is posited: $K\,x = y$ ...". Then, it is possible to schematize a direct problem, see Figure 6.

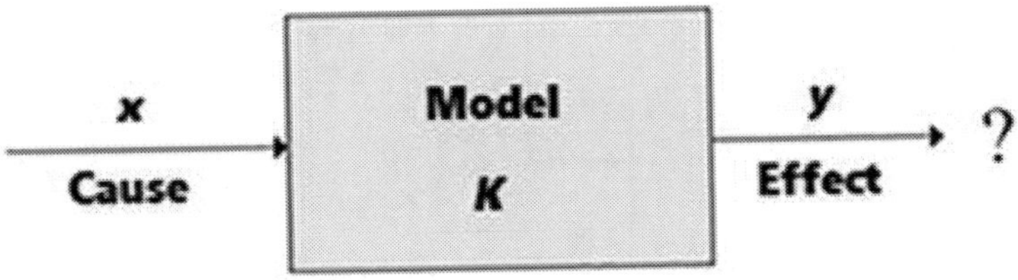

Figure 6. Scheme for a direct problem, following Groetsch ideas.

In a causation inverse problem, the model and the effect are well known, and the question is about the cause. This new problem can be schematized as shown in Figure 7.

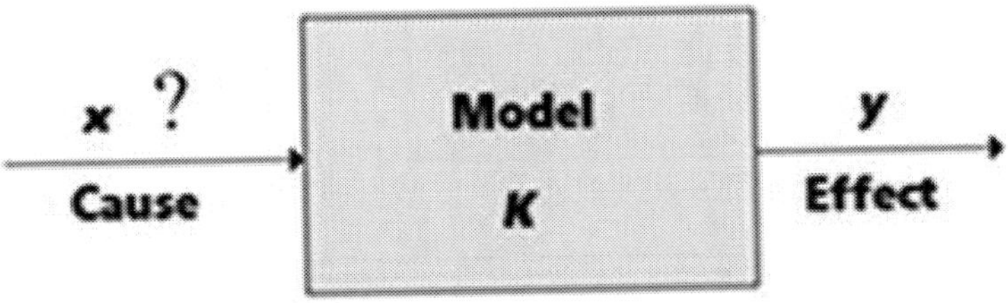

Figure 7. Scheme for causation inverse problems.

Thus, taking into account this theoretical framework, we consider that in the specific context of the topic dealt with in this chapter, the direct problem would be to determine the transfer function corresponding to a system of

CSTRs arranged in series and/or parallels. Consequently, the inverse causation problem consists in determining the system of CSTRs that corresponds to a given function $G(s)$. The mentioned causation inverse problem raises three fundamental questions in any mathematical problem: existence, stability and uniqueness [12].

Existence of the inverse problem.

It is obvious that in the inverse problem considered above –at least in its most general version– it is not possible to ensure the existence of a solution. In fact, a given function $G(s)$ not always corresponds to a system of CSTRs in series and/or parallels.

Even functions that might seem to correspond to a series of reactors, such as $(s) = \frac{s+1}{s^2+s+1}$, do not correspond to such a system of CSTRs. In fact, it is easy to note that the polynomial $s^2 + s + 1$ has a negative discriminant, which contradicts the previously prooved results.

A similar situation happens with the function $G(s) = \frac{3s+1}{(1+s)(1+2s)}$, which may correspond to a parallel of CSTRs, however, the numerator has a leading coefficient 3, which is not a convex linear combination of the residence times $\tau_1 = 1$ and $\tau_2 = 2$, once again contradicting previously prooved results.

Then, a quotient of polynomials such that the degree of the numerator is less than the degree of the denominator does not always correspond to a transfer function of a CSTRs system.

Stability of the inverse problem.

In order to study the stability of the causation inverse problem, let us consider the following function: $G(s) = \frac{As^2+4s+1}{(1+s)(1+2s)(1+3s)}$, and we are going to vary the parameter A in a neighborhood of 3.

If we consider $A = 2.99$, then: $G(s) = \frac{2.99s^2+4s+1}{(1+s)(1+2s)(1+3s)}$, and the corresponding partial fraction decomposition gives:

$G(s) = \frac{2.99s^2+4s+1}{(1+s)(1+2s)(1+3s)} = -\frac{0.005}{1+s} + \frac{1.01}{1+2s} - \frac{0.005}{1+3s}$, and obviously this function does not correspond to a CSTRs system.

Now, if we consider $A = 3$, then:

$G(s) = \frac{3\,s^2 + 4\,s + 1}{(1+s)(1+2\,s)(1+3\,s)} = \frac{(1+s)(1+3\,s)}{(1+s)(1+2\,s)(1+3\,s)} = \frac{1}{1+2\,s}$, which corresponds to a single CSTR with residence time $\tau = 2$.

Finally, if $A = 3.01$, then: $G(s) = \frac{3.01\,s^2 + 4\,s + 1}{(1+s)(1+2\,s)(1+3\,s)}$, and the corresponding partial fraction decomposition gives:

$G(s) = \frac{3.01\,s^2 + 4\,s + 1}{(1+s)(1+2\,s)(1+3\,s)} = \frac{0.005}{1+s} + \frac{0.99}{1+2\,s} + \frac{0.005}{1+3\,s}$, which is a parallel of three CSTRs with residence times $\tau_1 = 1$, $\tau_2 = 2$ and $\tau_3 = 3$ and flow fractions $f_1 = \frac{1}{200} = 0.005$, $f_2 = \frac{99}{100} = 0.99$ and $f_3 = \frac{1}{200} = 0.005$, respectively. Note that $f_1 + f_2 + f_3 = 1$, as expected.

So, as it can be observed, slight changes in one of the coefficients produce dramatic changes in the reactor system. Then, stability does not occur in the caustion inverse problem.

Uniqueness of the inverse problem.

In order to study the uniqueness of the causation inverse problem, let us consider the following reactor (Figure 8).

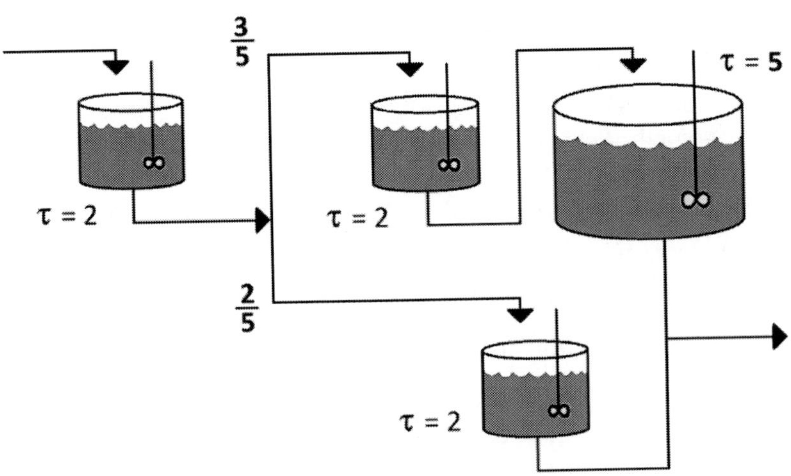

Figure 8. CSTRs system with series and parallels.

The transfer function for this reactor is:

$$G(s) = \frac{1}{1+2s}\left[\frac{3}{5}\,\frac{1}{(1+2s)(1+5s)} + \frac{2}{5}\,\frac{1}{1+2s}\right] = \frac{1}{1+2s}\,\frac{3+2(1+5s)}{5(1+2s)(1+5s)} = \frac{1}{1+2s}\,\frac{5(1+2s)}{5(1+2s)(1+5s)} = \frac{1}{1+2s}\,\frac{1}{1+5s} \tag{32}$$

This result corresponds to a series of two CSTRs with residence times $\tau_1 = 2$ and $\tau_2 = 5$.

Now, let us consider the following reactor (Figure 9).

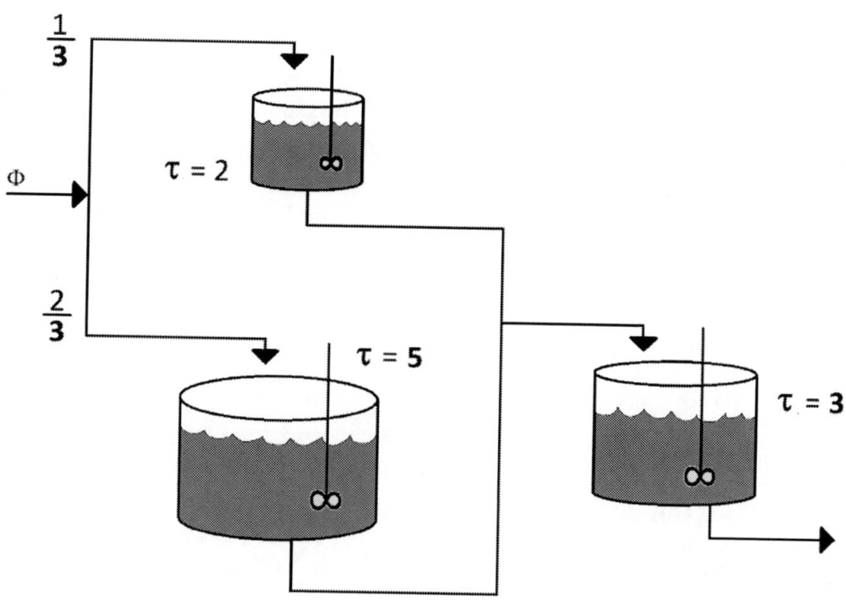

Figure 9. Another CSTRs system with series and parallels.

The transfer function for this reactor is:

$$G(s) = \left[\frac{1}{3}\,\frac{1}{(1+2s)} + \frac{2}{3}\,\frac{1}{1+5s}\right]\frac{1}{1+3s} = \frac{(1+5s)+2(1+2s)}{3(1+2s)(1+5s)}\,\frac{1}{1+3s} = \frac{3(1+3s)}{3(1+2s)(1+5s)}\,\frac{1}{1+3s} = \frac{1}{1+2s}\,\frac{1}{1+5s} \tag{33}$$

Once again, this function corresponds to a series of two CSTRs with residence times $\tau_1 = 2$ and $\tau_2 = 5$.

In other words, the uniqueness of the inverse problem does not occur since the same function $G(s) = \frac{1}{1+2s}\frac{1}{1+5s}$ corresponds to three different CSTRs systems: those of Figure 8, Figure 9 and a series of two CSTRs.

Conclusion

In this chapter, we analyze different reactors formed by series and parallels of CSTRs.

After exposing several theoretical results, we focus on the inverse problem, that is, to find a CSTRs system that corresponds to a certain rational function $G(s)$, with the degree of the numerator less than the degree of the denominator and constant terms equal to one. Moreover, we studied the existence, uniqueness and stability of the aforementioned inverse problem.

The result has been negative for all those problems, i.e., the reactor system does not always exist, if it exists it is not necessarily unique and finally small changes in the function $G(s)$, may produce signifincant changes in the reactor.

References

[1] Doggett, G.; Cockett, M.; Davies, A. G.; Phillips, D.; Derek-Woollins, J. *Maths for Chemists;* 2nd Ed.; Royal Society of Chemistry: London, 2015.

[2] Mickley, H.S., Sherwood, T.S. and Reed, C.E. (1975). *Applied Mathematics in Chemical Engineering,* 2º Ed., New Delhi: Tata Mc Graw-Hill Publ. Co. Ltd.

[3] Rice, R.G. and Do, D.D. (2012). *Applied Mathematics and Modeling for Chemical Engineers*, 2nd Ed. New Jersey: John Wiley & Sons.

[4] Steiner, E. *The Chemistry Maths Book.* Oxford University Press: Oxford, 2008.

[5] Martinez-Luaces, V., (2017). Laplace Transform in Chemistry degrees Mathematics courses, in: *Focus on Mathematics Educations Research*, Chapter 4. New York: Nova Publishers.

[6] Martinez-Luaces, V., (2007). Inverse-modelling problems in chemical engineering courses, in *Vision and change for a new century.* Proceedings of the Sixth Southern Hemisphere Conference on Matrices in Chemical

Problems Undergraduate Mathematics and Statistics Teaching and Learning. Montevideo, Uruguay: ISC-Delta.

[7] Martinez-Luaces, V., (2018). Square Matrices Associated to Mixing Problems, in: *Matrix Theory: Applications and Theorems*, Chapter 3, pp. 41-58. London, UK: In Tech Open Science.

[8] Hayes, R. E.; Mmbaga, J. P. *Introduction to Chemical Reactor Analysis*. CRC Press. Taylor & Francis Group: Boca Raton, FL, 2012.

[9] Westerterp, K. R., (1984). *Chemical Reactor Design and Operations*. Manchester: Wiley.

[10] Martinez-Luaces, Victor, José Antonio Fernández-Plaza, and Luis Rico. 2022. About the Notion of Inverse Problem in STEM Education. In: *Active Learning - Research and Practice for STEAM and Social Sciences Education*. IntechOpen. doi:10.5772/intechopen.106479.

[11] Groestch C. (1999). *Inverse Problems: Activities for Undergraduates*. Washington D.C.: Mathematical Association of America.

[12] Martinez-Luaces, V., (2009b). Modelling, applications and Inverse Modelling: Innovations in Differential Equations courses. In: D. Wessels y C. Snyman (Eds.), *Proceedings of Southern Right Delta '09, Seventh Southern Hemisphere Conference on Undergraduate Mathematics and Statistics Teaching and Learning*. South Africa: ISC-Delta.

Index

A

B

C

D

Q

R

S

T

U

V

W

About the Editor

Victor Martinez-Luaces, PhD

Victor Martinez-Luaces was born in Montevideo, Uruguay. He is a Chemical Engineer and Mathematician, and has a Master's degree and a PhD in Education. He is a Researcher and Mathematics Consultant at the Chemical Engineering Institute, UdelaR (Uruguay), and a PROFESTEAM Group member, University of Granada (Spain) Additionally, he was a Visiting Professor at LDAR, University of Paris VII (France). He was awarded twice with the first prize at RELME (2002, 2005) and received the Best Teacher Award at the Faculty of Engineering, UdelaR (1992, 1996). He was Chair, Convenor, and Keynote Speaker at International Scientific Conferences. He has published more than twenty books and book chapters (USA, UK, Germany, etc.) and more than fifty journal papers. He has served as a reviewer and guest editor in several international journals. He is currently working as a book editor.